Solar Energy Advancements in Agriculture and Food Production Systems

Solar Energy Advancements in Agriculture and Food Production Systems

Edited by

Shiva Gorjian

Biosystems Engineering Department, Faculty of Agriculture, Tarbiat Modares University (TMU), Tehran, Iran; Renewable Energy Department, Faculty of Interdisciplinary Science and Technology, Tarbiat Modares University (TMU), Tehran, Iran

Pietro Elia Campana

Mälardalen University, Future Energy Center, Västerås, Sweden

ACADEMIC PRESS

An imprint of Elsevier

Academic Press is an imprint of Elsevier
125 London Wall, London EC2Y 5AS, United Kingdom
525 B Street, Suite 1650, San Diego, CA 92101, United States
50 Hampshire Street, 5th Floor, Cambridge, MA 02139, United States
The Boulevard, Langford Lane, Kidlington, Oxford OX5 1GB, United Kingdom

ISBN: 978-0-323-89866-9

For Information on all Academic Press publications
visit our website at https://www.elsevier.com/books-and-journals

Publisher: Charlotte Cockle
Acquisitions Editor: Lisa Reading
Editorial Project Manager: Tim Eslava
Production Project Manager: Nirmala Arumugam
Cover Designer: Matthew Limbert
Cover Image Credit: Fraunhofer ISE

Typeset by MPS Limited, Chennai, India

Working together
to grow libraries in
developing countries

www.elsevier.com • www.bookaid.org

Contents

9.   **Solar applications for drying of agricultural and marine products**   313

Ankit Srivastava, Abhishek Anand, Amritanshu Shukla and Atul Sharma

10.   **Applications of robotic and solar energy in precision agriculture and smart farming**   351

Amir Ghalazman E., Gautham P. Das, Iain Gould, Payam Zarafshan, Vishnu Rajendran S., James Heselden, Amir Badiee, Isobel Wright and Simon Pearson

11. Economics and environmental impacts of solar energy technologies 391

Aneesh A. Chand, Prashant P. Lal, Kushal A. Prasad and Nallapaneni Manoj Kumar

12. Emerging applications of solar energy in agriculture and aquaculture systems 425

Shiva Gorjian, Fatemeh Kamrani, Omid Fakhraei, Haniyeh Samadi and Paria Emami

List of contributors

Mushtaque Ahmed Department of Soils, Water, and Agricultural Engineering, College of Agricultural and Marine Sciences, Sultan Qaboos University, Al-Khoudh, Muscat, Oman

Abhishek Anand Non-Conventional Energy Laboratory, Rajiv Gandhi Institute of Petroleum Technology, Jais, Amethi, India

Amir Badiee School of Engineering, University of Lincoln, Lincoln, United Kingdom

Pascal Biwole Université Clermont Auvergne, Clermont Auvergne INP, CNRS, Institut Pascal, Clermont-Ferrand, France; MINES Paris Tech, PSL Research University, PERSEE—Center for Processes, Renewable Energies and Energy Systems, Sophia Antipolis, France

Pietro Elia Campana Mälardalen University, Future Energy Center, Västerås, Sweden

Aneesh A. Chand School of Engineering and Physics, The University of the South Pacific, Suva, Fiji

Flemming Dahlke Thünen-Institute of Fisheries Ecology, Bremerhaven, Germany

Gautham P. Das Lincoln Institute for Agri-Food Technology, University of Lincoln, Lincoln, United Kingdom

Ipsa Sweta Dhal Fraunhofer Institute for Solar Energy Systems ISE, Freiburg im Breisgau, Germany

Hossein Ebadi MAHTEP Group, Department of Energy "Galileo Ferraris" (DENERG), Politecnico di Torino, Turin, Italy

Paria Emami Renewable Energy Department, Faculty of Interdisciplinary Science and Technology, Tarbiat Modares University (TMU), Tehran, Iran

Sina Eterafi Biosystems Engineering Department, Faculty of Agriculture, Tarbiat Modares University (TMU), Tehran, Iran

Omid Fakhraei Biosystems Engineering Department, Faculty of Agriculture, Tarbiat Modares University (TMU), Tehran, Iran

Ulfert Focken Thünen-Institute of Fisheries Ecology, Bremerhaven, Germany

Amir Ghalazman E. Lincoln Institute for Agri-Food Technology, University of Lincoln, Lincoln, United Kingdom

Shiva Gorjian Biosystems Engineering Department, Faculty of Agriculture, Tarbiat Modares University (TMU), Tehran, Iran; Renewable Energy Department, Faculty of Interdisciplinary Science and Technology, Tarbiat Modares University (TMU), Tehran, Iran

Iain Gould Lincoln Institute for Agri-Food Technology, University of Lincoln, Lincoln, United Kingdom

Charis Hermann Fraunhofer Institute for Solar Energy Systems ISE, Freiburg im Breisgau, Germany

James Heselden Lincoln Institute for Agri-Food Technology, University of Lincoln, Lincoln, United Kingdom

Simson Jakobsson KTH – Royal Institute of Technology, School of Industrial Engineering and Management, Stockholm, Sweden

Laxmikant D. Jathar Department of Mechanical Engineering, Imperial College of Engineering and Research, Pune, India

Fatemeh Kamrani Biosystems Engineering Department, Faculty of Agriculture, Tarbiat Modares University (TMU), Tehran, Iran

Karunesh Kant Université Clermont Auvergne, Clermont Auvergne INP, CNRS, Institut Pascal, Clermont-Ferrand, France; Advanced Materials and Technologies Laboratory, Department of Mechanical Engineering, Virginia Tech, Blacksburg, VA, United States

Daniel Ketzer Karlsruhe Institute of Technology (KIT), Karlsruhe, Germany

Nallapaneni Manoj Kumar School of Energy and Environment, City University of Hong Kong, Kowloon, Hong Kong, S.A.R. China

Prashant P. Lal School of Engineering and Physics, The University of the South Pacific, Suva, Fiji

Seyed Sina Mousavi Faculty of Civil Engineering, Babol Noshirvani University of Technology, Babol, Iran

Seyed Soheil Mousavi Ajarostaghi Faculty of Mechanical Engineering, Babol Noshirvani University of Technology, Babol, Iran

Özal Emre Özdemir Fraunhofer Institute for Solar Energy Systems ISE, Freiburg im Breisgau, Germany

Iva Papic KTH – Royal Institute of Technology, School of Industrial Engineering and Management, Stockholm, Sweden

Simon Pearson Lincoln Institute for Agri-Food Technology, University of Lincoln, Lincoln, United Kingdom

Fabienne Pennec Université Clermont Auvergne, Clermont Auvergne INP, CNRS, Institut Pascal, Clermont-Ferrand, France

Kushal A. Prasad School of Engineering and Physics, The University of the South Pacific, Suva, Fiji

Vishnu Rajendran S. Lincoln Institute for Agri-Food Technology, University of Lincoln, Lincoln, United Kingdom

Christine Rösch Karlsruhe Institute of Technology (KIT), Karlsruhe, Germany

Haniyeh Samadi Biosystems Engineering Department, Faculty of Agriculture, Tarbiat Modares University (TMU), Tehran, Iran

Laura Savoldi MAHTEP Group, Department of Energy "Galileo Ferraris" (DENERG), Politecnico di Torino, Turin, Italy

Ibrahim Shamseddine Université Clermont Auvergne, Clermont Auvergne INP, CNRS, Institut Pascal, Clermont-Ferrand, France; Université Libanaise, Centre de Modélisation, Ecole Doctorale des Sciences et Technologie, Hadath, Liban, Lebanon

Atul Sharma Non-Conventional Energy Laboratory, Rajiv Gandhi Institute of Petroleum Technology, Jais, Amethi, India

Amritanshu Shukla Non-Conventional Energy Laboratory, Rajiv Gandhi Institute of Petroleum Technology, Jais, Amethi, India

Ankit Srivastava Non-Conventional Energy Laboratory, Rajiv Gandhi Institute of Petroleum Technology, Jais, Amethi, India

Ghadie Tlaiji Université Clermont Auvergne, Clermont Auvergne INP, CNRS, Institut Pascal, Clermont-Ferrand, France

Max Trommsdorff Fraunhofer Institute for Solar Energy Systems ISE, Freiburg im Breisgau, Germany; Wilfried Guth Chair, Department of Economics, University of Freiburg, Wilhelmstr, Freiburg im Breisgau, Germany

Amir Vadiee Division of Sustainable Environment and Construction, School of Business, Society and Engineering, Mälardalen University, Västerås, Sweden

Nora Weinberger Karlsruhe Institute of Technology (KIT), Karlsruhe, Germany

Isobel Wright Lincoln Institute for Agri-Food Technology, University of Lincoln, Lincoln, United Kingdom

Jinyue Yan Mälardalen University, Future Energy Center, Västerås, Sweden

Payam Zarafshan Department of Agro-Technology, College of Aburaihan, University of Tehran, Tehran, Iran

Chapter 1

Solar energy for sustainable food and agriculture: developments, barriers, and policies

Shiva Gorjian[1,2], Hossein Ebadi[3], Laxmikant D. Jathar[4] and
Laura Savoldi[3]

[1]*Biosystems Engineering Department, Faculty of Agriculture, Tarbiat Modares University
(TMU), Tehran, Iran,* [2]*Renewable Energy Department, Faculty of Interdisciplinary Science and
Technology, Tarbiat Modares University (TMU), Tehran, Iran,* [3]*MAHTEP Group, Department of
Energy "Galileo Ferraris" (DENERG), Politecnico di Torino, Turin, Italy,* [4]*Department of
Mechanical Engineering, Imperial College of Engineering and Research, Pune, India*

1.1 Introduction

Since the early 1960s, the world's population has been doubled, and it is expected to exceed 9 billion people by 2050, worsening the global issue of "*Food Security*" as one of the most crucial facets of sustainability. According to the definition of the United Nation's Committee on World Food Security (CFS), food security means that: "all people, at all times, have physical, social, and economic access to sufficient, safe, and nutritious food that meets their food preferences and dietary needs for an active and healthy life" [1]. Food security is indeed a universal concept and its criteria's establishment is a major problem for the entire human race. As societies expand and mature, the concept of food security must be examined, amended, and redefined regularly [2]. Since 1961, food supply per capita has been increased by over 30%, accompanied by increased usage of nitrogen fertilizers (up to 800%) and irrigation water resources (over 100%) [3]. It has been proved that climate events are already affecting food systems, especially in countries with agricultural systems that are sensitive to weather instabilities. From another perspective, food systems themselves impact the state of the environment and therefore, they are considered as a driver of climate change [4]. New findings

Solar Energy Advancements in Agriculture and Food Production Systems.
DOI: https://doi.org/10.1016/B978-0-323-89866-9.00004-3

1

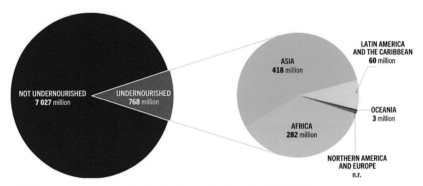

FIGURE 1.1   The number of global undernourished people in 2020 [9].

demonstrate that the current climate crisis has diminished global food production which is mainly due to anthropogenic activities. According to Ortiz-Bobea et al. [5], anthropogenic climate change has slumped the total factor productivity of global agriculture by 21% since 1961 which is equivalent to losing the last 7 years of productivity growth. This loss is even higher for less affluent countries, especially for the ones located in warmer regions such as Africa, Latin America, and the Caribbean. Therefore, the effects of the rising temperatures, carbon dioxide (CO_2) fertilization, and changes in annual precipitation could drive up some direct and indirect consequences on the agriculture sector ranging from reductions of the net revenue of farmers to fluctuations in national income due to low export volumes [6].

Climate change and nonclimate stressors (e.g., population and economic growth, demand for animal-sourced goods) are both putting strain on the food supply chain, affecting four pillars of food security including *availability, access, utilization,* and *stability* [7]. In 2020, after being essentially stable for 5 years, world hunger increased due to the COVID-19 pandemic. In this way, the prevalence of undernutrition jumped from 8.4% to roughly 9.9% in just 1 year (161 million more people in 2020) mainly due to the devastating impact on the world's economy, making the achievement of *Zero Hunger*[1] target by 2030 even more difficult [4,8]. Fig. 1.1 indicates that in 2020, more than half (418 million) of the worldwide population affected by hunger was in Asia and more than one-third (282 million) in Africa. Presenting figures in detail, in 2020, there were 46 million more people impacted by hunger in Africa, 57 million more in Asia, and 14 million more in Latin America and the Caribbean than in 2019. In this regard, it is crucial to recognize that the food security ecosystem is complex and multifaceted, requiring various *"players"* to secure its long-term viability [10]. Table 1.1 depicts the six most important trends affecting global food security in Asia.

---

1. Sustainable Development Goal 2 (SDG 2) aims to achieve "zero hunger."

TABLE 1.1 Trends affecting food security in Asia.

Factors	Remarks
Demographic trends	• Alongside urban population expansion, worldwide demographic trends are accelerating.
Transformation of agriculture	• The agricultural industry is undergoing transition, which is leading to anxiety over food security. • The rural population working in agriculture is reducing. • As a result of population expansion and inheritance-based fragmentation, farms are getting smaller.
Degradation of natural resources for food production	• This reality that the region's land and water resources are also under significant stress, adding to the pressure on agricultural sustainability.
Food price rise and volatility	• The costs of nearly all kinds of food have been increasing, counting basic food requirements. • In recent years, it was observed that food prices have been irregular.
Effect of fuels cost and energy security	• Higher oil prices may have an adverse impact on the cost of travel and shipment, influencing the cost of getting food from the producer to the consumer. • Increased food costs are a result of increased fuel costs, which have a significant impact on fertilizer costs.
Supply chain for commodities around the world	• To the displeasure of traditional shops and wet markets, supermarkets' share in the food retail sector has risen. • Farmworkers and small- to medium-sized suppliers face more hurdles as a result of the supply chain transformation, while urban consumers gain from the superstore revolutions in terms of enhanced financial and social access to food.

Source: Content is adapted from Ref. Barry Desker, Mely Caballero-Anthony, Paul Teng. Thought/Issues Paper on ASEAN Food Security: towards a more Comprehensive Framework. ERIA Discuss Pap Ser 2013;20.

1.1.1 Water, energy, and food security nexus

A framework to analyze the interconnection between water, energy, and food is called the WEF nexus, which includes the synergies, conflicts, and trade-offs among these resources. As depicted in Fig. 1.2, water is required to support livelihoods such as irrigated agriculture, fisheries, and food production, while at the same time, water is utilized to produce energy from hydropower

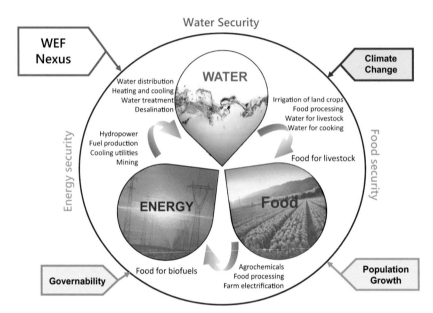

FIGURE 1.2   Representation of water−energy−food (WEF) nexus [11].

and biofuels. Then, the produced energy is employed for water distribution, plant cultivation, and food processing [11,12]. With the unprecedented increase in food demand by 60%, energy demand by 80%, and water demand by 50% [13], and also because of climate change consequences, the nexus objectives are only solutions that should be considered by governments and stakeholders for further developments. To battle these threats, the global community is investing in the "*energy-smart*" food systems where the access to new and modern energy systems increases, energy efficiency is enhanced, lower CO_2 is emitted, and the nexus approach is centered [14].

As a result, the role of renewable energy technologies can be promising in the WEF nexus, addressing some of the trade-offs between water, energy, and food. In this realm, the emerging clean systems alleviate the competition between the sectors, offering less resource-intensive processes than conventional energy systems. For example, the solar, wind, and tidal energy sources reduce the reliance on water demand and expand energy access to improve the security of supply across the WEF sectors [13]. Moreover, bioenergy technology has the potential to apply energy neutrality to the agriculture sector through a balance between energy production and consumption [15]. The nexus approach brings a novel perspective for the future policies in which sustainable development is pursued concerning simultaneous evaluations of water, energy, and food security. Since this concept is still nascent (initiated from the 2011 Nexus Conference [16]), it would be too early to list its outcomes, however, its potential effects must be taken into account to tackle current and future global challenges. What is also highlighted here is that since renewable technologies are localized and cannot be traded internationally, a country-specific policy analysis based on the WEF nexus is crucial to enhance environmental and energy security [17].

1.1.2 Agri-food supply chain

Globalization, urbanism, and agroindustrialization are all emerging developments that are putting pressure on agri-food chains and networks [18]. Agri-food supply chains (AFSCs) include all steps involved in the production, manufacturing, and distribution of food until its final consumption [19]. Generally, an AFSC is a complex network composed of a set of activities *"farm to the fork"* [20]. AFSCs frequently tackle major and complicated obstacles in attaining long-term sustainability, which include economic, environmental, and social factors [21]. Food safety has become a major issue for both consumers and producers as public awareness has grown. Improving food product quality is a major goal of agri-food cooperation, and customer satisfaction is critical for long-term profitability. In the agri-food industry, these supply chains involve postconsumption and preproduction operations [22]. The agri-food firms have a responsibility to offer food that is safe, secure, and long-lasting. Traceability is required in many nations to promote customer food security and confidence in the safety of their food supply [23].

Agricultural and horticultural products are often produced and distributed by AFSCs to end-users or consumers. Farmers, manufacturers, retailers, and consumers are the important participants in AFSCs who are directly involved in the logistics process [24]. In addition, several secondary stakeholders, such as government bodies, nonprofit organizations, food and industry groups, and financial institutions, serve as indirect partners. However, they may or may not participate in supply chain activities, but they frequently have a variety of effects on the processes that govern material, communications, and capital flow among all stakeholders [21]. The general structure of an AFSC is depicted in Fig. 1.3. AFSCs differ from other supply chains due to: (1) *nature of production*, which partially depends on biological processes, raising volatility and risk; (2) *type of product*, which also has particular attributes such as perishability and bulkiness that necessitate a specific type of supply chain; and (3) *public's and consumers' perceptions* on issues such as food hygiene, the welfare of animals, and environmental stress. Generally, there are two main types of AFSCs: (1) *fresh product supply chains*, including fresh fruits, vegetables, and flowers, and (2) *highly processed supply chains*, including canned food products, dessert products, etc. Growers, wholesalers, importers and exporters, and retailers make comprised the first supply chains. Producing, storage, packing, transporting, and selling of these products are the major procedures [25].

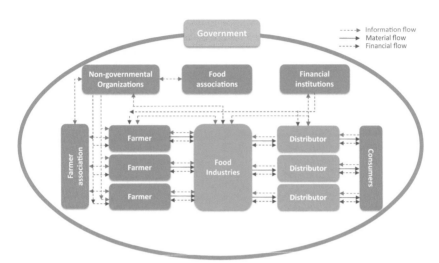

FIGURE 1.3 The general structure of an AFSC [21]. *AFSC*, Agri-food supply chain.

1.1.3 Energy supply and demand of agri-food sector

Energy is an essential component in all steps of the food chain including crops production, forestry, and dairy production, postharvest applications, and food storage, processing, transport, and distribution [26]. Traditionally, food production has relied on manual labor, animal power, and biomass consumption to supply the power demand of various agricultural production tasks including the production, storage, processing, transport, and distribution of food products. But over time, as agriculture has become more industrialized, these forms of energy inputs have been replaced by fossil fuels and make both farm production and food processing more intensive [27].

The energy utilized in the agri-food industry can be categorized as *direct energy* including electricity, mechanical power, solid, liquid, and gaseous fuels which are consumed for production, processing, and commercialization of products such as the energy required for irrigation, land preparation, and harvesting. While *indirect energy* refers to the energy required to manufacture inputs such as fertilizers, pesticides, as well as farm equipment and machinery [28]. The energy inputs in the agricultural value chains are depicted in Fig. 1.4 [29].

According to the study by Marshall and Brockway [30], global agriculture including, farm, aquaculture, fishing, and forestry (AAFF) energy systems consume nearly 27.9% of the total societies' energy supply in which the energy used for food supply accounts for 20.8% of this share by 2017. In detail, for the period between 1971 and 2017, the various energy inputs to AAFF as the source of power are illustrated in Fig. 1.5. Referring to this figure, the food required for human labor has increased by 15% which is

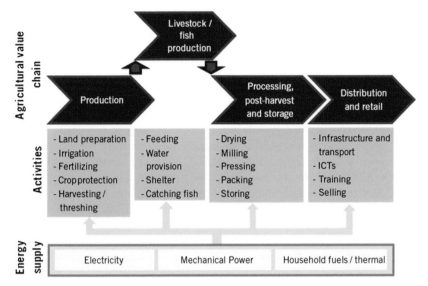

FIGURE 1.4 Energy inputs that enable various activities in agricultural value chains. *Adapted from Introducing the Energy-Agriculture Nexus - energypedia. info 2021. https://energypedia. info/wiki/Introducing_the_Energy-Agriculture_Nexus.*

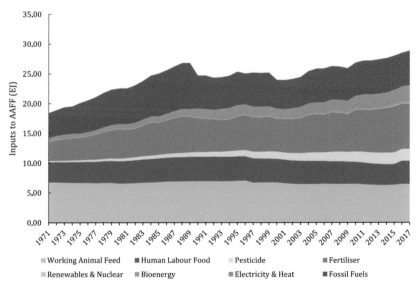

FIGURE 1.5 Various energy inputs to global agriculture, aquaculture, fishing, and forestry from 1971 to 2017 [30].

reflected in population growth balancing the decrease in the share of the global working population in AAFF as human labor has been replaced with work performed by fossil fuels [30]. The feed input required for working animal labor shows a 4% decrease which reflects that although agricultural mechanization has progressed in many parts of the world, this process is not homogenous and there are still a large number of countries that rely on animals to perform agricultural practices. What is more highlighted in this figure is the considerable growth in the use of fossil fuels, bioenergy, renewables, and nuclear power sources. A significant increase is also observed in the energy used for pesticides while the inputs for fertilizer application increased by 150% during this period. This positive trend mainly attributes to mechanization in agriculture and forestry, expansion of irrigated lands, growth in motorized fishing vessels, and total fishing efforts [31−33].

By the end of 2018, total final energy consumption (consumed by agriculture, fishery, and forestry) reached 16 EJ compared to 88, 121, and 119 EJ for residential, transport, and industry sectors, respectively [34]. However, the energy consumption in agriculture is geographically dissimilar and varies depending on the regional technology development. Rokicki et al. [35] found that EU countries are using less energy for agricultural activities and the form of the energy is shifting from crude oil with 60% share toward renewables with 10% from 2005 to 2018. Moreover, countries with developed economies have a higher tendency in this regard, where the implementation of renewable energy in agriculture has reached 20% by 2018 in the top five countries of Sweden, Austria, Finland, Germany, and Slovakia. Lin et al. [36] proved that farm management and structure can change the energy input, energy output, and energy-use efficiency (EUE). They claimed that mixed farming is the most recommended method of producing food with high EUE in the case of organic farming. Moreover, they concluded that as far as conventional arable farming is concerned, improved farm management and technologies can decrease the energy input from 14.0 to 12.2 GJ/ha/year, an increase the energy output from 155 to 179 GJ/ha/year, resulting in an improved EUE from 11.1 to 14.6. Wu and Ding [37] investigated the determinants of changes in agricultural energy intensity (CAEI) in China and found that countries like China that suffer from large agricultural energy intensity must take solutions to achieve sustainable development in the agricultural sector. For this purpose, governments can put some mechanisms into action such as regulations for removing old agricultural machinery and subsidies for purchasing energy-saving machinery and equipment to facilitate agricultural energy-related technology progress. They found that energy price and income significantly reduce CAEI, while labor has the inverse impact. Furthermore, they suggested improving people's environmental awareness to deepen agricultural capital accumulation and deregulate rural energy prices. However, according to Jiang et al. [38], agriculture mechanization and the increase in the number of machinery operated on the fields must be controlled based on the concept of green development such as environmental performance to avoid the over usage of fossil-based energy resources. Thus governments should support the development of low-carbon agricultural mechanization practices and provide technical training to farmers when an alternative machine is introduced to reduce CO_2 emissions.

1.1.4 Greenhouse gas emissions from agri-food sector

As described in the previous section, energy inputs play a crucial role in the productivity enhancement of agri-food systems and in meeting the global food demand of the growing world population. According to the report released by the Food and Agriculture Organization (FAO) in 2017 [39], the energy inputs of AFSCs, at all stages along different agri-food value chains, extremely relies on fossil fuels, resulting in the agri-food processing and production sector to be a significant source of greenhouse gas (GHG) emissions. Therefore, as long as the reduction of GHG emission is concerned, mitigation options are pronounced and this can be understood by reviewing the share of agricultural practices in current emission values. Fig. 1.6 indicates the major contributors of global GHG emissions in 2019, with a particular focus on agricultural activities. As shown in this figure, agriculture plays an important role among other sectors in terms of the emission of GHGs with the highest share of 55% and 45% for methane (CH_4) and nitrous oxide (N_2O), respectively. Moreover, enteric fermentation which is the digestion of carbohydrates by ruminant livestock is the largest source of CH_4 production in agricultural systems, while livestock manure is the second-largest driver for CH_4 and N_2O emission. Synthetic nitrogen fertilizers, as the third

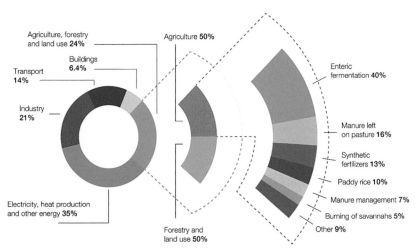

FIGURE 1.6 The share of agriculture among other sectors in global GHG emissions in 2019 [40].

contributor with a 13% share of GHG emissions, release N_2O gas when microbes start to process the nitrogen left by crops [40].

The emissions in food systems differ between continents and countries [41] owing to various production patterns, the distinction between unit emissions from plants and animals production, and the overall amount of agricultural intensity [42]. In this regard, Europe accounts for around 11% of worldwide GHG emissions from agricultural production, while Asia accounts for approximately 44%, followed by Africa with the share of 15%, Australia and Oceania with the share of 4%, and North and South America with the shares of 9% and 17%, respectively [43]. As shown in Fig. 1.6, agriculture, forestry, and land use account for 24% of total GHG emissions mainly owing to the widespread use of chemical fertilizers, pesticides, and animal manure. As a result of the rising world population, increasing demand for meat and dairy goods, and improvement of farming processes, this rate is expected to rise further. Representing more detailed shares, the livestock and fisheries account for 31% of global GHG emissions, following by crop production, land use, and supply chain with the total global share of 27%, 24%, and 18% respectively [44].

To accurately quantify the amount of GHG emissions from the agri-food sector, the following points should be considered:

1. *Livestock and fisheries account for 31% of food emissions:* This value only relates to the emissions of on-farm "production," excluding emissions from the land-use change or supply chain from the animal feed production;
2. *Crop production accounts for 27% of food emissions:* From this value, 21% comes from the production of food crops, while 6% comes from animal feed production;
3. *Land use accounts for 24% of food emissions:* From this value, 16% comes from livestock and 8% for food crops. Food production emissions from livestock and fisheries account for 31% of the total emissions;
4. *Supply chains account for 18% of food emissions:* From this value, a very small portion (6%) comes from transport, while food waste accounts for 3.3 billion tonnes of CO_2-eq (25%).

It is noteworthy to mention that effective mitigation strategies are those that do not affect the yields and are also cost-effective. Therefore, mitigation in agriculture and food production systems can be achieved under three main scopes as follows [45]:

• *Decreasing the emissions intensity through AFSCs, avoiding land-use change caused by agriculture:* It is expected that a 1.8 Gt CO_2-eq annual reduction can be achieved from enteric fermentation and manure management from all crops in addition to the enhanced fertilizer production with current technologies. This could lead to a 30% decrease in agricultural GHG emissions by 2030.

- *Sequestering additional carbon in agricultural systems:* Carbon sequestration can bring a 0.7 and 1.6 Gt CO_2-eq annual reduction in agricultural GHG emission by 2030 through a series of soil, crop, and livestock management practices. However, this technique still poses a wide range of doubts on the economy and availability for farmers.
- *Increasing overall agricultural productivity with decreasing food losses, and wastes or reducing demands for biofuels:* It has been predicted that there is a 3 Gt CO_2-eq mitigation potential from the changes in diets and by cutting the current wastes from foods. In the case of diet, turning from high-carbon intensity agricultural products such as meat from ruminants could result in 75% mitigation while the rest of 25% can be achieved by reductions in food wastes.

1.2   Sustainable food and agriculture

Sustainable agriculture consists of three main dimensions; (1) *economy*, (2) *society*, and (3) *environment*. To be sustainable, the agriculture sector must meet the food demand of present and future generations, while ensuring profitability, environmental health, and social and economic equity [46]. Therefore economic concerns over economic justice are considered to support local and small-scale agricultural businesses while guaranteeing long-term profitability. Additionally, environmental issues are mainly associated with the adverse impacts of agriculture on land, water, and wildlife resources. Last but not least, accepting the public welfare concerns revolving around food quality and the needs of present and future generations [47]. Sustainable food and agriculture (SFA) contributes to all four pillars of food security and three dimensions of sustainability. Globally, FAO promotes SFA to help countries achieve "*Zero Hunger*" and "*Sustainable Development Goals*" (SDGs) [46]. Surveying a group of farmers, Laurett et al. [48] concluded that natural agriculture, investment in innovation and technology, and environmental aspects can define sustainable development in agriculture. However, factors of lack of information and knowledge, and lack of planning and support are the two main barriers in the dissemination of this concept. Additionally, they asserted that family farmers can be motivated by external influencers and through engagement with the sustainability movement.

Sustainable development in agriculture is still a challenge where the growing food demand must be met by more sustainable agricultural activities [39], also given the competition for land use which is rising rapidly and requires international frameworks to protect food production. This means that not only a revolution in energy efficiency but also a shift in energy resources utilized in the agricultural sector is required [49]. Carbon footprint (CF) is a good indicator representing the status of sustainable development in the agriculture and food production sector. Adewale et al. [50] stated that agricultural CF is farm-specific and the value ranges from 7144 to 3410 kg

CO_2-eq/ha/year for a typical small-scale and a large-scale farm, respectively. They also found that the share of contributors varies so that for small-scale farms, the share of 47.6% for the fuel use is dominant, followed by the share of 11.3% for greenhouse facilities, and the net soil's emission of 10.3%. However, in the case of large-scale farms, the electricity used for irrigation with 47.5%, fuel use with 26.1%, and soil amendments with 20.1% are the major contributors. In another study, Mantoam et al. [51] evaluated various types of farm machinery and investigated their agricultural CF. They revealed that the typical CF for tractors lies between 11 and 30 t CO_2-eq, while this value ranges from 27 to 176 t CO_2-eq for other types of machinery such as sugarcane harvesters, coffee harvesters, sprayers, planters, and combiners. It is worthy to be noted that on-farm emissions due to burning fossil fuels are a function of cropping practices, type of the utilized machinery, level of mechanization, and the production scale [52]. Over time, the use of low-carbon energy sources as an alternative to fossil fuels in the agri-food sector has been rapidly increased. Sustainable agriculture production systems and energy-smart AFSCs with higher access to modern energy services can be pragmatic and cost-effective solutions to ensure energy security and achieve sustainable development [28]. In this regard, employing renewable energy sources (RESs) to meet the energy demand of the entire AFSC can assist in improving the access to energy sources, mitigating energy security concerns, and reducing the reliance on fossil fuels [27]. Hence, GHG emissions will consequently be reduced as an ultimate goal to achieve SFA development.

1.2.1   Global potential and development of solar energy technology

The energy transition is not a new topic among the researchers, but today its importance is gaining attention more than ever due to the need for urgent application, especially with an average increase of 1.3% in energy-related (CO_2) emissions for the period of 2014 to 2019. Although 2020 was an exception due to the worldwide pandemic where the emission decreased by 7%, the ascending trend is anticipated to rebound in short term [53]. To alleviate the COVID-19 crisis, over US$12 trillion as fiscal stimulus including at least US$732.5 billion for the energy sector was announced by governments around the world. As of April 2021, from the total value globally supplied by governments, nearly US$264 billion in the form of stimulus packages were allocated to renewables as incentives while, at the same time, more than US$309 billion was given to fossil fuels [54]. Considering the 1.5°C climate pathway defined by the *Intergovernmental Panel on Climate Change* (IPCC), nations are forced to reach net-zero global emissions through a sort of strategies such as investment shifts to clean energy, decarbonization of electricity generation, electrification of energy end-uses, energy savings, and development of anthropogenic CO_2 removal techniques [49]. Despite the efforts put in this way, the current trend could not meet the goals, and therefore a portfolio of mitigation options contains the deployment of clean energy supply and a spectrum of end-use technologies is required. In this way, the energy demand will be diminished while low-carbon fuels such as electricity, hydrogen, and biofuels are letting to be in the demand sector [55]. For this reason, 2020 was one of the brightest years for renewables with the highest growth. In this year, the total capacity increased by nearly 10.3% compared to the value in 2019, reaching a global capacity of almost 2.8 TW [56]. Taking the ambitions for renewable energy integration into consideration, it is expected that by 2030s, low-carbon technologies will take the lead in the power sector while the CCUS (carbon capture, utilization, and storage)-equipped coal and gas power plants will assist to guarantee the use of remaining fossil fuel be carbon-neutral or -negative. As shown in Fig. 1.7, direct CO_2 emissions from end-use sectors face a 65% decrease by 2050 based on the Faster Transition Scenario [57]. This can be attributed to rapid improvements in energy efficiency, as well as cut-offs on the use of fossil fuels due to the sharp increase in the penetration of renewable technologies such as solar and wind. Cost reductions coupled with increasing incentives for investments in renewable energy are the main drivers in the future energy mix which are expected to enhance the share of renewables in the total primary energy supply from 14% in 2015 to 63% in 2050 [58].

It has been estimated that solar energy, as one of the key energy sources in the future energy mix, will allocate the total final energy consumption of 25% [15% solar PV, 7% solar thermal, 3% concentrated solar power (CSP)] by 2050 [58]. Additionally, solar energy is considered a promising option due to its extensive availability, cheapness, and versatility especially for millions of underprivileged people in developing countries [59]. During recent years, solar photovoltaics (PVs) has shown the highest cost reduction among

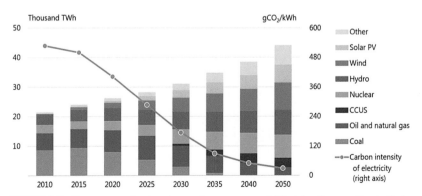

FIGURE 1.7 Power generation based on various sources and CO_2 intensity according to Faster Transition Scenario from 2010 to 2050 [57]. Source: *IEA, All rights reserved.*

all renewable technologies where its wholesale selling price has been reduced 15 times from 2000 to 2019, and it is expected to be further decreased in the near future [60]. Statistics indicate a capacity addition of 115 GW for solar PV in 2019, which is expected to reach over 145 GW (8% increase) in 2021. There is also a positive sign with a rise in the number of countries entering the PV market, which will make this industry more robust and thrive [61]. It has also been estimated that the share of utility-scale applications in annual PV additions to be increased over 55% in 2020 to nearly 70% in 2022. Fig. 1.8 shows the annual additions of solar PV capacity in different types of applications. As shown in this figure, the overall PV deployment has increased from 25% in 2016 to almost 45% in 2018, owing to the attractive support scheme of China. While in 2019, this trend was reversed due to the reduced feed-in-tariffs (FITs) of China for commercial and industrial PV projects. However, sustained support resulted in doubled deployment of residential applications from 2019 to 2020.

Although system prices for large-scale PV systems have reached below US$0.6/W, two factors of the regulatory framework and its further evolution towards market mechanisms remain important for further development of worldwide PV markets. The advent of new market segments, including floating PV (FPV), agri-PV (the combination of PV with agriculture), and off-grid PV is also advancing the penetration of PV technology in various countries [62]. In the case of solar thermal systems with direct heat generation, an accumulated capacity of 479 GW_{th} with a global turnover of US$16.1 billion was obtained by the end of 2019 which is equal to the reduction of 41.9 million tons of oil and 135.1 million tons of CO_2. Despite the growth in some emerging local markets, in 2019, the worldwide solar thermal market shrank by 6% compared to that of 2018. The reason for such a decrease refers to the fact that most traditional markets are still focused on small-scale solar water heating systems for single-family houses while heat pumps and PV systems are becoming better alternatives [63]. As shown in Fig. 1.9, the global solar

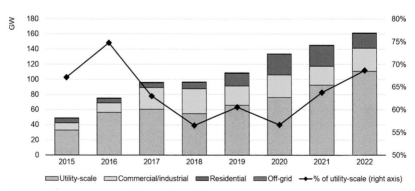

FIGURE 1.8 Annual solar PV capacity additions by application segment, 2015−2022 [61]. Source: *IEA, All rights reserved.*

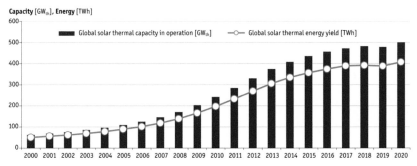

FIGURE 1.9   Global solar thermal capacity in operation and annual energy, 2000−20 [64].

thermal capacity reached 501 GW$_{th}$ (715 million m^2) in 2020 compared to the total global capacity of 62 GW$_{th}$ (89 million m^2) in 2000, for glazed and unglazed solar water collectors^2 in operation. While the corresponding annual solar thermal energy yields amounted to 407 TWh in 2020 from 51 TWh in 2000 [64]. The global solar thermal market was gradually decreased in 2020 with the global added capacity of 25.2 GW$_{th}$ in comparison with 26.1 GW$_{th}$ in 2019, mainly driven by the constraints associated with the COVID-19 pandemic, causing the investment decisions to be postponed [54].

The rapid growth of demand for cooling and refrigeration will continue, especially in emerging countries (with an estimated several hundred million sold alternating current (AC) units per year by 2050), representing a huge potential for solar-powered cooling systems using thermal and PV systems. This is mainly owing to consuming less conventional energy sources and employing natural refrigerants of water and ammonia [64]. This suggests that the emergence of new markets such as agriculture concerning the applicability of solar heating and cooling systems and supports from governmental programs could give an impetus to the total market while addressing SDGs.

The deployment of solar CSP with a total capacity of 4.9 GW in 2017 has been accelerated with thermal energy storage (TES) systems, leading to poly-generation technologies as a unique feature of CSP systems [65]. The global CSP capacity experienced 1.6% growth in 2020, reaching a total global capacity of 6.2 GW with a 100 MW parabolic trough project coming into operation in China which was commissioned as 600 MW in 2019. The reason behind this reduced market growth was many challenges that the CSP sector has been faced in recent years mostly include the cost competition increase from solar PV, the CSP incentive programs expiration, and several operational issues at facilities in operation (Fig. 1.10A). Additionally, this sector was affected by delays and stoppages in construction that occurred in China, Chile, and India [54]. According to Islam et al. [66], CSP is expected

---

2. Not include concentrating, air or hybrid collectors.

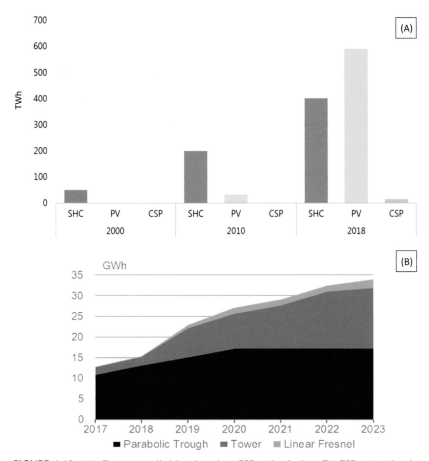

FIGURE 1.10 (A) Energy supplied by three key CSP technologies. (B) CSP generation by technology (2017−2023) [67]. Source: *IEA, All rights reserved.*

to supply 5% and 12% of the global electricity in 2050, referring to the moderate and advanced scenarios, respectively. Moreover, market projections indicate that the number of people to be employed in this sector will significantly rise until 2050, reaching the maximum employment of 1.4 million people under the advanced CSP market growth. Fig. 1.10B indicates the generation capacity of different CSP technologies from 2017 to 2023.

1.2.2 Advent of solar energy technology in agriculture sector

Different RESs including solar, wind, biomass, hydro, geothermal, and ocean energy can be converted to a wide range of energy carriers including electricity, heat, as well as liquid and gases biofuels to supply the energy demand of different applications involved in the agri-food sector [68]. Among different RESs, solar energy is the most abundant renewable source with the highest adaptability with conventional and modern agricultural operations [69]. Solar energy can be converted into both heat and electricity,[3] providing the power requirements of several agricultural applications. In this regard, by using solar thermal collectors, solar energy can be converted into heat, while using PV technology, solar radiation can be directly converted into electricity [70]. The utilization of solar energy in agriculture can increase reliability by eliminating the heavy reliance of agricultural operations on fossil fuels, reducing GHG emissions to a large extent. On the other hand, since mechanization in agriculture has bounded with digitalization and utilization of smart technologies for more precise field operations, electricity is becoming a prime demand in agricultural energy contexts. However, the mobile nature and harsh environment of agricultural activities reflect the need for a robust source of electricity such as solar PV systems [71].

1.2.2.1   Challenges and barriers

The developing markets of solar technologies imply that supports must be continued with focuses on research and development for cost reduction and efficiency improvement. The high investment cost is found to be the most common economical barrier in the dissemination of solar systems in developing countries, and therefore policies that reduce the financial risks should be implemented to attract the lower-medium and low-income classes. Additionally, policies taking the solar-derived wastes into account are also scarce and limited, while if solar energy is deployed in industries dealing with human health such as agri-food, the adverse effects of this technology must be well known [72].

Other challenges that have delayed the deployment of solar energy technologies are the instability in the performance due to their heavy reliance on the availability of solar radiation [73]. To overcome this issue, solar energy systems can be integrated with other RESs as auxiliary energy sources, increasing reliability and supporting a steady energy supply to end-users. Such systems are known as hybrid renewable energy systems (HRESs) [74]. In this case, solar-biomass systems can be employed to supply the power demand of water heating, space heating and cooling, power generation, and hydrogen production applications; solar-wind systems can be utilized to supply electricity; and solar-geothermal systems can be employed to supply the energy demands of heating and cooling, power generation, and water desalination. Additionally, in more advanced configurations, the solar-based HRESs may compose of solar-wind-geothermal or solar-wind-biomass power generation systems. Note

3. Solar energy technologies are presented in Chapter 2.

that solar-based HRESs are not necessarily composed of two or more RESs as fossil fuel-based systems can also be integrated.

In some cases, TES units are employed to increase the reliability of solar systems, extending their working hours over the sunshine hours. In terms of solar PV systems, the main challenge is still the limited efficiency for commercially available modules, so that thin-film PV modules with 3%−14% electric efficiency and crystalline-based modules with efficiency values below 21% are currently available on the market [75]. On the other side, concerns regarding the sustainable supply of cadmium and tellurium materials used in the fabrication of thin-film solar cells have overshadowed the development of this industry. While, low thermal efficiency, as well as limited heat carrying capacity of heat transfer fluids coupled with the need for TES with efficiency constraints, are among the problems associated with solar thermal systems [76].

Solar energy can be utilized to supply the power requirement of several conventional agricultural applications in the form of solar-powered crop drying systems, solar-powered desalination technologies, solar-powered greenhouse cultivation systems, solar-powered heating and cooling systems, and solar-powered water pumping and irrigation systems, as well as several innovative smart farming applications such as solar-powered farm robots, solar-powered communication networks, and solar-powered advanced electric machinery and tractors. Fig. 1.11 indicates an overview of the most common applications of solar energy in agriculture and food production

FIGURE 1.11 An overview of applications of solar energy in agriculture and food production systems: (A) solar-powered agricultural greenhouses [77], (B) solar-powered irrigation system [77], (C) an installed agrivoltaic system [78], (D) aquavoltaic system using FPV modules [79], (E) a central solar heating system [80], (F) a solar water heating system [81], (G) a solar-powered desalination system [82], (H) a solar-powered crop dryer [77], (I) a solar-powered crop protection system [77], (J) a solar-powered autonomous robot for use in vineyards [83], (K) an off-board solar-powered electric tractor [71], (L) a solar-powered weather station [84].

systems. As shown in this figure, two nearly innovative applications of solar PV systems for the coproduction of food and electricity have been emerged, known as agrivoltaic and aquavoltaic systems, where the first is the cogeneration of crops and electricity on the same farmland, while the second uses FPV modules as a structure for aquaculture systems. Additionally, solar CSP systems can also be employed to provide the heat and electricity demands of agricultural and food production systems, mainly on large scales such as solar-powered commercial seawater green-houses (SWGHs) [77]. As a result, the simplicity of the most renewable energy systems in addition to the eco-friendly nature of these systems will trigger sustainable rural development and raise levels of agriculture productivity worldwide.

1.2.2.2 Global policies and regulations

Although there is significant progress for solar energy systems, the presence of some major challenges and barriers hinders the deeper penetration of such eco-friendly technologies. Considering the global market, PV has still a policy-driven market, and the rest of the world should follow the policies established by leading countries such as China and the USA in terms of yearly installation as well as cumulative installed capacity. The current PV market is driven by two major policies: (1) *FIT* and (2) *renewable obligation certificates (ROCs)* [75]. Moreover, the PV market has always been a play-ground for many political interventions which all lack the persistence to construct the desired market conditions. However, to solve this problem, governmental support is required to assist the PV market to serve as a tool to achieve a long-term policy goal. This can be applied through fewer external interventions while the market for renewable must be self-sustaining [85].

Zakeri et al. [86] demonstrated that by using three alternative policies to support small-scale PV systems, such technology will become profitable for the users. The proposed incentives include a PV self-consumption FIT bonus; *energy storage policies* for rewarding discharge of electricity from home bat-teries at times the grid needs most; and dynamic retail pricing mechanisms for enhancing the arbitrage value of residential electricity storage. According to the study by Aprà et al. [87], since CSP technology is less known among policymakers, these systems have not been perceived as much as their values. From the academic point of view, CSP still suffers from a lack of knowledge in terms of environmental impacts, indicating the challenges for animal habitat, water use, visual impacts, and the effects on endangered spe-cies. Furthermore, one of the facts that make CSP a favorable option in the energy markets is that the levelized cost of electricity could not be the only factor comparing CSP with other renewable technologies for power genera-tion purposes. But other parameters such as the flexibility added by CSP, amount of CO_2 reductions, and the hidden externalized cost must be taken into account. To promote the CSP market in developing countries, Labordena et al. [88] suggested that new policies to de-risk CSP finance are necessary to make CSP competitive with the coal power industry, which can be achieved through certain regulations such as long-term power purchase agreement, concessional loans, and/or loan guarantees.

Based on a literature review, most of the policies taken for the solar energy industry consist of tax exemption, subsidies, FIT, formation incentives, and renewable portfolio standards (RPSs) while FIT and RPS are the two successful models implemented so far. What can improve the addressed policies is the combination of laws, regulations, economic encouragement, technical research and development, and industrialized support as well as renewable energy model projects coordinated in a structural framework for each country. Considering agriculture-based solar systems, policies are still limited and the previous regulations and incentives may need some revisions according to the economic, ecological, and social problems of the target populations, which are mostly rural societies in the agricultural context. In this way, one of the primary and effective steps is the fair transfer of financial and technological resources from the developed countries to the poorer nations to improve the sustainable development status in these areas [89]. Hopefully, the number of successful carbon development mechanism projects in agriculture is rising all over the world, which is based on sustainable and renewable energy technologies. However, referring to the high capital cost investments of these projects, higher governmental supports are needed to pave the way for development in the agriculture sector [68]. It is notable to mention that incentive policies for the use of renewables in agriculture must be well-analyzed before the incorporation since there are some examples in the literature that the overattention to these policies have resulted in several negative side effects such as the increase in agricultural land prices and direct and indirect land-use changes [90]. One of the productive programs in action is the *Adaptation for Smallholder Agriculture Programme* introduced by *International Fund for Agricultural Development*. This program aims to link directly the climate finance to smallholder farmers and help them to build resilience to climate change by employing renewable technologies [91].

1.3 Aims and framework of the book

There is a global will to battle the current humanity challenges associated with food and energy security. Mitigating climate change and guaranteeing sustainable food supply for the growing global population are the two main difficulties to be solved with the nexus approach. Increasing the global population on one hand, and decreasing the availability of fossil fuels with growing prices, on the other hand, have drastically increased the pressure on the agriculture and food production sectors. Additionally, destructive environmental consequences due to the use of fossil fuels have made the situation worse.

Solar energy with the fast-growing associated technologies and a nascent market has shown promising potential for integration with a wide range of agricultural activities to offer an alternative sustainable solution for current practices. With the flexibility and sustainability added by solar systems to AFSCs, it is expected that both mitigation and adaptation strategies will be met in coping with climate change impacts. The interventions between solar energy and agriculture must be analyzed in-depth, especially in terms of global markets and implemented policies while considering the economic aspects is also crucial to address and support this integration. In general, solar energy could open new doors to agricultural technologies, giving birth to new and novel systems that form future activities. As an example, the synergy coming from solar electrification and digitalization in agriculture offers a promising solution for acceleration in the developments of precision farming. Additionally, solar energy is potentially making farm operations safer, reducing the destructive impacts causing by the use of conventional fuels which harm human lives with a large risk. In this regard, researchers, applied scientists, engineers, and technicians in the agriculture sector need to be updated with theories and principles to develop new, low-cost, and sustainable energy technologies. Governments and industries must also increase research and development activities to reduce costs and ensure solar technology readiness for rapid deployment while supporting longer-term technology innovations. The use of solar energy has been growing exponentially in the last decade and it is expected to follow a similar trend in the future. For farmers, this means that the cost to install solar technologies is likely to be decreased and reach grid-parity.

This book aims to provide up-to-date knowledge and information on recent advances in solar energy technologies and their specific applications in the agriculture and food production sectors. Following the introductory chapter (*Chapter 1*), which sets out the aims and structure of the book, *Chapter 2* provides an overview of solar energy technologies, discussing their principles, advancements, and applications. In this context, a wide range of solar energy conversion technologies is presented and discussed and recent advancements are highlighted. *Chapter 3*, represents developments of agricultural greenhouses integrated with various solar energy technologies and discusses the possible employment of thermal energy storage systems, while thermal modeling of solar greenhouses is also provided. In *Chapter 4*, the principles and advances of PV-powered water pumping systems for irrigation are presented by highlighting the water-food-energy nexus and reviewing the current and future challenges. Two emerging concepts of agrivoltaics and aquavoltaics are introduced in *Chapters 5* and *6*, respectively. In these chapters, both concepts are introduced, and different designs and configurations along with the challenges and barriers are described. *Chapter 7* discusses solar heating and cooling applications in agricultural operations and food processing systems, introducing some innovative active and passive integrated solar systems applications. In *Chapter 8*, solar desalination, one of the most important aspects in the agriculture sector, with the main target of supplying water for irrigation is presented and discussed. In this chapter, different solar desalination technologies employed in agricultural applications are introduced, highlighting their technical, economic, and environmental aspects. In *Chapter 9*, one of the most common applications of solar energy which is solar drying is presented and discussed. In this chapter, different types of solar drying technologies for drying agricultural and marine products are studied and some commercial solar drying projects are introduced. *Chapter 10* represents the novel integration of solar energy with precision agriculture and smart farming applications. This chapter presents an overview of robotic technologies for agriculture workspaces and describes the role of solar energy in novel agricultural practices. In *Chapter 11*, different solar energy technologies that could potentially be used in the agriculture and food sectors are discussed, evaluating both their economic and environmental aspects. Additionally, several tools employing to model and investigate the techno-economic and environmental impacts of solar energy technologies are introduced and discussed. *Chapter 12* provides some emerging applications of solar energy in agriculture and aquaculture systems, describing their potentials for global deployment. In this chapter, both active and passive utilization of solar energy in some specific applications are explored.

As a whole, this book provides a technical and scientific endeavor to assist society and farming communities in different regions and scales to improve their productivity and sustainability. To fulfill the future needs of modern sustainable agriculture, this book addresses highly actual topics providing innovative, effective, and more sustainable solutions for agriculture by using sustainable, environmentally friendly, modern energy-efficient, and cost-improved solar energy technologies. This comprehensive book is expected to be served as a practical guide to scientists, engineers, decision-makers, and stakeholders who are involved in agriculture and related primary industries, sustainable energy development, and climate change mitigation projects. By including real-scale solar-based agricultural projects implemented around the world in each chapter, highlighting their main associated challenges and benefits, it is expected that the knowledge gap between market/real-world applications and research in this field will be bridged.

Acknowledgment

The authors would like to thank Tarbiat Modares University (http://www.modares.ac.ir) for the received financial support (grant number IG/39705) for the "Renewable Energies Research Group."

References

[1] Food Security | IFPRI: International Food Policy Research Institute; 2021. Available from: https://www.ifpri.org/topic/food-security [accessed 04.08.21].

[2] Islam MS, Kieu E. Tackling regional climate change impacts and food security issues: a critical analysis across ASEAN, PIF, and SAARC. Sustainaibility 2020;12. Available from: https://doi.org/10.3390/su12030883.

[3] Mbow, C., C. Rosenzweig, L.G. Barioni, T.G. Benton, M. Herrero, M. Krishnapillai, et al. Food security Climate change and land: an IPCC special report on climate change, desertification, land degradation, sustainable land management, food security, and greenhouse gas fluxes in terrestrial ecosystems, 2019, pp. 437–550.

[4] The State of Food Security and Nutrition in the World 2021. Transforming food systems for food security, improved nutrition and affordable healthy diets for all; 2021.

[5] Ortiz-Bobea A, Ault TR, Carrillo CM, Chambers RG, Lobell DB. Anthropogenic climate change has slowed global agricultural productivity growth. Nat Clim Chang 2021; 11:306–12. Available from: https://doi.org/10.1038/s41558-021-01000-1.

[6] Molua EL, Lambi CM. The economic impact of climate change on agriculture in Cameroon; 2007.

[7] Sadeg SA, Al-samarrai K. Securing foods in Libya concepts, challenges and strategies. Conf Pap; 2019.

[8] Food and Agriculture Organisation of the United Nations. Food security and nutrition in the world the state of transforming food systems for affordable healthy diets; 2020.

[9] FAO Report on World Hunger: more than 820 million people are hungry; 2021. Available from: https://farmpolicynews.illinois.edu/2019/07/fao-report-on-world-hunger-more-than-820-million-people-are-hungry/.

[10] Barry Desker, Caballero-Anthony Mely, Paul Teng. Thought/issues paper on ASEAN food security: towards a more comprehensive framework. ERIA Discuss Pap Ser 2013;20.

[11] Mahlknecht J, González-Bravo R, Loge FJ. Water-energy-food security: a Nexus perspective of the current situation in Latin America and the Caribbean. Energy 2020; 194:116824. Available from: https://doi.org/10.1016/j.energy.2019.116824.

[12] Lee L-C, Wang Y, Zuo J. The nexus of water-energy-food in China's tourism industry. Resour Conserv Recycl 2021;164:105157. Available from: https://doi.org/10.1016/j.resconrec.2020.105157.

[13] IRENA. Renewable energy in the water, energy and food nexus. Int Renew Energy Agency 2015;1–125.

[14] Dubois O, Flammini A, Kojakovic A, Maltsoglou I, Puri M, Rincon L, et al. Energy Access Food Agriculture 2017;1–15.

[15] Harchaoui S, Chatzimpiros P. Can agriculture balance its energy consumption and continue to produce food? a framework for assessing energy neutrality applied to French agriculture. Sustain 2018;10. Available from: https://doi.org/10.3390/su10124624.

[16] Hoff H. Understanding the Nexus. Background paper for the Bonn2011 Nexus conference: The Water, Energy and Food Security Nexus. In: Background paper for the Bonn2011 Nexus conference: The Water, Energy and Food Security Nexus). Stockholm; 2011.

[17] Sarkodie SA, Owusu PA. Bibliometric analysis of water–energy–food nexus: sustainability assessment of renewable energy. Curr Opin Env Sci Heal 2020;13:29–34. Available from: https://doi.org/10.1016/j.coesh.2019.10.008.

[18]   Da Silva CA, Baker D, Shepherd AW, Jenane C, Miranda-da-Cruz S. Agro-industries for development. Wallingford: CABI; 2009. Available from: https://doi.org/10.1079/9781845935764.0000.

[19]   Hu J, Zhang J, Mei M, Yang Wmin, Shen Q. Quality control of a four-echelon agri-food supply chain with multiple strategies. Inf Process Agric 2019;6:425−37. Available from: https://doi.org/10.1016/j.inpa.2019.05.002.

[20]   Sid S, Mor RS, Panghal A, Kumar D, Gahlawat VK. Agri-food supply chain and disruptions due to COVID-19: effects and strategies. Braz J Oper Prod Manag 2021;18: e20211148. Available from: https://doi.org/10.14488/BJOPM.2021.031.

[21]   Dania WAP, Xing K, Amer Y. Collaboration behavioural factors for sustainable agri-food supply chains:a systematic review. J Clean Prod 2018;186:851−64. Available from: https://doi.org/10.1016/j.jclepro.2018.03.148.

[22]   Tsolakis NK, Keramydas CA, Toka AK, Aidonis DA, Iakovou ET. Agrifood supply chain management: a comprehensive hierarchical decision-making framework and a critical taxonomy. Biosyst Eng 2014;120:47−64. Available from: https://doi.org/10.1016/j.biosystemseng.2013.10.014.

[23]   Oltra-Mestre MJ, Hargaden V, Coughlan P, Segura-García, del Río B. Innovation in the Agri-Food sector: exploiting opportunities for Industry 4.0. Creat Innov Manag 2021;30:198−210. Available from: https://doi.org/10.1111/caim.12418.

[24]   Pérez Perales D, Verdecho M-J, Alarcón-Valero F. Enhancing the sustainability performance of agri-food supply chains by implementing industry 4.0. Collab Netw Digit Transform 2019;496−503. Available from: https://doi.org/10.1007/978-3-030-28464-0_43.

[25]   Aramyan L, Ondersteijn CJM, KOOTEN OVAN, Oude Lansink A. Performance indicators in agri-food production chains. Quantifying agri-food supply chain. the Netherlands: Springer; 2006. p. 49−66. Available from: https://doi.org/10.1007/1-4020-4693-6_5.

[26]   Becerril H, de los Rios I. Energy efficiency strategies for ecological greenhouses: experiences from Murcia (Spain). Energies 2016;9:866. Available from: https://doi.org/10.3390/en9110866.

[27]   Flammini A, Bracco S, Sims R, Cooke J, Elia A. Costs and benefits of clean energy technologies in the milk, vegetable and rice value chains: intervention level. Food and Agriculture Organization of the United Nations, Deutsche Gesellschaft für Internationale Zusammenarbeit (GIZ) GmbH; 2018.

[28]   Sims R, Flammini A, Puri M, Bracco S. Opportunities agri-food chains become energy-smart 2016;43.

[29]   Introducing the Energy-Agriculture Nexus - energypedia. Info; 2021. Available from: https://energypedia.info/wiki/Introducing_the_Energy-Agriculture_Nexus [accessed 03.09.2021].

[30]   Marshall Z, Brockway PE. A net energy analysis of the global agriculture, aquaculture, fishing and forestry system. Biophys Econ Sustain 2020;5:9. Available from: https://doi.org/10.1007/s41247-020-00074-3.

[31]   Pellegrini P, Fernández RJ. Crop intensification, land use, and on-farm energy-use efficiency during the worldwide spread of the green revolution. Proc Natl Acad Sci USA 2018;115:2335−40. Available from: https://doi.org/10.1073/pnas.1717072115.

[32]   Sauer T, Havlík P, Schneider UA, Schmid E, Kindermann G, Obersteiner M. Agriculture and resource availability in a changing world: the role of irrigation. Water Resour Res 2010;46. Available from: https://doi.org/10.1029/2009WR007729.

[33]   Anticamara JA, Watson R, Gelchu A, Pauly D. Global fishing effort (1950−2010): trends, gaps, and implications. Fish Res 2011;107:131−6. Available from: https://doi.org/10.1016/j.fishres.2010.10.016.

[34] Data tables − Data & Statistics - IEA n.d.

[35] Rokicki T, Perkowska A, Klepacki B, Bórawski P, Bełdycka-Bórawska A, Michalski K. Changes in energy consumption in agriculture in the EU countries. Energies 2021;14:1570. Available from: https://doi.org/10.3390/en14061570.

[36] Lin H-C, Huber JA, Gerl G, Hülsbergen K-J. Effects of changing farm management and farm structure on energy balance and energy-use efficiency—a case study of organic and conventional farming systems in southern Germany. Eur J Agron 2017;82:242−53. Available from: https://doi.org/10.1016/j.eja.2016.06.003.

[37] Wu S, Ding S. Efficiency improvement, structural change, and energy intensity reduction: evidence from Chinese agricultural sector. Energy Econ 2021;99:105313. Available from: https://doi.org/10.1016/j.eneco.2021.105313.

[38] Jiang M, Hu X, Chunga J, Lin Z, Fei R. Does the popularization of agricultural mechanization improve energy-environment performance in China's agricultural sector? J Clean Prod 2020;276:124210. Available from: https://doi.org/10.1016/j.jclepro.2020.124210.

[39] The future of food and agriculture−Trends and challenges. Rome; 2017.

[40] Richards M, Arslan A, Cavatassi R, Rosenstock T. Climate change mitigation potential of IFAD investments. IFAD Research Series 35. Rome; 2019.

[41] Vermeulen SJ, Campbell BM, Ingram JSI. Climate change and food systems. Annu Rev Env Resour 2012;37:195−222. Available from: https://doi.org/10.1146/annurev-environ-020411-130608.

[42] Rojas-Downing MM, Nejadhashemi AP, Harrigan T, Woznicki SA. Climate change and livestock: impacts, adaptation, and mitigation. Clim Risk Manag 2017;16:145−63. Available from: https://doi.org/10.1016/j.crm.2017.02.001.

[43] Mrówczyńska-Kamińska A, Bajan B, Pawłowski KP, Genstwa N, Zmyślona J. Greenhouse gas emissions intensity of food production systems and its determinants. PLoS One 2021;16:1−20. Available from: https://doi.org/10.1371/journal.pone.0250995.

[44] Food production is responsible for one-quarter of the world's greenhouse gas emissions - Our World in Data; 2021. Available from: https://ourworldindata.org/food-ghg-emissions [accessed 12.08.21].

[45] Dickie A, Streck C, Roe S, Zurek M, Haupt F, Dolginow A. Strategies for mitigating climate change in agriculture: abridged report; 2014.

[46] Sustainable Food and Agriculture | Food and Agriculture Organization of the United Nations. Food Agric Organ United Nations; 2021. Available from: http://www.fao.org/sustainability/en/ [accessed 05.08.21].

[47] Weil RR. Defining and using the concept of sustainable agriculture. J Agron Educ 1990;19:126−30. Available from: https://doi.org/10.2134/jae1990.0126.

[48] Laurett R, Paço A, Mainardes EW. Antecedents and consequences of sustainable development in agriculture and the moderator role of the barriers: Proposal and test of a structural model. J Rural Stud 2021;. Available from: https://doi.org/10.1016/j.jrurstud.2021.06.014.

[49] Rogelj J, den Elzen M, Höhne N, Fransen T, Fekete H, Winkler H, et al. Paris Agreement climate proposals need a boost to keep warming well below 2 °C. Nature 2016;534:631−9. Available from: https://doi.org/10.1038/nature18307.

[50] Adewale C, Reganold JP, Higgins S, Evans RD, Carpenter-Boggs L. Agricultural carbon footprint is farm specific: case study of two organic farms. J Clean Prod 2019;229:795−805. Available from: https://doi.org/10.1016/j.jclepro.2019.04.253.

[51] Mantoam EJ, Angnes G, Mekonnen MM, Romanelli TL. Energy, carbon and water footprints on agricultural machinery. Biosyst Eng 2020;198:304−22. Available from: https://doi.org/10.1016/j.biosystemseng.2020.08.019.

[52] FAO. Global database of GHG emissions related to feed crops: a life cycle inventory; 2017.

[53] IRENA. World energy transitions outlook; 2021.

[54] Renewables 2021, Global Status Report. Paris, France: REN21; 2021.

[55] Fragkos P. Global energy system transformations to 1.5°C: the impact of revised intergovernmental panel on climate change carbon budgets. Energy Technol 2020;8:2000395. Available from: https://doi.org/10.1002/ente.202000395.

[56] IRENA. Renewable capacity statistics 2021. International Renewable Energy Agency (IRENA). Abu Dhabi; 2021.

[57] Perspectives for a clean energy transition. The critical role of buildings; 2019.

[58] Gielen D, Boshell F, Saygin D, Bazilian MD, Wagner N, Gorini R. The role of renewable energy in the global energy transformation. Energy Strateg Rev 2019;24:38−50. Available from: https://doi.org/10.1016/j.esr.2019.01.006.

[59] Gorjian S, Ebadi H. Introduction. In: Shukla SGA, editor. Photovoltaic solar energy conversion. Elsevier; 2020. p. 1−26. Available from: https://doi.org/10.1016/B978-0-12-819610-6.00001-6.

[60] Green MA. How did solar cells get so cheap? Joule 2019;3:631−3. Available from: https://doi.org/10.1016/j.joule.2019.02.010.

[61] Renewable energy market update. OECD; 2020. Available from: https://doi.org/10.1787/afbc8c1d-en.

[62] Masson G, Kaizuka I. Trends in photovoltaic applications 2020. Report IEA-PVPS T1-38:2020; 2020.

[63] Weiss W, Monika Spörk-Dür. Solar heat worldwide 2020—global market development and trends in 2019. vol. 24. Gleisdorf, Austria; 2020.

[64] Weiss W, Spörk-Dür M. Solar heat worldwide, global global market development and trends in 2020, Detailed Market Data 2019. AEE—Institute for Sustainable Technologies 8200 Gleisdorf, Austria; 2021.

[65] Răboacă MS, Badea G, Enache A, Filote C, Răsoi G, Rata M, et al. Concentrating solar power technologies. Energies 2019;12. Available from: https://doi.org/10.3390/en12061048.

[66] Islam MT, Huda N, Abdullah AB, Saidur R. A comprehensive review of state-of-the-art concentrating solar power (CSP) technologies: current status and research trends. Renew Sustain Energy Rev 2018;91:987−1018. Available from: https://doi.org/10.1016/j.rser.2018.04.097.

[67] Solar energy mapping the road ahead; 2019.

[68] Chel A, Kaushik G. Renewable energy for sustainable agriculture. Agron Sustain Dev 2011;31:91−118. Available from: https://doi.org/10.1051/agro/2010029.

[69] Gorjian S, Zadeh BN, Eltrop L, Shamshiri RR, Amanlou Y. Solar photovoltaic power generation in Iran: development, policies, and barriers. Renew Sustain Energy Rev 2019;106:110−23. Available from: https://doi.org/10.1016/j.rser.2019.02.025.

[70] Shakouri M, Ebadi H, Gorjian S. Solar photovoltaic thermal (PVT) module technologies. In: Gorjian S, Shukla A, editors. Photovolatic solar energy conversion. 1st (ed.) Elsevier; 2020. p. 79−116. Available from: https://doi.org/10.1016/B978-0-12-819610-6.00004-1.

[71] Gorjian S, Ebadi H, Trommsdorff M, Sharon H, Demant M, Schindele S. The advent of modern solar-powered electric agricultural machinery: a solution for sustainable farm operations. J Clean Prod 2021;292:126030. Available from: https://doi.org/10.1016/j.jclepro.2021.126030.

[72] Ozoegwu CG, Akpan PU. Solar energy policy directions for safer and cleaner development in Nigeria. Energy Policy 2021;150:112141. Available from: https://doi.org/10.1016/j.enpol.2021.112141.

[73] Guo S, Liu Q, Sun J, Jin H. A review on the utilization of hybrid renewable energy. Renew Sustain Energy Rev 2018;91:1121−47. Available from: https://doi.org/10.1016/j.rser.2018.04.105.

[74] Upadhyay S, Sharma MP. A review on configurations, control and sizing methodologies of hybrid energy systems. Renew Sustain Energy Rev 2014;38:47−63. Available from: https://doi.org/10.1016/j.rser.2014.05.057.

[75] Gul M, Kotak Y, Muneer T. Review on recent trend of solar photovoltaic technology. Energy Explor Exploit 2016;34:485−526. Available from: https://doi.org/10.1177/0144598716650552.

[76] Timilsina GR, Kurdgelashvili L, Narbel PA. Solar energy: markets, economics and policies. Renew Sustain Energy Rev 2012;16:449−65. Available from: https://doi.org/10.1016/j.rser.2011.08.009.

[77] Gorjian S, Singh R, Shukla A, Mazhar AR. On-farm applications of solar PV systems. In: Gorjian S, Shukla ABT-PSEC, editors. Photovolatic solar energy conversion. Elsevier; 2020. p. 147−90. Available from: https://doi.org/10.1016/b978-0-12-819610-6.00006-5.

[78] Toledo C, Scognamiglio A. Agrivoltaic systems design and assessment: a critical review, and a descriptive model towards a sustainable landscape vision (three-dimensional agrivoltaic patterns). Sustainability 2021;13:6871. Available from: https://doi.org/10.3390/su13126871.

[79] Pringle AM, Handler RM, Pearce JM. Aquavoltaics: synergies for dual use of water area for solar photovoltaic electricity generation and aquaculture. Renew Sustain Energy Rev 2017;80:572−84. Available from: https://doi.org/10.1016/j.rser.2017.05.191.

[80] Faninger G. Solar hot water heating systems. Comprehensive renewable energy, vol. 3. Elsevier; 2012. p. 419−47. Available from: https://doi.org/10.1016/B978-0-08-087872-0.00312-7.

[81] Zheng R, Yin Z. Solar thermal heating and cooling in China. Renewable heating and cooling. Elsevier; 2016. p. 221−39. Available from: https://doi.org/10.1016/B978-1-78242-213-6.00010-2.

[82] Gorjian S, Ghobadian B, Ebadi H, Ketabchi F, Khanmohammadi S. Applications of solar PV systems in desalination technologies. Photovolatic solar energy conversion. Elsevier; 2020. p. 237−74. Available from: https://doi.org/10.1016/B978-0-12-819610-6.00008-9.

[83] Gorjian S, Minaei S, MalehMirchegini L, Trommsdorff M, Shamshiri RR. Applications of solar PV systems in agricultural automation and robotics. In: Gorjian S, Shukla A, editors. Photovolatic solar energy conversion First. London: Elsevier; 2020. p. 191−235. Available from: https://doi.org/10.1016/B978-0-12-819610-6.00007-7.

[84] Karami E, Rafi M, Haibaoui A, Ridah A, Hartiti B, Thevenin P. Performance analysis and comparison of different photovoltaic modules technologies under different climatic conditions in Casablanca. J Fundam Renew Energy Appl 2017;07. Available from: https://doi.org/10.4172/2090-4541.1000231.

[85] Kihlström V, Elbe J. Constructing markets for solar energy—a review of literature about market barriers and government responses. Sustainability 2021;13:3273. Available from: https://doi.org/10.3390/su13063273.

[86] Zakeri B, Cross S, Dodds PE, Gissey GC. Policy options for enhancing economic profitability of residential solar photovoltaic with battery energy storage. Appl Energy 2021;290:116697. Available from: https://doi.org/10.1016/j.apenergy.2021.116697.

[87] Aprà FM, Smit S, Sterling R, Loureiro T. Overview of the enablers and barriers for a wider deployment of CSP tower technology in Europe. Clean Technol 2021;3. Available from: https://doi.org/10.3390/cleantechnol3020021.

[88] Labordena M, Patt A, Bazilian M, Howells M, Lilliestam J. Impact of political and economic barriers for concentrating solar power in sub-Saharan Africa. Energy Policy 2017;102:52−72. Available from: https://doi.org/10.1016/j.enpol.2016.12.008.

[89] Del Río P. Encouraging the implementation of small renewable electricity CDM projects: an economic analysis of different options. Renew Sustain Energy Rev 2007;11:1361−87. Available from: https://doi.org/10.1016/j.rser.2005.12.006.

[90] Troost C, Walter T, Berger T. Climate, energy and environmental policies in agriculture: simulating likely farmer responses in Southwest Germany. Land Use Policy 2015;46:50−64. Available from: https://doi.org/10.1016/j.landusepol.2015.01.028.

[91] Renewable energy for smallholder agriculture (RESA). Rome, Italy; 2020.

Chapter 2

Solar energy conversion technologies: principles and advancements

Seyed Soheil Mousavi Ajarostaghi[1] and Seyed Sina Mousavi[2]
[1]*Faculty of Mechanical Engineering, Babol Noshirvani University of Technology, Babol, Iran,*
[2]*Faculty of Civil Engineering, Babol Noshirvani University of Technology, Babol, Iran*

2.1 Sustainable energy technologies

The employment of energy in human societies is growing to improve the quality of life and economic elements. Utilizing energy resources is an important topic in the present era. Today, approximately 3.5 billion people worldwide do not have access to electricity [1]. Depend on technology, increasing the quality of life and the continuous growth in the global population, especially in developing countries, has dramatically increased the energy demand. In this regard, the consumption of fossil fuels is rising, leading to destructive environmental impacts and increasing health risks to living organisms on Earth [2]. Greenhouse gases (GHGs) in the atmosphere absorb and reflect infrared rays (IRs), resulting in the lower part of the atmosphere and the surface of the earth are maintained warm. The global average atmospheric carbon dioxide (CO_2) has been increased from 280 ppm in the preindustrial period to 409.8 ppm in 2019 [3].

To keep the earth safe and deal with potential environmental threats, sustainable and pollution-free technologies have been introduced, known as renewable energy technologies (RETs). Energy sources can be divided into three main categories: (1) *fossil fuels*, (2) *nuclear energy*, and (3) *renewable energies* [4]. Renewable energy sources (RESs) refer to the types of energy that, unlike nonrenewable energies (fossil and nuclear), can be re-created or renewed by nature in a short period [5]. The leading types of renewable energies are solar [6], wind [7], geothermal [8,9], marine energy [10,11], biomass [12], and biofuels [13]. RESs can provide zero or almost zero percent pollution. RETs are reliable, cost-effective, and environment-friendly methods to meet the energy requirements of rural areas on a small scale. Some of the

Solar Energy Advancements in Agriculture and Food Production Systems.
DOI: https://doi.org/10.1016/B978-0-323-89866-9.00005-5
29

RETs are currently applicable, but many of them are still under development. In 2021 RESs account for about 7.23% of the world's total energy supply compared to the share of 2% in 1998 [14].

2.1.1 Solar energy harvesting

The importance of the sun in providing life is probably recognized to humans in all ancient civilizations so that the Babylonians, the ancient Hindus, the Persians, and the Egyptians worshiped the sun. According to the historical archives, the ancient Greeks were pioneers in using passive solar devices in their homes and undoubtedly experimented with using solar energy in various methods. In the 2nd century BC, there is a story that Archimedes reflected the sun's rays from the shiny bronze shields to a focal point and thus could set fire to enemy ships. The Romans maintain the tradition of using the sun to allow the sun's heat to be trapped by a kind of glass. The Romans even passed a law criminalizing obscuring a neighbor's access to sunlight [15].

In early 1818, scientists discovered that the electrical conductivity of some substances, such as selenium, is increased several times once exposed to sunlight. However, in the 1950s, researchers working on transistors in *Bell Telephone Laboratories* found that silicon could be used as an effective solar cell. This shortly led to the use of silicon solar cells in spacecraft. In 1958 "Vanguard 1" was the initial satellite to use this new development. This program paved the way for more research on better and cheaper solar cells. In the 1970s, after rapid growth in oil prices, working and focusing on this field was significantly improved. In 1977 the US government established the *National Renewable Energy Laboratory (NREL)*. Another sign of the rapid growth of silicon solar cell technology was the construction of the first solar park in California in 1982, which could produce 1 MW of power. One year later, another larger Canadian sun park with a total capacity of 5.2 MW was established. The United States has built several photovoltaic (PV) power plants in the range of 250−550 MW. After 34 years passed from the time that first solar farm was built in California, China made a solar farm with a capacity of 850 MW. Additionally, at the end of 2016, the global capacity of solar PV production was more than 300 GW. In 2017 Chinese companies planned to spend US$1 billion to build a large solar farm (1 GW) on 2500 ha in Ukraine, in a deprived area located in the south of the land and affected by the 1986 nuclear explosions [15].

The received solar energy by the earth (wavelengths between 0.38 and 250 μm) warms the atmosphere and earth's surface, providing energy for every climate zone and ecosystem. This energy heats the molecules of GHGs [such as CO_2 and methane (CH_4)] and water contained in the atmosphere, where most of this thermal energy is emitted into space at different wavelengths between 5 and 50 μm at night [16]. The Earth receives 174 petawatt

(PW) of incoming solar radiation in the upper atmosphere. About 30% of this amount is reflected in space, while the rest is absorbed by the oceans, land, and bodies on Earth. The whole received solar energy by the Earth's atmosphere, oceans, and land masses is approximately 3.85 Yotta joules per annum (YJ/a). Currently, total energy consumption in the world is less than 0.02% of the total solar energy that shines on the earth. Most people in the world live in zones with an insolation level of $3.5-7.0$ kWh/m^2 day. This amount of available solar energy makes it an attractive source for electricity generation [17].

Solar energy can be harnessed using mature, evolving, and innovative technologies including solar heating systems, PV technology, concentrated solar power (CSP), solar ponds, solar cooling systems, solar lighting, and photosynthesis. Some of these technologies have been developed in the last 30 years to reduce the accumulation of atmospheric CO_2 caused by burning fossil fuels. The significance of solar energy is owing to its being inexhaustible and its capability to provide the energy demand of a wide range of small- to large-scale applications [18]. In 2016 renewable energy share in electricity generation was less than a quarter of the world's electricity demand. The available data proves that among different RESs, the growth rate of wind and solar energy in providing the global electricity demand is fast. In 2017 Germany supplied almost all of its electricity requirements through renewable energy, and for 4 days, Portugal runs only on renewable energies [19].

In 2014 solar energy provided about 1% of the total primary energy which was much less than the share of traditional forms of energy or other sources of renewable energy (see Fig. 2.1). According to the report released by the *International Energy Agency* (*IEA*) in 2018, by 2050, the sun will be the largest source of electricity generation in the world considering up to

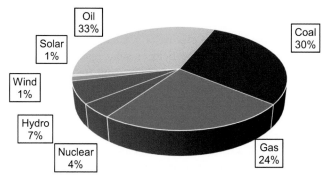

FIGURE 2.1 World's primary energy consumption in 2014. *Data adapted from News − IEA. How solar energy could be the largest source of electricity by mid-century. https://www.iea.org/ news/how-solar-energy-could-be-the-largest-source-of-electricity-by-mid-century; 2021 [accessed 19.08.2021].*

16% of the world's electricity generation by solar PV and 11% by the solar CSP [20]. Hence, achieving this goal entails large investments in currently existing solar technologies and those novel ones that will be emerged in the future.

The world continues to burn more fossil fuels each year, especially for electricity generation. This is mainly due to the overall export of coal from Australia and the United States to Japan and China [16,21]. Instead of importing expensive fuels, investments can be made in local developments of RETs. In addition, new research and development (R&D) in the field of solar energy and other RESs can improve international trade [22]. With more than a third of the world's population (3.5 billion people, mainly in Asia-Pacific, and sub-Saharan Africa) without access to electricity, solar energy holds good promise for improving living standards and reducing GHG emissions [23].

2.1.2 Benefits and challenges of using solar energy

There are three main motivations for humans to be adapted to using renewable energies. First, burning large amounts of coal in power plants results in significant fogging problems, causing smoke and health issues (especially in China because of its heavy reliance on coal power plants), forcing governments around the world (especially Europe and China) to think seriously about reducing the reliance on coal-fired power generation and increasing the development of RETs. Second, pumping oil or gas from both offshore and onshore wells will become more expensive. And the last reason is that the price of renewable energies, especially solar PV, has been significantly dropped as in 2017, it reached less than US$1.56/W (for roof installation) and US$0.86/W (for solar farm installations) compared to US$10/W in 2007. Therefore solar energy has started to offer competitive prices compared to the energy derived from coal, gas, and oil [24,25]. The rapid decline in solar cell prices has led many researchers to confirm that solar PV will shortly play an important role in global electricity generation. Additionally, involved companies which are active in the field of solar energy should make extra efforts to make solar technologies compatible with fossil-based technologies and precisely plan for their future market [26]. It is worthy to be noted that fossil fuels are associated with a high energy density that makes them more convenient to be used while solar energy is a diluted and unequally distributed source. Although solar energy is a clean source with fewer environmental impacts, replacing fossil fuels with renewable energies is not an easy task and requires a precise strategic management program. The main reason is that the fossil fuel industry has been well established so that most human societies entirely depend on it. Additionally, as mentioned before, fossil fuels are highly concentrated energy sources, in all their forms, and can release large amounts of energy compared to renewables. However, concerning environmental impacts and limited remained reservoirs, humans have to be adapted, albeit slowly [27]. Generally, solar energy with the main benefits of being abundant, carbon-free, and capable of providing the highest level of energy security creates no pollution and waste after consumption and is predictable [28]. It has been estimated that only one 50-MW CSP plant based on parabolic collectors can prevent the annual consumption of 30 million liters of fossil fuel and release 90,000 tons of CO_2.

The two main drawbacks of using solar energy are as follows [29]:

1. Solar energy is a diluted source of energy and for instance, producing an average amount of 1 GW electricity from PV under a warm climate, where the peak mid-day available solar energy is 1200 W/m^2 requires a solar PV farm with an area of about $20-25$ km^2, including PV arrays, the proper distance between them, and access roads. In the United Kingdom, each PV solar farm must have an area twofold of the one mentioned above for the same capacity. However, if the area occupied by coal mining is considered, the coal life cycle will occupy more land in comparison with PVs.

2. In many countries, especially those located in temperate regions, the electricity demand is low when the sun shines. Therefore the peak demand usually occurs in the early morning and evening when either solar radiation is not available or the generated power is not sufficient. Under temperate climates, a solar farm is estimated to generate almost 20% of its total capacity, while in areas with intense sunlight (e.g., the Atacama Desert in Chile), the electricity generation exceeds 35%.

Considering the abovementioned issues, supplying solar electricity and injecting the generated power into a national grid becomes a challenging topic. This is because power generation and consumption must be always in balance and therefore, in most cases, energy storage systems at a lower cost through hydraulics, compressed energy storage, and thermal energy storage (TES) will be required to support high sun penetration. In terms of electricity storage, the price of batteries (especially lithium-ion) is also declining mainly because of the growth of electric vehicles (EVs) [18].

The efficiency of solar energy systems is graded according to their performance under the standard test condition with the radiation amount of 1000 W/m^2, corresponding to the maximum radiation expected on a clear summer day at moderate latitudes. The actual level of solar radiation depends on latitude and local climatic conditions, but the average annual density of solar energy in most areas is from 100 to 350 W/m^2. Therefore the capacity factor for solar PV (actual output power/DC rated output power) is between 10% and 35%, depending on the location. The price of generating electricity from solar PV compared to coal was estimated by Lazard [28] (see Fig. 2.2). The analyses demonstrated that PV solar farming costs (predicted to be as low as US$0.86/W in 2017) are lower than coal (estimated to

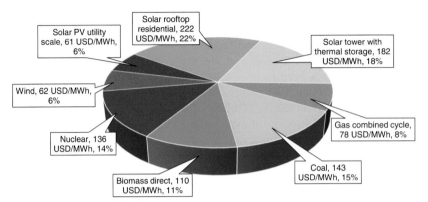

FIGURE 2.2 The maximum cost for electricity generation by different technologies in late 2014. *Data adapted from Energy Innvoation. Comparing the levelized cost of energy technologies; 2021. https://energyinnovation.org/2015/02/07/levelized-cost-of-energy/ [accessed 19.08.2021].*

be in the range of US$1.5−2.5/W) in the same range of gas-fired power stations. However, residential rooftop PV (estimated to be about US$1.56/W in 2017) is not cost-effective for feed into national grids without government subsidies [24].

To absorb enough energy from the sun, the collectors or other solar equipment should be installed on a large area to provide a more exposed surface to the incoming solar irradiation [5]. The crucial drawback of solar energy is the inherent intermittent characteristic of solar radiation which makes its availability climate-, site-, and time-specific. Hence, the amount of available solar radiation depends on the location with specific meteorological features, time of the day, and season of the year [30]. For instance, cloudy weather reduces the electric efficiency of solar PV modules and the thermal performance of solar collectors due to the low amount of available solar radiation. Since solar-based technologies have not been widely commercialized, precise control and management of solar power plants remain unsolved. However, some potential solutions to overcome these limitations are the employment of fossil-based systems and energy storage units as backup facilities.

2.2 Solar energy technologies

Solar energy applications are divided into two main categories of the power plant and nonpower plant applications as presented in Fig. 2.3. Accordingly, it can be seen that the nonpower plant solar energy applications include six main categories of SWH, solar space heating and cooling, solar desalination, solar dryer, solar cocker, and solar furnace (see Section 2.2.1). Moreover,

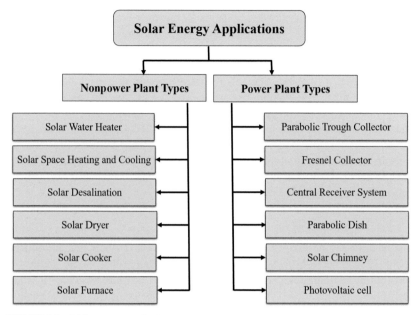

FIGURE 2.3 Different types of solar energy applications [31].

according to this figure, there are six types of power plant solar energy applications including power plants equipped with parabolic trough collectors (PTCs), power plants equipped with Fresnel collectors, central receiver systems, power plants equipped with solar parabolic dishes, solar chimney, and PV power plants (see Section 2.2.2).

2.2.1 Nonpower plant applications

2.2.1.1 Solar water heating

Solar water heaters (SWHs), so-called solar domestic hot water (SDHW) systems, are cost-effective facilities to provide hot water for any applications from domestic to industrial. Solar water heating systems mainly include storage tanks and solar collectors. There are two types of solar water heating systems as *active systems* which have circulating pumps and controls, and *passive systems* contain no additional components [32]. Active solar water heating systems are also categorized based on using direct and indirect circulation systems. In the former type, pumps circulate household water through the collectors and transfer it to the target consumer. They can work appropriately under climates that freezing hardly occurs. In the latter type, pumps circulate a nonfreezing heat transfer fluid (HTF) through the collectors and a heat exchanger (HE), heating the water that runs into the home. This type is widely used in climates that water is disposed to freeze. Passive solar water heating systems are usually less expensive, but they are not generally effective, instead, they offer more reliability and may last longer [33]. There are two basic passive systems, including (1) *integral collector-storage* and (2) *thermosyphon* systems. The first one works best in zones where temperatures hardly fall below freezing. They also work efficiently in households with significant daytime and evening hot water needs. In thermosyphon systems, water flows when warm water increases as cool water sinks. In these systems, the collector needs to be installed below the storage tank to increase the amount of warm water in the tank. Although these systems are reliable, contractors should pay careful attention to the roof design due to the heavy storage tank. They are usually more expensive as compared to integral collector-storage passive systems [15].

2.2.1.2 Solar space heating and cooling

The heating and cooling of buildings using solar energy was first introduced in the 1930s and significantly improved in less than a decade. By adding an absorption refrigeration system to solar water heating systems, a solar absorption refrigeration system can be achieved, providing water heating application, as well as space heating and cooling in buildings during hot seasons [15]. Solar space heating systems are analogous to solar water heating systems but commonly include larger solar collectors' areas and storage units, and a more sophisticated design. These heating systems can use a non-toxic liquid, water, or air as a HTF. The heated liquid or air is then circulated within residential areas to provide space heating [34]. Another solar space heating technology utilizes transpired solar collectors (TSCs) installed at the exterior south-facing wall of buildings. The perforations in these collectors permit air to cross and be heated. This solar-heated air is then directed to the building's ventilation system [35]. There are two types of solar cooling systems: (1) *desiccant* and (2) *absorption chiller*. In desiccant systems, air passes over a common desiccant or "drying material" such as silica gel to attract moisture from the air and make it more pleasant. The desiccant is renewed by using solar heat to dry out the moist air. The absorption chiller uses solar water heating collectors and a thermal-chemical sorption process to produce air-conditioning without electricity. The process is almost similar to that of a refrigerator, but no compressor is utilized. As a substitute, the absorption cycle is driven by a heated fluid from solar collectors [15].

2.2.1.3 Solar desalination

In desalination systems, freshwater is produced from saltwater through solar-powered thermal evaporation processes or membrane methods. Solar stills are also simple and low-cost devices that work based on evaporation and distillation with the main drawback of low productivity [36]. The desalination technique has been first used in the 1950s in the Middle East and tropical regions around the world. The most effective method to drive desalination plants is using solar energy to supply their heat and electricity demands. In this way, the water can be supplied in locations with no accessibility to freshwater for drinking or other applications. Solar desalination systems are available on both domestic and industrial scales [37].

2.2.1.4 Solar drying

Drying is the removal of some moisture content from food and products to raise their shelf life by preventing the growth of bacteria and mold, reducing the waste of products. Solar dryers use solar energy directly or indirectly to dry materials and also require air flows that are provided naturally or by force to accelerate the drying process. Solar dryers are designed and manufactured in different scales and designs for various products and applications [38].

2.2.1.5 Solar cooking

There is a wide range of solar cookers with different designs. In a typical box-type solar cooker, the sun's rays enter the box through the glass cover and are absorbed by the black coating surface, increasing the box temperature to 88°C. The principle of operation of a solar cooker is to collect direct rays of the sun at a focal point and increase the temperature at that point [39]. Solar cooking is performed by utilizing the sun's ultraviolet (UV) rays, allowing them to enter the cooker where they are then converted into Infrared rays that cannot escape. Infrared radiation energy vibrates vigorously and heats the water, fat, and protein molecules of the food. Solar cooking provides a quick and healthy cooking process and it is easy to operate, without noise, environmental-adapted, and in most cases it is portable. Three main types of solar cookers are *box-type cookers, panel cookers,* and *parabolic cookers* [40].

2.2.1.6 Solar furnace

In the early 18th century, Notora built the first solar furnace in France, setting fire to a wood mound 60 m away. Bessemer, the father of world steel, also provided the required heat in his furnace from solar energy [41]. The most common solar furnace systems utilized flat or curved mirrors to concentrate solar radiation into a focal point. Whenever solar beam rays strike parallelly to the mirrors, they are concentrated at the focal point, accumulating the sun's vast thermal energy at one point and reaching high temperatures. Currently, several solar furnace projects are in operation around the world and due to their ability to provide high-temperature values up to 3500°C are mainly utilized in industrial applications [42]. The main benefits of solar furnaces are their immense heat capabilities, lack of required fuel, and ease of use, and some of their main drawbacks are being expensive with their reliance on the intermittent energy source of sunshine [43].

2.2.2 Power plant applications

A solar thermal power plant is a facility composed of high-temperature solar concentrators that convert absorbed thermal energy into electricity using power generation cycles. In solar thermal power plants, the primary function of solar concentrators is generating the steam required to drive turbines that are connected to generators. Solar thermal power plants consist of two main units as follows [44]:

1. a solar field composed of concentrators that absorbs the sun's rays and produces steam by the absorbed heat;
2. a traditional system that converts steam into electricity using turbine-generators (similar to the operating principle of conventional power plants).

According to the type of employed concentrators and their geometric shapes, solar thermal power plants are divided into several categories as follows [44]:

- solar thermal power plants based on PTCs
- solar thermal power plants based on linear Fresnel reflectors (LFRs)
- solar thermal power plants based on parabolic dish collectors (PDCs)
- solar towers/heliostats
- solar chimney power plants

Where the first two types using line-focus reflectors, while the second two types employ point-focus concentrators.

2.3 Solar energy technology subsystems

2.3.1 Solar thermal collectors

Solar thermal collectors are devices that absorb solar radiation and convert it into heat. Then, the generated heat is transferred by a HTF to provide the heat demand of a specific application [45]. A comprehensive description of solar thermal collectors is provided by Kalogirou [46]. Fig. 2.4 shows a classification of solar thermal collectors. Typical examples of these collectors are also presented in Fig. 2.5. Depending on the temperature range, solar collectors are classified as (1) *low-temperature collectors* (<100°C) which are used in SDHW systems and air heaters; (2) *medium-temperature collectors* (100°C−300°C) which are utilized for air heating applications in offices, hospitals, and food industry; and (3) *high-temperature collectors* (> 300°C) which are mostly employed in high-temperature industrial applications and solar thermal power plants [53]. Low-temperature solar thermal collectors are typically in the form of unglazed flat plates coated with black color which are commonly used to heat swimming pools. While glazed flat-plate collectors (FPCs) and evacuated tube collectors (ETCs) are categorized as

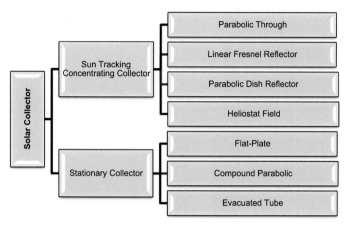

FIGURE 2.4   Classification of solar thermal collectors [31].

medium-temperature collectors. Medium-temperature collectors are employed in a wide range of applications including water/air heating applications for residential and commercial uses [35].

In CSP plants, usually, mirrors or lenses are utilized to concentrate a large area of sunlight onto a relatively small collection area. In this way, the light-collecting area is exposed to sunlight where its temperature rises. Therefore solar energy which is converted into heat is used to drive a heat engine (usually a steam turbine engine) to generate electricity. Generally, solar collectors utilized the absorbed thermal energy to generate steam and then drive the steam turbine to produce electricity [44].

2.3.2   Photovoltaic technology

PV technology converts sunlight directly into electricity using semiconductor materials. The PV effect is a process in which two dissimilar materials in close contact produce an electrical voltage when struck by light or other radiant energy [54]. It was first observed by Henri Becquerel in 1839 when he immersed a sheet of platinum (Pt) coated with a thin layer of silver chloride in an electrolytic solution, and then illuminated the sheet while it was connected to a counter electrode [5]. The primary device for photo-electrical conversion is a solar cell. A solar cell is a semiconductor device that directly converts solar energy into electricity through the PV effect. In PV electricity generation when the sun illuminates a solar cell, the electrons present in the valence band absorb energy, being excited and jump to the conduction band. These highly excited electrons diffuse, generating an electromotive force, and thus some of the light energy is converted into electricity [55]. The schematic view of a solar cell is shown in Fig. 2.6.

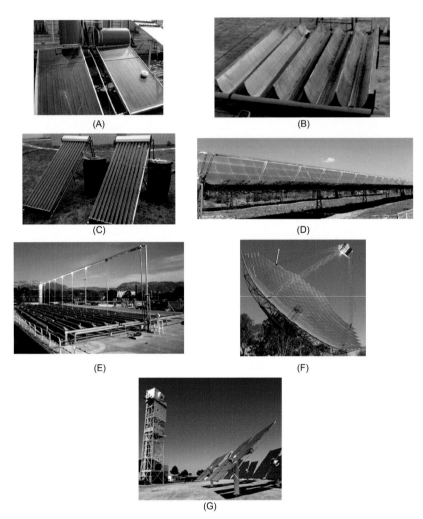

FIGURE 2.5 Photos of various types of solar thermal collectors: (A) flat-plate collectors (FPCs) [31], (B) compound parabolic concentrator (CPCs) [47], (C) evacuated tube collector (ETCs) [48], (D) parabolic through concentrators (PTCs) [49], (E) linear Fresnel reflectors (LFRs) [50], (F) parabolic dish concentrator (PDC) [51], and (G) heliostat field (solar tower) [52].

Since the awareness of the PV effect in the first half of the 19th century, various technologies were developed for the fabrication of solar cells. Before the 1980s, most of these methods, due to their low efficiency, remained at the laboratory scale and were not emerged in the industrial arena except for specific devices such as satellites and limited capacities such as multikilowatt home PV systems. From the mid-1980s, fabrication of the first solar

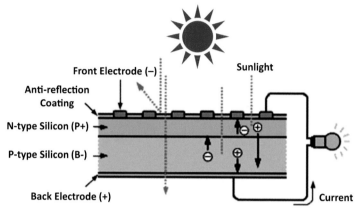

FIGURE 2.6 Schematic structure of a solar cell [5].

cells with an efficiency of 20% and rising fuel prices at one time made entering this technology into the commercial scale more realistic, paving the way to justify the employment of this technology in large-scale applications [56,57].

2.3.2.1 Solar cell materials

The most commonly used materials to fabricate commercial solar cells are silicon (Si), cadmium telluride (CdTe), copper indium gallium selenide (CIGS), and gallium arsenide (GaAs) [2,58,59]. Solar cells which are made of crystalline silicon (c-Si) are commercially available on the market are classified into two main categories of crystalline and amorphous silicon (a-Si) (thin-film) solar cells. The c-Si solar cells are also classified as monocrystalline (mono-Si) and poly-crystalline (poly-Si). The manufacturing process of mono-Si is more complex than poly-Si, causing a relatively high price of PV modules about US$0.29−0.38/Wp compared to poly-Si with the price range of US$0.18−0.19/Wp [60]. The highest confirmed the efficiency of PV cells is 26.7% for mono-Si technology, 22.3% for poly-Si cells, and from 10.2% to 21.7% for thin-film technology [61]. The a-Si solar cells have a lightweight and flexible structure that makes them more suitable to be installed on curved surfaces such as the roof of greenhouses or building facades. Although c-Si solar cells offer higher values of efficiency, their main drawback is that if their surface is partially blocked from reaching solar radiation, their efficiency and consequently their output power will sharply be reduced. While some researches have shown better performance for a-Si solar cells under low-irradiance and high-temperature conditions. Recently, novel solar cell technologies with higher values of efficiency including semitransparent PV (STPV) cells based on recently developed dye synthesized solar cells (DSSCs) and organic PV (OPV) cells have emerged that can

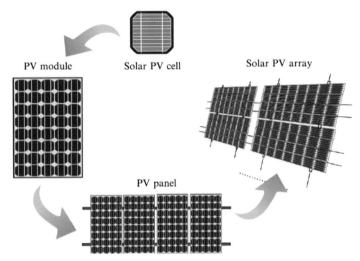

FIGURE 2.7 From a PV cell to a PV array [64]. *Adapted from Wikimedia Commons. File:From a solar cell to a PV system.svg. https://commons.wikimedia.org/wiki/File:From_a_solar_cell_to_a_PV_system.svg; 2021 [accessed 19.08.2021].*

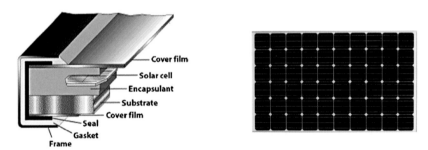

FIGURE 2.8 Typical view of a photovoltaic module [65].

mitigate undesirable shading effects of traditional opaque PV modules and manipulate the light spectrum [62,63].

2.3.2.2 Solar photovoltaic modules

A solar PV module is a collection of solar cells which are mainly connected in series. A single solar cell can generate a very small amount of power in the range of a fraction of 0.1 to 2–3 W. Therefore, to generate electricity in large amounts to fulfill high power requirements, several solar cells are connected to make a solar PV module. In practice, maybe power in large quantities such as kW and MW is required. For this purpose, solar PV modules are connected in series and/or parallel combinations to create arrays (see Fig. 2.7). The main layer of a typical PV module is shown in Fig. 2.8. The main environmental parameters that significantly affect the performance of PV modules are; solar intensity, ambient temperature, wind speed, and dust accumulation [66,67].

In addition to conventional modules, there are concentrating PV (CPV) modules that use curved mirrors or lenses to focus sunlight onto small, highly efficient, multijunction (MJ) solar cells. CPV modules often use sun trackers and sometimes a cooling system to further increase their efficiency. Depending on the concentration ratio, CPV modules are classified as high-concentration CPV (HCPV) and low-concentration CPV (LCPV) modules [68]. Systems using high-concentration PV (HCPV) modules possess the highest value of efficiency among all existing PV technologies, reaching nearly 40% for modules and 30% for systems. They enable the employment of a smaller PV array, reducing land use, waste heat and material, and balance of system (BOS) costs. The rate of annual CPV installations peaked in 2012 and has fallen to near zero since 2018 with the faster price drop in crystalline silicon PVs [69].

2.3.2.3   Components of a solar photovoltaic system

A solar PV system consists of solar PV modules (and in large scales PV arrays) and several other components such as power converters (DC−AC and DC−DC converters), AC and DC isolators, charge controllers, and in some cases battery energy storage systems [70]. In solar PV systems with battery storage, a charge controller is used that regulates the charging and discharging process of batteries. Once required, under cloudy conditions or at nighttime, the energy stored in the battery can be consumed. When AC loads are integrated, DC-generated electricity is converted to AC through an inverter while it can be directly consumed by DC loads. A power meter is also used to record and measure the electricity flow feeding the load [71]. Fig. 2.9 represents a schematic view of a typical solar PV system and its main components. The various components of a typical solar PV system are briefly introduced as follows [70].

2.3.2.3.1   Photovoltaic module

PV modules contains different numbers of solar cells that convert directly solar energy into electricity.

2.3.2.3.2   Inverter or electronic power converter

An inverter is a component that converts DC power generated by PV module (s) to the standard AC power on a voltage and frequency value required by the load(s).

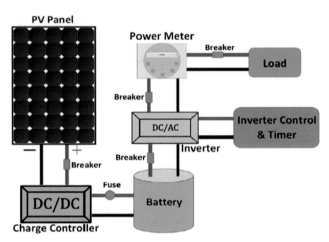

FIGURE 2.9   Schematic view of a typical PV system using battery storage [70].

2.3.2.3.3   Maximum power point tracker

To make the best use of the power produced by PV modules, the point with the highest current and voltage must be chosen. Achieving the optimal point is done by the maximum power point tracker (MPPT). Since the PV output power depends on environmental parameters (i.e., solar irradiance and ambient temperature), an MPPT should be employed to optimize the energy capture.

2.3.2.3.4   Other components

Other components of a PV system are called the BOS. The BOS includes installation and wiring systems so that PV modules can be installed in different places. Additional components will be required for specific installations of PV modules such as roof mounting and floating PV systems.

2.3.2.4   *Different configurations of photovoltaic systems*

PV systems are installed into two main configurations according to their application, limitations, and installation conditions: (1) *on-grid* (grid-connected) solar PV systems, and (2) *off-grid* (stand-alone) solar PV systems as shown in Fig. 2.10A and B. There are also other configurations which are known as hybrid solar PV systems in which the PV system is integrated with other power generation devices such as wind turbines or diesel generators [72]. In grid-connected PV systems, generated DC electricity by PV modules is connected to an inverter via junction boxes, converting DC into AC. On-grid PV systems generally fall into two main groups of *distributed* and *central* PV systems. The former types are connected to a low-voltage distribution grid for small industrial or household applications, while

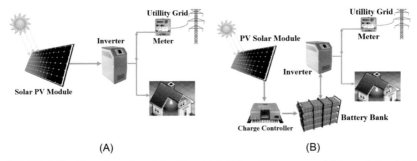

FIGURE 2.10 The schematics of the (A) grid-tied/direct PV system and (B) stand-alone/off-grid PV system [72].

central types are usually large-scale ground-mounted PV systems that are connected to medium- or high-voltage transmission grids. Solar off-grid PV systems are more suitable to be used in locations with no available electricity grid. They usually incorporate battery storage to supply electricity when the sun is not shining. Solar water pumping systems are the major exception to this type [71].

Grid-connected PV systems are installed in many parts of the world in small capacities from 1 to 5 kW on roofs of residential houses and in larger capacities (MW and GW) as PV power plants also known as solar farms or solar parks. The main advantages of on-grid PV systems are as follows [72]:

- easy installation and commissioning
- high efficiency and no need for complex accessories
- no need for batteries to store electrical energy

Grid-connected PV systems equipped with storage systems are especially suitable for residential applications and small businesses because these systems use stored electricity for sensitive loads such as refrigerators, lighting, elevators, water pumps, etc. Under normal conditions, when the grid supplies electricity, the power generated by the PV system is used to charge batteries, while the excess production is injected into the local grid [71]. In case of power surplus, the required power is provided by the grid. In each case, the embedded batteries or other energy storage devices are fully charged. The off-grid PV systems usually provide electricity demand for telecommunication and television stations, residential houses, nomadic tents, and rural cottages [72].

2.3.3 Energy storage technologies

2.3.3.1 Thermal energy storage systems

TES systems are key components especially for large-scale solar CSP plants, mostly, to balance the mismatch between supply and demand but also for other motivations including the difference between electricity price during valley and peak times, and the possibility to shave the consumption peaks. Similarly, if the primary commodity is thermal energy, TES systems are fundamental to balance mismatch between supply and demand, for dealing with price arbitrage and peak shaving strategies, and if there is a mismatch between production and consumption sites. The ways TES systems are operated provide both economic and environmental benefits [73]. A TES system is integrated into the CSP plants to meet the following criteria:

- Reduce sudden fluctuations due to intermittent weather. Although HTFs' thermal inertia can keep the plant going through short-term weather fluctuations such as clouds passing over the solar field, large-scale CSP plants have shown that this thermal inertia may not be sufficient to prevent the turbine from shutting down [74].
- Changing the power generation interval from peak hours of production to peak hours of consumption. The TES system can control the solar CSP plant by storing thermal energy during off-peak consumption hours and using the energy stored during peak consumption hours [75].
- Extend the power generation after the sunset to increase the solar plant's capacity. The solar field with TES is larger than the case without TES [75].

TES systems are divided into three types: (1) *sensible heat*, (2) *latent heat*, and (3) *sorption and chemical energy storage* (*also known as thermochemical*). Typically, TES systems have higher performance values and have lower capital costs as compared to other energy storage solutions [76]. The TES system of the *Solar Two project* [77] showed a thermal efficiency of more than 98% and enable the solar plant to produce power during the night. Typically, TES in commercial CSP plants is molten salt and consists of two storage tanks. The heat is stored and directly extracted from the heat storage tank. Increasing the storage capacity of TES systems in CSP plants can significantly increase the costs, making them unsuitable for long-term energy storage solutions [78].

2.3.3.1.1 Thermal energy storage materials

The TES materials must have excellent thermal and physical properties such as high thermal conductivity, high specific and latent heat, a suitable melting point for the required application, low vapor pressure, thermal and chemical stability, easy availability, and low cost [79]. Based on the state of storage material, TES systems can be divided into two systems: (1) *sensible TES* (STES) systems and (2) *latent TES* (LTES) systems. In STES, no phase change happens during the heat absorption process. The quantity of thermal energy stored depends on the specific heat, volume, temperature variation, and density of TES material. The utilized heat exchange materials used in this type are water [74], molten salts [80,81], thermal oils [80,82], liquid metals [80,82], rocks [83], sand [84], and concrete blocks [85].

In the LTES, phase change materials (PCMs) are utilized to store thermal energy. The PCM is heated until it changes its phase. During the phase change process, the material absorbs a huge amount of thermal energy called the latent heat of fusion. The phase change can be solid-to-liquid, solid-to-gas, or solid-to-solid, nevertheless, latent heat is smaller in solid-to-solid phase change [86]. Before reaching the phase change point, the temperature of the PCM increases, representing the STES. A wide variety of PCMs for different operating temperature ranges are available. Fig. 2.11 shows the classification of different PCMs for LTES. Both inorganic and organic materials can be used in LTES systems. The inorganic materials include metal alloys, saline composites, and salt hydrates, while organic materials comprise paraffin and nonparaffin materials [88]. Organic materials are used for low- and medium-temperature ranges but they are not suitable for high-temperature applications. Generally, inorganic materials can operate at higher temperatures, and their energy storage potential is almost twice the organic material. In this kind of TES system, PCMs include salts (for instance, $LiNO_3$ and KNO_3), salt eutectics (composition of two or more

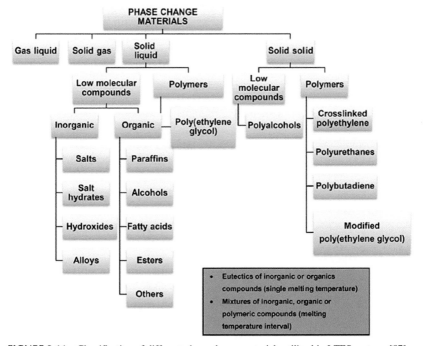

FIGURE 2.11 Classification of different phase change materials utilized in LTES systems [87].

salts), salt hydrates (low-temperature applications), and metals and alloys (costly) [76,79]. In some cases, the water/ice is also utilized as PCM because of its very low cost [89−91].

Table 2.1 presents a list of TES systems being used in large commercial CSP plants (mostly PTC-based). Forty-five out of one hundred existing and under construction PTC-CSP plants are utilizing TES units. Forty-four among them is using a two-tank indirect TES system with molten salt as the storage medium. Only one small plant (3 MW) is using a packed bed TES system with rocks as a storage medium where air at ambient pressure is being used as HTF.

2.3.3.1.2 Different types of thermal energy storage systems

TES systems can also be classified as passive and active [93]. Active TES systems are charged or discharged by forced convection in a HE. Active TES systems can additionally be divided into indirect and direct systems. An indirect system uses two storage mediums, while in a direct TES system, the same material is employed as a storage medium and HTF. In direct TES systems, a proper material with particular properties simultaneously performs as a storage medium and HTF should be chosen. In the passive type, the HTF flows in the TES system to charge or discharge the material used as a storage medium. The HTF transfers the thermal energy captured in the solar field to the storage in the charging process and collects the stored thermal energy from the storage medium during the discharge process. Active types are mostly employed in the current commercial CSP plants [74].

1. Active thermal energy storage systems

Fig. 2.12 shows the schematic diagrams of various types of solar-TES systems in active mode. It can be concluded that generally, there are three configurations, including (1) *dual-tank indirect type*, (2) *dual-tank direct type*, and (3) *single-tank indirect* (thermocline). In the dual-tank indirect type (see Fig. 2.12A), the HTF does not directly store thermal energy. The HTF transfers thermal energy to another fluid as a TES medium. In this type, generally molten salt is employed as a TES material. This indirect type contains two storage tanks, one for the storage medium at higher temperature and another for the lower temperature. The cost of this type is about US $50−80 kWh_{th} [92].

In the dual-tank direct type (see Fig. 2.12B), the hot HTF is directly stored in the hot storage tank to be used after sunset or during cloudy periods. The low-temperature HTF, flowing from the power block, is stored in the cold storage tank before being heated in the solar field [94]. Direct steam and molten salts are the two active direct TES choices for commercial applications [74]. In a direct storage system, when molten salt is also used as HTF, a costly drain-back system is vital throughout the nighttime and cloudy

TABLE 2.1 CSP plants integrated with thermal energy storage implemented around the world till 2019 [92].

Project	Country	Capacity (MW)	Storage material (%)	Storage type	HTF type
Andasol-1,2	Spain	50	*Ms 60—40	*TTI	Dowtherm A
Andasol-3	Spain	50	MS	TTI	Thermal Oil
Arcosol 50	Spain	49.9	Ms 60—40	TTI	*DDO
Arenales	Spain	50	Ms 60—40	TTI	Diphyl
Aste 1A, 1B	Spain	50	Ms 60—40	TTI	Dowtherm A
Astexol II	Spain	50	Ms 60—40	TTI	Thermal Oil
Casablanca	Spain	50	Ms 60—40	TTI	DDO
Extresol-1,2,3 (EX-1,2,3)	Spain	50	Ms 60—40	TTI	DDO
La Dehesa, La Florida, Manchasol-1,2, Thermesol 50	Spain	50	Ms 60—40	TTI	DDO
Solana	United States	280	MS	TTI	Therminol VP-1
SEGS I	United States	13.8	Storage system damaged	TTI	–
Bokpoort	South Africa	55	MS	TTI	Dowtherm A
Ilanga I, Kathu Solar Park, KaXu Solar One, Xina Solar One	South Africa	100	MS	TTI	Thermal Oil

(Continued)

TABLE 2.1 (Continued)

Project	Country	Capacity (MW)	Storage material (%)	Storage type	HTF type
Chabei	China	64	MS	TTI	MS
Delingha and Rayspower Yumen	China	50	MS	TTI	Thermal Oil
Gansu Akesai	China	50	MS	TTI	MS
Gulang Project	China	100	MS	TTI	Thermal Oil
Urat Middle Banner Project	China	100	MS	TTI	Thermal Oil
Yumen CSP Project	China	50	MS	TTI	Thermal Oil
Diwakar	India	100	MS	TTI	Synthetic Oil
KVK Energy Solar	India	100	MS	TTI	Synthetic Oil
Airlight Energy Ait-Baha	Morocco	3	Rocks	Packed bed of rocks	Air at ambient pressure
NOOR I	Morocco	160	MS	TTI	Dowtherm A
NOOR II	Morocco	200	MS	TTI	Thermal Oil
DEWA	UAE	600	MS	TTI	Thermal Oil
Archimede	Italy	5	Ms 60—40	TTI	MS

*Ms, molten salt; TTI, two-tank indirect; DDO, diphenyl/diphenyl oxide.

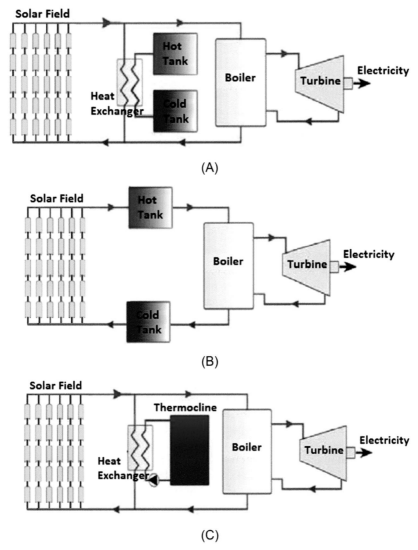

FIGURE 2.12   Various solar thermal energy storage systems in active mode [92]: (A) dual-tank indirect type, (B) dual-tank direct type, (C) single-tank indirect type.

days. However, the experience with large commercial plants demonstrates that using molten salt for a dual-tank direct type is promising [94].

In the single-tank indirect type (see Fig. 2.12C), the same tank contains both hot and cold fluids. The HTF transfers its thermal energy to the TES fluid during the charging cycle. The hot and cold fluids in the single-tank TES can be separated due to temperature layers with cold and hot fluid on the lower and upper levels, respectively. These hot and cold fluid layers are called the thermocline. The formation of thermocline generally needs fillers materials like gravel and quartzite.

2. Passive thermal energy storage systems

The schematic diagrams of passive TES systems are depicted in Fig. 2.13. There are two types of passive TES systems which are explained in the following sections.

Concrete block-based thermal energy storage system*

Fig. 2.13A shows a concrete blocks TES system. Concrete has many advantages, including easy construction, nonflammable, nontoxic, superior mechanical properties, and is low cost [95]. The superior mechanical properties of the concrete eliminate the requirement of the tank, thereby further reducing the storage cost. The heat exchange tubes are permanently immersed in the concrete block and the HTF flows in these tubes to charge and discharge the concrete block. An insulating cover helps to reduce the outer surface heat losses. The volume changes in each charge/discharge cycle adversely affect the heat exchanger pipes, causing gaps ruptures. Moreover, in the case of corrosion, it is extremely difficult to replace the embedded pipes [96].

Packed bed thermal energy storage system

The packed bed TES system consists of an insulated tank containing a packed bed of filler material for heat storage as shown in Fig. 2.13B. It is charged/discharged through the HTF, which transfers its heat by direct contact with the surface not requiring a HE. The HTF fluid can be liquid or gaseous. When the liquid HTF is used, a part of the total thermal energy is also stored by that liquid, and this packed bed system is named a dual storage system. The packed bed system has the advantage of the thermal stratification effect along the bed. When a gaseous HTF, such as air, is employed as HTF, the filler material stores the total heat of the packed bed TES system. Typical HTFs are thermal oils, water, supercritical CO_2 gas, and air. The common bed fillers are gravel, soil, gypsum, rocks, sand, small concrete cubes and spheres, metal blocks, and pebbles [97]. The thermal stability of

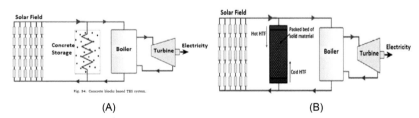

(A)   (B)

FIGURE 2.13 Various solar thermal energy storage systems in passive mode [73]: (A) concrete block thermal energy storage system, (B) packed bed thermal energy storage system.

the filler substance limits the temperature range of the packed bed system. The packed bed systems using earth materials such as sand and gravel have an operating temperature range between 50°C and 300°C [79]. The factors that affect the packed bed system are the ratio of gravel to container size, bed dimensions and porosity, the mass flow rate of HTF, and pressure drop of HTF across the bed [97].

2.3.3.2 Battery storage

With all the new advancements, solar PV is rapidly closing the gap and it is becoming one of the cheapest sources of electricity in most of the regions around the world. The great advantage of CSP plants over PV systems is that the solar energy is converted into heat, which can be rather easily stored. Electricity is more difficult and expensive to store making solar PV electricity not dispatchable as much as CSP. However, battery storage systems are rapidly improving in terms of efficiencies and costs [98].

When integrated with PV, batteries can perform many different services behind and in front of the meter in the form of ancillary services. The most common technologies for stationary battery applications are lead-acid batteries and Li-ion batteries. Concerning the EV applications, nickel−cadmium (NiCd) and nickel−metal hydride (NiMh) batteries are also utilized. However, due to their limited energy density and low autonomy, they have been substituted by Li-ion batteries since 2009 [99,100]. Lead-acid batteries are in use since the late 1800s, being the oldest technology among the others referred to in this section. Since the first appearance of sealed batteries in 1957, known as Valve Regulated Lead-acid (VRLA) batteries, the technology has no much evolution as expected [101]. This type of battery has some limitations, such as (1) a useable capacity between 30% and 50% of the nominal capacity to avoid excessive battery degradation; (2) a very low lifetime of 3−5 years if compared to the PV average lifetime of 25 years; (3) a limited number of cycles during lifetime between 300 and 500 with an 80% depth-of-discharge (DOD); (4) high sensitivity to Peukert's loss showing that the power output required is higher than the one specified by the manufacturer; and (5) the delivered capacity is lower than expected [102]. Despite their limitations, VRLA batteries still dominate the market for PV off-grid applications due to their affordable costs for large installed capacities. Still, they are the main weakness of the system and in recent years start to be substituted by more promising technologies like Li-ion batteries [102]. At present, Li-ion batteries are the major technology applied in EVs and also a perfect candidate for stationary applications. They can be used with a DOD up to 80% of their total capacity, with a large number of cycles vary between 2000 and 5000 according to some manufacturers and higher efficiency with large loads as their power inverters have almost no Peukert's loss [103]. Regarding chemistry, four different cathode cell technology can be highlighted: (1) lithium nickel-cobalt-aluminum (NCA), (2) lithium iron phosphate (LFP), (3) lithium nickel-manganese-cobalt (NMC), and (4) lithium-manganese-oxide−nickel-manganese-cobalt (LMO-NMC).

NCA batteries present a superior behavior concerning calendar life, the highest specific capacity vs cell potential ratio, and low capacity degradation concerning changes in temperature and state of charge (SOC) [104]. LFP degradation is more temperature-driven as compared to SOC-driven, resulting in noticeable power and capacity losses at high temperatures. LFP is the safest regarding the possible thermal release, however, this comes with a trade-off of reducing specific energy and specific capacity against cell potential ratio [105]. The key properties of NMC and LMO-NMC batteries come specifically from the NMC technology. However, LMO is used to further improve safety and increase specific capacity. NMC and LMO-NMC show comparable behavior in capacity loss degradation, and the power loss is highly affected by high SOCs and temperatures (above 50°C). However, the influence of the SOC tends to be less present in pure NCM technology. Compared to NCA and LFP, the NMC and LMO-NCM have the second-best specific capacity versus cell potential ratio and the second-worst thermal characteristics [106]. Table 2.2 provides an overview of the key properties of the reviewed battery technologies.

2.4 Applications of nanofluids in solar systems

In recent years, there has been a growing tendency to use nanomaterials in engineering applications [109]. If the nanoparticles are uniformly scattered, and stably floated in base fluids, resulting in the base fluid's thermal properties being significantly increased. The nanoparticles ranging from 1 to 100 nm in colloidal combination with a base liquid (nanoparticle fluid suspensions) are called nanofluid, and they were primarily invented by Choi in 1995 [110] to develop HTFs with supreme thermal properties as compared to conventional particle fluid suspensions [111,112].

Depending on the application, nanoparticles are being produced with different substances, including carbide ceramics, semiconductors, oxide ceramics, metals, nitride ceramics, carbon nanotubes, and compound materials such as nanoparticle core polymer shell composites, and alloyed nanoparticles [113]. Apart from nonmetallic, metallic, and other materials to produce nanoparticles, entirely new substances and structures are being applied as "doped." The goal is to achieve the most outstanding thermal properties correlated with minimum probable volume fraction ($\varphi < 1\%$). Hence, the suspension of approximately nonaccumulated or mono-dispersed nanoparticles in liquids is vital to enhance heat transfer [31].

Nanofluids have been widely employed in several applications. They are becoming the next generation of HTFs since they present new possibilities to

TABLE 2.2 An overview of battery technologies [102,107,108].

Specifications	Lead-acid	Lithium-ion			Lithium-air
		NCA	LFP	NCM	
Specific energy/capacity (Wh/kg)	30–50	200–250	90–140	140–200	3,500
Cycle life	300–500	1000–1500	2000	1000–2000	10–300
Charge/discharge efficiency (%)	50–95	80–90	80–90	80–90	60–90
Installation Cost (US $/kWh)	200	350	580	420	Under research
Nominal cell voltage (V)	2	3.6	3.3	3.8	2.96
Maintenance	3–6 Months	Not required	Not required	Not required	Under research
Thermal protection complexity	Thermally Stable	Very High	Low	High	High
First appearance in	Late 1800s	1999	1999	2006	1996
Applications	Power Grid	EVs and power grid	EVs and power grid	EVs, electronic devices, and power grid	EVs, electronic devices, and power grid

improve the heat transfer efficiency compared to pure fluids [31,114]. Nanofluids applications among different fields are shown in Fig. 2.14.

Fig. 2.15 shows the percentage contributions of several nanofluids employed by researchers for solar applications. TiO_2, CuO, and Al_2O_3 were marked as the most used nanofluids in evacuated collectors, while CeO_2, WO_3, Silver, Ag, CeO_2, GNP, and Cu have been rarely applied. Many researchers have presented that utilizing nanofluids could enhance solar collectors' performances. Sabiha et al. [115] asserted the maximum efficiency improvement of 93.43% for heat-pipe HE owing to the employment of SWCNTs (single surface carbon nanotubes) nanofluids with a 0.025 kg/s mass flow rate. In addition, Gan et al. [116] confirmed that the minimum thermal efficiency improvement in solar ETCs is 16.5% when nanofluids are employed.

One of the most significant challenges that researchers are facing in nanofluids applications is the size of nanoparticles. In most cases, the size of nanofluids has ranged from 1 to 100 nm as shown in Fig. 2.16. Over 40% of studies on solar systems have employed nanofluids with size ranges from 1 to 25 nm, while about 34% of the studies have employed nanofluids with the size between 25 and 50 nm, and the remaining utilized nanofluids with the size between 50 and 100 nm. Moreover, according to the investigations based on nanofluids volume fraction counted as an effective parameter in collector efficiency, it can be seen that 75% of the studied nanofluids volume fraction ranges from 0% to 1%, and 25% from 1% to 4%. The highest

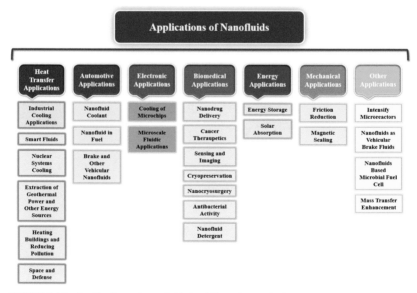

FIGURE 2.14 Applications of nanofluids in different industrial fields [31].

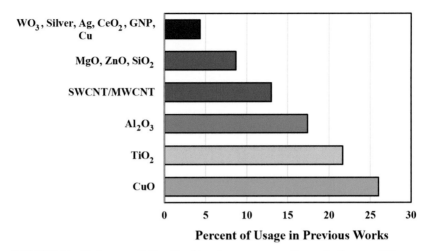

FIGURE 2.15 Various nanofluids utilized in solar systems according to the literature [31].

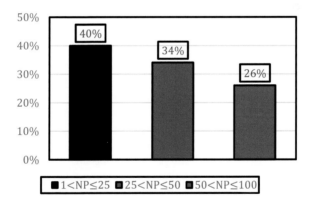

FIGURE 2.16 Classification of nanoparticle size (nm) reported in the literature related to solar systems [31].

performance was found with nanoparticles ranging from 1.0 to 25 nm and a volume fraction of 0%−1% [31].

2.4.1 Benefits and challenges of using nanofluids

The benefits of utilizing nanofluids in solar systems are as follows [31,45]:

- **Superior heat transfer properties:** Nanofluids have some unique features such as high density, increased conductivity, increased heat transfer, and low specific heat capacity that result in their effectiveness to enhance the thermal performance of solar systems.

- **Small size and large surface area:** Nanoparticles are featured by tiny particle size and an extensive surface area, leading to a remarkable rise in the absorption of solar energy.
- **Superior optical characteristics:** Optical characteristics of nanofluids such as high absorption and extinction coefficients make them superior to other fluids. Nanofluids show low emittance in the infrared spectrum and high absorption in the solar spectrum range.
- **Reduction of required heat transfer area:** The use of nanofluids leads to a substantial reduction of the required heat transfer surface of thermal systems.
- **Privileges of nanofluids over microsuspension:** The addition of nanofluids leads to enhanced stability and a longer lifetime in pumps and pipes due to mitigating the problems of clogging and sedimentation compared to systems employing suspensions with micro or millimeter-sized particles.

There are several factors of interest when considering a given synthetic approach. Some of these reasons express why a theoretical definition of nanofluids has not been organized [31,117].

- **Thermal instability:** Thermal behavior of nanofluids is disparate from standard solid-liquid suspensions or solid-solid composites.
- **Chemical compatibility:** They often cause limitations in the system's adaptability to various chemicals and conditions.
- **High cost:** The major reason for the limited use of nanofluids in the solar industry is their high production cost. In general, the production of nanofluids requires sophisticated and advanced equipment.
- **Manufacture process difficulty:** The propensity to agglomerate into larger particles is one of the main challenges of nanofluids during the production process, limiting their advantage of high surface area. Therefore particle scattering additives are often combined with pure fluids contain nanoparticles.

2.5 Photovoltaic–thermal technology

During the process of electricity generation using a solar cell, only a small fraction of received solar radiation is converted to electricity while a large portion of solar energy is getting dumped as heat. Every 1°C increase in the surface temperature of a PV cell causes a 0.5% decrease in efficiency [118,119]. The heat generated in a solar PV module can be extracted and used in several other applications, resulting in the improvement of its performance. Therefore, by cooling the surface of PV modules utilizing the air/water as HTF, the electrical performance of the PV module can significantly be enhanced [120]. In recent decades, special attention has been paid to photovoltaic–thermal (PVT) technology because of its advantages over PV

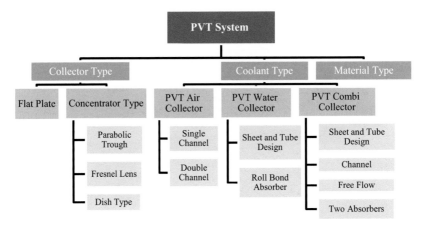

FIGURE 2.17 Broad classification of photovoltaic−thermal module technologies [123].

modules and solar thermal collectors. PVT is a combination of PV and solar thermal technologies that simultaneously converts solar radiation to power and low-temperature heat [121].

A PVT module usually contains a solar PV module, a duct, coolant (air/water), a DC fan, and a thermal collector [122]. A broad classification of PVT modules is shown in Fig. 2.17. The coolant in a PVT module is mainly utilized for drying applications, space, and water heating [124]. Ibrahim et al. [125] categorized PVT systems as natural and forced flow types according to fluid flow inside the thermal module under the PV module. Although the simplest and most economical technique for PV cell cooling is natural air circulation (convection), this technique is less effective in regions where the ambient temperature is more than 20°C [126]. The performance of a PVT module is analyzed considering the main factors of thermal and electrical efficiencies. The electrical efficiency of a PV module is mostly contingent on the cell temperature since the materials utilized in the fabrication of PV cells are sensitive to temperature variations.

2.5.1 Air-based photovoltaic−thermal modules

The extracted heat from the PV cell can be utilized in various applications such as space or floor heating, water heating, and materials drying. The detailed schematic of an air-based PVT module is depicted in Fig. 2.18. The impact of applying some design modifications has been analyzed by researchers to improve the performance of the air-based PVT modules. The investigated improvements include utilizing various designs of fins, thin metallic sheets (TMS), hexagonal honeycomb HE, and v-grooved absorbers. Tonui and Tripanagnostopoulos [127] employed aluminum fins in the air channel to enhance the heat transfer rate between the PV surface and air-

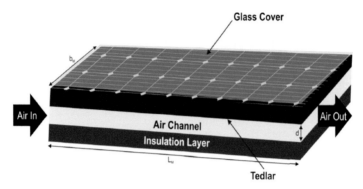

FIGURE 2.18 Schematic view of an air-based photovoltaic–thermal collector [123].

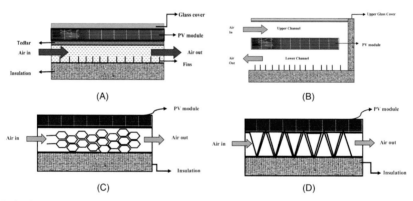

FIGURE 2.19 Schematic of air-based photovoltaic–thermal modules equipped with [123]: (A) fins, (B) double-pass air channel with fins in the lower part, (C) honeycomb-shaped air channel, and (D) V-grooved absorber plate.

fluid flow. In another work, Mojumder et al. [128] experimentally evaluated the effect of utilizing rectangular fins in the air channel on the thermal and electrical efficiencies of the PVT module. The results indicated that the maximum thermal and electric efficiencies of the modified system are 56.2% and 13.8%, respectively. The schematic of the proposed system is illustrated in Fig. 2.19A.

Kumar and Rosen [129] investigated the heat transfer rate in an air-based PVT module with a double-pass air channel in the presence of the fins (see Fig. 2.19B). In the proposed PVT module, the thermal and electric efficiencies were enhanced by 15.5% and 10.5%, respectively in comparison with the conventional PVT module. Hussain et al. [130] experimentally analyzed the air-cooling process of a PV cell by using a honeycomb-shaped air channel. The schematic of the proposed PVT system is shown in Fig. 2.16C. According to the results, the maximum achieved thermal efficiency was 87%

at solar irradiance of 828 W/m^2 and $\dot{m}_{air} = 0.11$ kg/s. Yu et al. [131] evaluated the thermal efficiency of an air-based PVT module equipped with a v-grooved aluminum absorber plate as shown in Fig. 2.19D. Accordingly, it was concluded that the electric and thermal efficiencies of the proposed model are significantly increased in comparison with traditional systems.

An efficient method for heat exchange enhancement in a PVT module and accordingly increasing the thermal efficiency was proposed by Ahmed and Mohammad [132]. They evaluated the thermal performance of a double-pass air-based PVT module in the presence of the porous media. Obtained results revealed that the thermal efficiency is increased by 80.23% and the electrical efficiency is decreased by 8.7% compared to the case without porous media. In another study by Dhiman et al. [133] an experimental analysis was performed on the thermal and electrical performance of a double-pass air-based PVT collector equipped with porous media inside the air channel. The thermal efficiency of the proposed system was enhanced by around 10%−12%. Youssef and Adam [134] employed a porous medium in both single and double-pass air-based PVT air modules. It was concluded that the thermal efficiency of the proposed model is enhanced by nearly 10% in comparison with the same cases without a porous medium.

Tahmasbi et al. [135] used porous metal foams in the PVT module to improve the PV cell's cooling process and consequently its thermal and electrical efficiencies. The impacts of some efficient parameters including porous zone thickness, solar heat flux, and inlet flow's Reynolds number on the PV cell's efficiency were evaluated. In the proposed system, the thermal and electrical efficiencies were enhanced by almost 85% and 3%, respectively in comparison with the case without porous media. Ahmed and Mohammed [132] investigated the impact of utilizing porous media on the thermal performance of a double-pass PVT air collector by performing experimental tests. Results indicated that the presence of porous media in the lower part of the proposed collector leads to enhancement in the efficiency by nearly 3%.

As an efficient method, a thermoelectric (TE) module was employed by Yousef and Adam [134] in a PVT module to directly convert the recovered thermal energy into electricity. A TE module works based on the principles of Seebeck and Peltier's effects. A potential difference is formed among the hot and cold junctions because of the temperature difference [136]. Babu and Ponnambalam [122] evaluated the performance of a PVT module by using a thermoelectric generator (TEG), as shown in Fig. 2.20A. Different effective factors such as the method of integration of TEG, location, properties of TEG, and its thermal resistance have been investigated by researchers. Moreover, some comprehensive studies have evaluated the impact of utilizing TEG on the thermal performance of the PVT module [137,138]. According to the obtained results, the highest achieved thermal efficiency of the PVT-TEG system has been around 30%. In another study, Tiwari et al. [139] experimentally evaluated the performance of an air-based PVT module

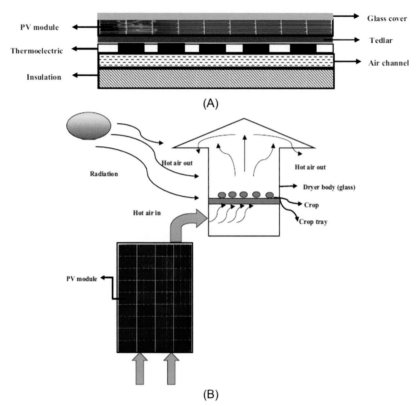

FIGURE 2.20 (A) Schematic of an air-based PVT-TEG module [123], and (B) an air-based PVT module integrated with a crop dryer [123].

integrated with a drying system which is beneficial in protecting the crops (see Fig. 2.20B). Experimental results indicated that a better drying performance is obtained by forced convection for high-moisture corps.

2.5.2 Water-based photovoltaic−thermal modules

The output of a water-based PVT module is electricity and hot water. The main difference between air-based and water-based PVT modules is in the structure of the HTF's channel. In a water-based PVT module, the coolant water is pumped in some tubes (or ducts) placed in the air channel, resulting in a reduction in the temperature of the PV module and consequently enhancement in the electrical efficiency of the module. Diwania et al. [123] presented different designs of water-based PVT modules considering the water flow patterns inside the PV module as shown in Fig. 2.21.

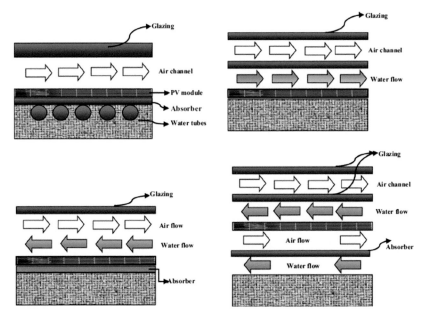

FIGURE 2.21   Various designs of water-based photovoltaic−thermal modules [123].

2.5.3   Recent developments in water-based photovoltaic−thermal modules

2.5.3.1   Utilizing water-based nanofluid

Recently, there has been a growing tendency to use different nanomaterials in coolant applications to improve thermal conductivity and fluid stability. Utilizing pure water as coolant fluid in PVT modules leads to 13% of incident solar radiation absorption. Hence, researchers concentrated on using nanofluids with direct sunlight absorbers and direct solar absorption collectors [140]. Considering developments in production methods, nanofluids with direct absorption solar collectors have expected much attention due to more efficient thermal energy use than conventional solar thermal collectors. The values of thermal and electrical efficiency of PVT modules are considerably enhanced when nanofluids are utilized as coolant fluid in direct absorption solar collectors [141−143]. Different researchers stated the enhanced electrical and thermal efficiency of PVT modules employing various water-based nanofluids in the channel as listed in Table 2.3.

2.5.3.2   Utilizing phase change material

Another new efficient method that has been recently introduced to decline the PV cell's temperature is using PCMs. In the phase change process of the PCM, the material interchanges among the solid and liquid phases. The

TABLE 2.3 The list of some studies utilized water-based nanofluid.

Author	Year	Water-based nanofluid	Maximum electrical efficiency (%)	Maximum thermal efficiency (%)
Shamani et al. [144]	2017	SiO_2	12.70	64.4
Sardarabadi and Farad [145]	2016	Al_2O_3, ZnO, and TiO	23.5 (ZnO and TiO)	110 (ZnO)
Ghadiri et al. [146]	2015	Fe_2O_3	17	33
Radwan et al. [147]	2016	Al_2O_3	19	62
Al-Waeli et al. [148]	2017	SiC, Al_2O_3, and CuO	16.5 (SiC)	49 (SiC)

difference between the ambient and PCM temperature causes heat transfer between them, and the phase of the PCM is changed from solid to liquid or vice versa [149,150]. Preet et al. [151] experimentally evaluated the thermal and electrical performance of a PVT-PCM module. The schematic of the proposed system is shown in Fig. 2.22A. Diallo et al. [152] proposed a novel PVT module integrated with heat pipes and the PCM. In their design, it was assumed that as the fluid of heat pipes absorbs heat, the refrigerant in the evaporation section is evaporated. Then, the heat is transferred from the central tube (contains refrigerant) to the water, while the excess heat reaches the PCM in the outer tube and is stored as the PCM starts melting/fusion. Consequently, when heat flux from the heat pipe decreases, the PCM releases the heat to the water and keeps the hot water's temperature constant. Results reported the thermal and electric efficiencies as 55.6% and 12.2%, respectively.

2.5.3.3   Utilizing graphite layer

Liang et al. [153] presented a new model in which the graphite layer was utilized below the water channel (see Fig. 2.22B). The stated average electrical efficiency of the proposed PVT model and conventional type was around 6.46 and 5.15%, respectively.

In addition to conventional flat-plate PVT modules, the simultaneous generation of electricity and heat can be performed by employing concentrating reflectors and tracking devices, enhancing the overall efficiency of the whole device. CPVT modules are mainly utilized in large-scale or high-temperature facilities employing concentrating collectors such as PTCs, PDC, LFRs, and CPCs integrated with single- or two-axis tracking systems.

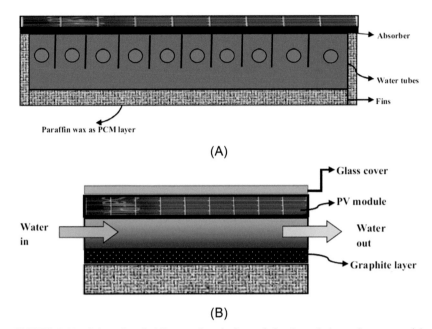

FIGURE 2.22 Schematic of: (A) water-based photovoltaic—thermal-phase change material module [123], and (B) water-based photovoltaic—thermal-graphite layer module [123].

A comprehensive study on the recent advancements in PVT technology can be found in Ref. [121].

2.6 Conclusions and prospects

Efforts should be made to focus on the proper utilization of abundant potential natural resources by effective technological advancement that will be led to encouraging sustainable development. The energy deficit worldwide, especially in some countries including India, China, and South Africa, plays an important role in the development of solar energy to overcome the energy demand in the context of the global energy mix. While other energy surplus countries such as Brazil and Russia can export alternative renewable energies to other countries to meet their energy demand. This chapter presents principles and advancements of solar energy technology considering both power plant and nonpower plant applications. In this regard, different techniques to harness solar energy include thermal methods and direct electricity generation using PV technology along with energy storage methods are presented and discussed. Additionally, the employment of nonfluids in solar systems and progressive PVT technology considering their benefits and challenges are described. The results of this study indicate that recent advancements made in solar energy technologies have made the future of this renewable technology more promising in both domestic and industrial applications. These innovations are eminent in five main areas as follows:

1. The material used in the structure of solar cells significantly affects the performance of PV cells. The new materials used to fabricate solar cells have significantly reduced their cost. Additionally, these newly used materials have also increased the efficiency of solar cells to a favorable extend, making this technology more competitive with fossil-based conventional technologies.

2. The second area considers the main developments made in the field of solar thermal collectors. Since these collectors are mainly utilized in residential and industrial applications, any progress in their performance can increase the production capacity of solar power plants, making them a suitable alternative to traditional plants.

3. The third area is related to integrating energy storage technologies into solar systems which is considered one of the most critical challenges in this field. With the integration of energy storage systems, performing solar systems during periods with no sufficient radiation (night, rainy weather, etc.) becomes possible. Solar energy can be stored as thermal energy in TES systems or electricity in storage batteries. Significant advances in technology and reductions in costs can make both technologies more feasible to be utilized in solar systems.

4. The most efficient area of developing solar systems is utilizing nanofluids in these systems. The impact of using nanofluids, as a reliable technique, to enhance the thermal performance of solar systems has already been proved in various applications. The most critical challenge in this field is to solve the problem of nanoparticles' sedimentation which is one of the most recent topics of investigation.

5. PVT technology is another innovation area in the field of solar energy systems. In PVT modules, the heat received due to the cooling of PV cells can be utilized in other applications, resulting in appropriate operating temperatures of PV cells and assisting for their safe operation. Till now, several cooling methods have been proposed by researchers, and many efficient viable models have been proposed and implemented.

References

[1] Ayaburi J, Bazilian M, Kincer J, Moss T. Measuring "reasonably reliable" access to electricity services. Electr J 2020;33:106828. Available from: https://doi.org/10.1016/j.tej.2020.106828.

[2] Gorjian S, Zadeh BN, Eltrop L, Shamshiri RR, Amanlou Y. Solar photovoltaic power generation in Iran:d, policies, and barriers. Renew Sustain Energy Rev 2019;106:110−23. Available from: https://doi.org/10.1016/j.rser.2019.02.025.

[3] Jogdand OK. Study on the effect of global warming and greenhouse gases on environmental system. Green Chemistry and Sustainable Technology. Apple Academic Press; 2020. p. 275−306. Available from: https://doi.org/10.1201/9780367808310-12.

[4] Yamakawa CK, Qin F, Mussatto SI. Advances and opportunities in biomass conversion technologies and biorefineries for the development of a bio-based economy. Biomass Bioenergy 2018;119:54−60. Available from: https://doi.org/10.1016/j. biombioe.2018.09.007.

[5] Gorjian S, Ebadi H. Introduction. In: Shukla SGA, editor. Photovoltaic solar energy conversion. Elsevier; 2020. p. 1−26. Available from: https://doi.org/10.1016/B978-0-12-819610-6.00001-6.

[6] Dimitriev O, Yoshida T, Sun H. Principles of solar energy storage. Energy Storage 2020;2:e96. Available from: https://doi.org/10.1002/est2.96.

[7] Roy S, Saha UK. Review on the numerical investigations into the design and development of Savonius wind rotors. Renew Sustain Energy Rev 2013;24:73−83. Available from: https://doi.org/10.1016/j.rser.2013.03.060.

[8] Javadi H, Ajarostaghi SSM, Rosen MA, Pourfallah M. A comprehensive review of backfill materials and their effects on ground heat exchanger performance. Sustain 2018;10. Available from: https://doi.org/10.3390/su10124486.

[9] Javadi H, Mousavi Ajarostaghi SS, Rosen MA, Pourfallah M. Performance of ground heat exchangers: a comprehensive review of recent advances. Energy 2019;178:207−33. Available from: https://doi.org/10.1016/j.energy.2019.04.094.

[10] Khan K, Ahmed S, Akhter M, Alam R, Hossen M. Wave and tidal power generation. Encycl Hydrocarb Vol III/ N Dev Energy, Transp Sustain 2018;(4):575−94.

[11] Sleiti AK. Tidal power technology review with potential applications in Gulf Stream. Renew Sustain Energy Rev 2017;69:435−41. Available from: https://doi.org/10.1016/j. rser.2016.11.150.

[12] Bilandzija N, Voca N, Jelcic B, Jurisic V, Matin A, Grubor M, et al. Evaluation of Croatian agricultural solid biomass energy potential. Renew Sustain Energy Rev 2018;93:225−30. Available from: https://doi.org/10.1016/j.rser.2018.05.040.

[13] Tursi A. A review on biomass: importance, chemistry, classification, and conversion. Biofuel Res J 2019;6:962−79. Available from: https://doi.org/10.18331/BRJ2019.6.2.3.

[14] Renewables. Global status report. Paris, France: REN21; 2021.

[15] Letcher TM, Trevor M., Fthenakis VM. A comprehensive guide to solar energy systems: with special focus on photovoltaic systems; 2018.

[16] Letcher TM. Climate change: observed impacts on the planet Earth. Elsevier; 2021.

[17] World Insolation Map; 2021. https://www.solar-facts.com/world-solar/world-insolation. php [accessed 19.08.2021].

[18] Escombe F. Novel hydroelectric storage concepts. Storing energy with spec. ref. to renew. Energy Sources. Elsevier Inc; 2016. p. 39−67.

[19] The Guardian. Portugal runs for four days straight on renewable energy alone. Renewable energy; 2021. https://www.theguardian.com/environment/2016/may/18/portugal-runs-for-four-days-straight-on-renewable-energy-alone [accessed 19.08.2021].

[20] News − IEA. How solar energy could be the largest source of electricity by mid-century, https://www.iea.org/news/how-solar-energy-could-be-the-largest-source-of-electricity-by-mid-century; 2021 [accessed 19.08.2021].

[21] Letcher TM. Future energy: improved, sustainable and clean options for our planet. Elsevier; 2020. Available from: https://doi.org/10.1016/C2018-0-01500-5.

[22] Briefing CC. Solar energy for heat and electricity: the potential for mitigating climate change. 2009.

[23] Streatfeild JEJ. Low electricity supply in sub-Saharan Africa: causes, implications, and remedies. J Int Commer Econ 2018;2018:1.

[24]   Jäger-Waldau A. Snapshot of photovoltaics—March 2017. Sustainability 2017;9:783. Available from: https://doi.org/10.3390/su9050783.

[25]   Pvinsights. http://pvinsights.com/; 2021 [accessed 19.08.2021].

[26]   Zeng K, Gauthier D, Minh DP, Weiss-Hortala E, Nzihou A, Flamant G. Characterization of solar fuels obtained from beech wood solar pyrolysis. Fuel 2017;188:285−93. Available from: https://doi.org/10.1016/j.fuel.2016.10.036.

[27]   Ediger VŞ. An integrated review and analysis of multi-energy transition from fossil fuels to renewables. Energy Procedia 2019;156:2−6.

[28]   Energy Innvoation. Comparing the levelized cost of energy technologies; 2021. https://energyinnovation.org/2015/02/07/levelized-cost-of-energy/ [accessed 19.08.2021].

[29]   Fthenakis V, Kim HC. Land use and electricity generation: a life-cycle analysis. Renew Sustain Energy Rev 2009;13:1465−74. Available from: https://doi.org/10.1016/j.rser.2008.09.017.

[30]   Duffie JA, Beckman WA, Blair N. Solar engineering of thermal processes, photovoltaics and wind. John Wiley & Sons; 2020.

[31]   Olfian H, Ajarostaghi SSM, Ebrahimnataj M. Development on evacuated tube solar collectors: a review of the last decade results of using nanofluids. Sol Energy 2020;211:265−82. Available from: https://doi.org/10.1016/j.solener.2020.09.056.

[32]   Chan H-Y, Riffat SB, Zhu J. Review of passive solar heating and cooling technologies. Renew Sustain Energy Rev 2010;14:781−9. Available from: https://doi.org/10.1016/j.rser.2009.10.030.

[33]   Sadhishkumar S, Balusamy T. Performance improvement in solar water heating systems—a review. Renew Sustain Energy Rev 2014;37:191−8. Available from: https://doi.org/10.1016/j.rser.2014.04.072.

[34]   Khalifa AJN, Abbas EF. A comparative performance study of some thermal storage materials used for solar space heating. Energy Build 2009;41:407−15.

[35]   Gorjian S, Ebadi H, Calise F, Shukla A, Ingrao C. A review on recent advancements in performance enhancement techniques for low-temperature solar collectors. Energy Convers Manag 2020;222:113246. Available from: https://doi.org/10.1016/j.enconman.2020.113246.

[36]   Gorjian S, Ghobadian B. Solar desalination: a sustainable solution to water crisis in Iran. Renew Sustain Energy Rev 2015;48:571−84. Available from: https://doi.org/10.1016/j.rser.2015.04.009.

[37]   Gorjian S, Ghobadian B, Ebadi H, Ketabchi F, Khanmohammadi S. Applications of solar PV systems in desalination technologies. In: Gorjian S, Shukla A, editors. Photovolt. Sol. Energy Convers. 1st (ed.) London: Elsevier; 2020. p. 237−74. Available from: https://doi.org/10.1016/B978-0-12-819610-6.00008-9.

[38]   Gorjian S, Hosseingholilou B, Jathar LD, Samadi H, Samanta S, Sagade AA, et al. Recent advancements in technical design and thermal performance enhancement of solar greenhouse dryers. Sustainability 2021;13:7025. Available from: https://doi.org/10.3390/su13137025.

[39]   Palanikumar G, Shanmugan S, Chithambaram V, Gorjian S, Pruncu CI, Essa FA, et al. Thermal investigation of a solar box-type cooker with nanocomposite phase change materials using flexible thermography. Renew Energy 2021;178:260−82. Available from: https://doi.org/10.1016/j.renene.2021.06.022.

[40]   Cuce E, Cuce PM. A comprehensive review on solar cookers. Appl Energy 2013;102:1399−421. Available from: https://doi.org/10.1016/j.apenergy.2012.09.002.

[41] Heywood H. Solar energy: a challenge to the future. Nature 1957;180:115−18. Available from: https://doi.org/10.1038/180115a0.

[42] Jafrancesco D, Sansoni P, Francini F, Contento G, Cancro C, Privato C, et al. Mirrors array for a solar furnace: optical analysis and simulation results. Renew Energy 2014;63:263−71. Available from: https://doi.org/10.1016/j.renene.2013.09.006.

[43] Tiwari DR, Kumar D, Kumar A, Kumar R, Ojha A, Raj A. Design and fabrication of solar furnace. Invertis J Renew Energy 2019;9:1. Available from: https://doi.org/10.5958/2454-7611.2019.00001.8.

[44] Behar O. Solar thermal power plants − a review of configurations and performance comparison. Renew Sustain Energy Rev 2018;92:608−27. Available from: https://doi.org/10.1016/j.rser.2018.04.102.

[45] Elsheikh AH, Sharshir SW, Mostafa ME, Essa FA, Ahmed, Ali MK. Applications of nanofluids in solar energy: a review of recent advances. Renew Sustain Energy Rev 2018;82:3483−502. Available from: https://doi.org/10.1016/j.rser.2017.10.108.

[46] Kalogirou SA. Solar thermal collectors and applications. Prog Energy Combust Sci 2004;30:231−95. Available from: https://doi.org/10.1016/j.pecs.2004.02.001.

[47] Arunkumar T, Velraj R, Denkenberger DC, Sathyamurthy R, Kumar KV, Ahsan A. Productivity enhancements of compound parabolic concentrator tubular solar stills. Renew Energy 2016;88:391−400. Available from: https://doi.org/10.1016/j.renene.2015.11.051.

[48] Papadimitratos A, Sobhansarbandi S, Pozdin V, Zakhidov A, Hassanipour F. Evacuated tube solar collectors integrated with phase change materials. Sol Energy 2016;129:10−19. Available from: https://doi.org/10.1016/j.solener.2015.12.040.

[49] Li X, Xu E, Ma L, Song S, Xu L. Modeling and dynamic simulation of a steam generation system for a parabolic trough solar power plant. Renew Energy 2019;132:998−1017. Available from: https://doi.org/10.1016/j.renene.2018.06.094.

[50] Beltagy H, Semmar D, Lehaut C, Said N. Theoretical and experimental performance analysis of a Fresnel type solar concentrator. Renew Energy 2017;101:782−93. Available from: https://doi.org/10.1016/j.renene.2016.09.038.

[51] Li L, Dubowsky S. A new design approach for solar concentrating parabolic dish based on optimized flexible petals. Mech Mach Theory 2011;46:1536−48. Available from: https://doi.org/10.1016/j.mechmachtheory.2011.04.012.

[52] Roca L, de la Calle A, Yebra LJ. Heliostat-field gain-scheduling control applied to a two-step solar hydrogen production plant. Appl Energy 2013;103:298−305. Available from: https://doi.org/10.1016/j.apenergy.2012.09.047.

[53] Alfaro-Ayala JA, Martínez-Rodríguez G, Picón-Núñez M, Uribe-Ramírez AR, Gallegos-Muñoz A. Numerical study of a low temperature water-in-glass evacuated tube solar collector. Energy Convers Manag 2015;94:472−81. Available from: https://doi.org/10.1016/j.enconman.2015.01.091.

[54] Gray JL. The physics of the solar cell. Handbook of. photovoltaic science and engineering. Chichester, UK: John Wiley & Sons, Ltd; 2011. p. 82−129. Available from: https://doi.org/10.1002/9780470974704.ch3.

[55] Plante RH, Russell H. Solar energy, photovoltaics, and domestic hot water: a technical and economic guide for project planners, builders, and property owners. Academic Press; 2014.

[56] Mukashev BN, Betekbaev AA, Kalygulov DA, Pavlov AA, Skakov DM. Study of silicon production processes and development of solar-cell fabrication technologies. Semiconductors 2015;49:1375−82. Available from: https://doi.org/10.1134/S1063782615100164.

[57] Lee TD, Ebong AU. A review of thin film solar cell technologies and challenges. Renew Sustain Energy Rev 2017;70:1286−97. Available from: https://doi.org/10.1016/j.rser.2016.12.028.

[58] Badescu V, Lazaroiu GC, Barelli L. Power engineering advances and challenges part B: electrical power 2018;. Available from: https://doi.org/10.1201/9780429453717.

[59] Gorjian S, Ebadi H, Trommsdorff M, Sharon H, Demant M, Schindele S. The advent of modern solar-powered electric agricultural machinery: a solution for sustainable farm operations. J Clean Prod 2021;292:126030. Available from: https://doi.org/10.1016/j.jclepro.2021.126030.

[60] EQ International. The price difference between mono- and multi-si widens while the overseas polysilicon rises slowly. The Leading Solar Magazine In India. https://www.eqmagpro.com/the-price-difference-between-mono-and-multi-si-widens-while-the-overseas-poly-silicon-rises-slowly/; n.d. [accessed 30.3. 2021].

[61] Green MA, Hishikawa Y, Dunlop ED, Levi DH, Hohl-Ebinger J, Ho-Baillie AWY. Solar cell efficiency tables (version 51). Prog Photovolt Res Appl 2018;26:3−12. Available from: https://doi.org/10.1002/pip.2978.

[62] Parida B, Iniyan S, Goic R. A review of solar photovoltaic technologies. Renew Sustain Energy Rev 2011;15:1625−36. Available from: https://doi.org/10.1016/j.rser.2010.11.032.

[63] Gul M, Kotak Y, Muneer T. Review on recent trend of solar photovoltaic technology. Energy Explor Exploit 2016;34:485−526. Available from: https://doi.org/10.1177/0144598716650552.

[64] Wikimedia Commons. File: from a solar cell to a PV system.svg. https://commons.wikimedia.org/wiki/File:From_a_solar_cell_to_a_PV_system.svg; 2021 [accessed 19.08.2021].

[65] Plante RH. Solar photovoltaic systems. Solar.energy, photovoltaics, domestic hot water. Elsevier; 2014. p. 75−92. Available from: https://doi.org/10.1016/B978-0-12-420155-2.00005-0.

[66] Jordehi AR. Parameter estimation of solar photovoltaic (PV) cells: a review. Renew Sustain Energy Rev 2016;61:354−71. Available from: https://doi.org/10.1016/j.rser.2016.03.049.

[67] Maghami MR, Hizam H, Gomes C, Radzi MA, Rezadad MI, Hajighorbani S. Power loss due to soiling on solar panel: a review. Renew Sustain Energy Rev 2016;59:1307−16. Available from: https://doi.org/10.1016/j.rser.2016.01.044.

[68] Wiesenfarth M, Philipps SP, Bett AW, Horowitz K, Kurtz S. Current status of concentrator photovoltaic (CPV). Technology. 2017;. Available from: https://www.ise.fraunhofer.de/.

[69] Philipps SP, Bett AW, Horowitz K, Kurtz S. Current status of concentrator photovoltaic (CPV) technology; 2015.

[70] Mahmud M, Huda N, Farjana S, Lang C. Environmental impacts of solar-photovoltaic and solar-thermal systems with life-cycle assessment. Energies 2018;11:2346. Available from: https://doi.org/10.3390/en11092346.

[71] Solanki CS. Solar photovoltaics: fundamentals, technologies and applications. Phi learning pvt. Ltd; 2015.

[72] Awasthi A, Shukla AK, S.R. MM, Dondariya C, Shukla KN, Porwal D, et al. Review on sun tracking technology in solar PV system. Energy Rep 2020;6:392−405. Available from: https://doi.org/10.1016/j.egyr.2020.02.004.

[73] Mohammadi S, Mohammadi A. Stochastic scenario-based model and investigating size of battery energy storage and thermal energy storage for micro-grid. Int J Electr Power Energy Syst 2014;61:531−46. Available from: https://doi.org/10.1016/j.ijepes.2014.03.041.

[74] Akinyele DO, Rayudu RK. Review of energy storage technologies for sustainable power networks. Sustain Energy Technol Assess 2014;8:74−91. Available from: https://doi.org/10.1016/j.seta.2014.07.004.

[75] Kuravi S, Trahan J, Goswami DY, Rahman MM, Stefanakos EK. Thermal energy storage technologies and systems for concentrating solar power plants. Prog Energy Combust Sci 2013;39:285−319. Available from: https://doi.org/10.1016/j.pecs.2013.02.001.

[76] González-Roubaud E, Pérez-Osorio D, Prieto C. Review of commercial thermal energy storage in concentrated solar power plants: steam vs. molten salts. Renew Sustain Energy Rev 2017;80:133−48. Available from: https://doi.org/10.1016/j.rser.2017.05.084.

[77] Pacheco JE, Bradshaw RW, Dawson DB, De la Rosa W, Gilbert R, Goods SH. Final test and evaluation results from the solar two project. SAND2002-0120 2002:1−294.

[78] Wang Z. Thermal storage systems. Des. solar and thermal power plants. Elsevier; 2019. p. 387−415. Available from: https://doi.org/10.1016/B978-0-12-815613-1.00006-7.

[79] Alva G, Lin Y, Fang G. An overview of thermal energy storage systems. Energy 2018;144:341−78. Available from: https://doi.org/10.1016/j.energy.2017.12.037.

[80] Benoit H, Spreafico L, Gauthier D, Flamant G. Review of heat transfer fluids in tube-receivers used in concentrating solar thermal systems: properties and heat transfer coefficients. Renew Sustain Energy Rev 2016;55:298−315. Available from: https://doi.org/10.1016/j.rser.2015.10.059.

[81] Jacob R, Belusko M, Inés Fernández A, Cabeza LF, Saman W, Bruno F. Embodied energy and cost of high temperature thermal energy storage systems for use with concentrated solar power plants. Appl Energy 2016;180:586−97. Available from: https://doi.org/10.1016/j.apenergy.2016.08.027.

[82] Kenda ES, N'Tsoukpoe KE, Ouédraogo IWK, Coulibaly Y, Py X, Ouédraogo FMAW. *Jatropha curcas* crude oil as heat transfer fluid or thermal energy storage material for concentrating solar power plants. Energy Sustain Dev 2017;40:59−67. Available from: https://doi.org/10.1016/j.esd.2017.07.003.

[83] Zanganeh G, Commerford M, Haselbacher A, Pedretti A, Steinfeld A. Stabilization of the outflow temperature of a packed-bed thermal energy storage by combining rocks with phase change materials. Appl Therm Eng 2014;70:316−20. Available from: https://doi.org/10.1016/j.applthermaleng.2014.05.020.

[84] Schlipf D, Schicktanz P, Maier H, Schneider G. Using sand and other small grained materials as heat storage medium in a packed bed HTTESS. Energy Procedia 2015;69:1029−38. Available from: https://doi.org/10.1016/j.egypro.2015.03.202.

[85] Alonso MC, Vera-Agullo J, Guerreiro L, Flor-Laguna V, Sanchez M, Collares-Pereira M. Calcium aluminate based cement for concrete to be used as thermal energy storage in solar thermal electricity plants. Cem Concr Res 2016;82:74−86. Available from: https://doi.org/10.1016/j.cemconres.2015.12.013.

[86] Cárdenas B, León N. High temperature latent heat thermal energy storage: phase change materials, design considerations and performance enhancement techniques. Renew Sustain Energy Rev 2013;27:724−37. Available from: https://doi.org/10.1016/j.rser.2013.07.028.

[87] Pielichowska K, Pielichowski K. Phase change materials for thermal energy storage. Prog Mater Sci 2014;65:67−123. Available from: https://doi.org/10.1016/j.pmatsci.2014.03.005.

[88] Teggar M, Arıcı M, Mert MS, Mousavi Ajarostaghi SS, Niyas H, Tunçbilek E, et al. A comprehensive review of micro/nano enhanced phase change materials. J Therm Anal Calorim 2021;1−28. Available from: https://doi.org/10.1007/s10973-021-10808-0.

[89] Pakzad K, Mousavi Ajarostaghi SS, Sedighi K. Numerical simulation of solidification process in an ice-on-coil ice storage system with serpentine tubes. SN Appl Sci 2019;1:1258. Available from: https://doi.org/10.1007/s42452-019-1316-4.

[90] Mousavi Ajarostaghi SS, Sedighi K, Aghajani Delavar M, Poncet S. Numerical study of a horizontal and vertical shell and tube ice storage systems considering three types of tube. Appl Sci 2020;10:1059. Available from: https://doi.org/10.3390/app10031059.

[91] Mousavi Ajarostaghi SS, Sedighi K, Delavar MA, Poncet S. Influence of geometrical parameters arrangement on solidification process of ice-on-coil storage system. SN Appl Sci 2020;2:109. Available from: https://doi.org/10.1007/s42452-019-1912-3.

[92] Bilal Awan A, Khan MN, Zubair M, Bellos E. Commercial parabolic trough CSP plants: research trends and technological advancements. Sol Energy 2020;211:1422−58. Available from: https://doi.org/10.1016/j.solener.2020.09.072.

[93] Agalit H, Zari N, Maaroufi M. Suitability of industrial wastes for application as high temperature thermal energy storage (TES) materials in solar tower power plants − a comprehensive review. Sol Energy 2020;208:1151−65. Available from: https://doi.org/ 10.1016/j.solener.2020.08.055.

[94] Yang H, Wang Q, Huang Y, Gao G, Feng J, Li J, et al. Novel parabolic trough power system integrating direct steam generation and molten salt systems: preliminary thermo-dynamic study. Energy Convers Manag 2019;195:909−26. Available from: https://doi. org/10.1016/j.enconman.2019.05.072.

[95] Mousavi SS, Ouellet-Plamondon CM, Guizani L, Bhojaraju C, Brial V. On mitigating rebar−concrete interface damages due to the pre-cracking phenomena using superabsor-bent polymers. Constr Build Mater 2020;253:119181. Available from: https://doi.org/ 10.1016/j.conbuildmat.2020.119181.

[96] Pelay U, Luo L, Fan Y, Stitou D, Rood M. Thermal energy storage systems for concen-trated solar power plants. Renew Sustain Energy Rev 2017;79:82−100. Available from: https://doi.org/10.1016/j.rser.2017.03.139.

[97] Singh H, Saini RP, Saini JS. A review on packed bed solar energy storage systems. Renew Sustain Energy Rev 2010;14:1059−69. Available from: https://doi.org/10.1016/j. rser.2009.10.022.

[98] Faraz T. Benefits of concentrating solar power over solar photovoltaic for power genera-tion in Bangladesh. Proc 2nd Int Conf Dev Renew Energy Technol ICDRET 2012;2012:183−7.

[99] Armand M, Tarascon J-M. Building better batteries. Nature 2008;451:652−7. Available from: https://doi.org/10.1038/451652a.

[100] Yang P, Tarascon J-M. Towards systems materials engineering. Nat Mater 2012;11:560−3. Available from: https://doi.org/10.1038/nmat3367.

[101] Diouf B, Pode R. Potential of lithium-ion batteries in renewable energy. Renew Energy 2015;76:375−80. Available from: https://doi.org/10.1016/j.renene.2014.11.058.

[102] Diouf B, Avis C. The potential of Li-ion batteries in ECOWAS solar home systems. J Energy Storage 2019;22:295−301. Available from: https://doi.org/10.1016/j. est.2019.02.021.

[103] Gyan P. Calendar ageing modeling of lithium-ion batteries. Mat4Bat Summer Sch 2015;1−49.

[104] Keil P, Schuster SF, Wilhelm J, Travi J, Hauser A, Karl RC, et al. Calendar aging of lithium-ion batteries. J Electrochem Soc 2016;163:A1872−80. Available from: https:// doi.org/10.1149/2.0411609jes.

[105] Thompson AW. Economic implications of lithium ion battery degradation for vehicle-to-grid (V2X) services. J Power Sources 2018;396:691−709. Available from: https://doi.org/10.1016/j.jpowsour.2018.06.053.

[106] Krieger EM, Cannarella J, Arnold CB. A comparison of lead-acid and lithium-based battery behavior and capacity fade in off-grid renewable charging applications. Energy 2013;60:492−500. Available from: https://doi.org/10.1016/j.energy.2013.08.029.

[107] Freitas Gomes IS, Perez Y, Suomalainen E. Coupling small batteries and PV generation: a review. Renew Sustain Energy Rev 2020;126:109835. Available from: https://doi.org/10.1016/j.rser.2020.109835.

[108] IRENA. Electricity storage and renewables: costs and markets to 2030; 2017.

[109] Mousavi SS, Mousavi Ajarostaghi SS, Bhojaraju C. A critical review of the effect of concrete composition on rebar−concrete interface (RCI) bond strength: a case study of nanoparticles. SN Appl Sci 2020;2:1−23. Available from: https://doi.org/10.1007/S42452-020-2681-8.

[110] Choi SUS, Eastman JA. Enhancing thermal conductivity of fluids with nanoparticles. IL (United States): Argonne National Lab; 1995.

[111] Alsarraf J, Shahsavar A, Babaei Mahani R, Talebizadehsardari P. Turbulent forced convection and entropy production of a nanofluid in a solar collector considering various shapes for nanoparticles. Int Commun Heat Mass Transf 2020;117:104804. Available from: https://doi.org/10.1016/j.icheatmasstransfer.2020.104804.

[112] Sheremet MA, Pop I, Mahian O. Natural convection in an inclined cavity with time-periodic temperature boundary conditions using nanofluids: application in solar collectors. Int J Heat Mass Transf 2018;116:751−61. Available from: https://doi.org/10.1016/j.ijheatmasstransfer.2017.09.070.

[113] Wen D, Lin G, Vafaei S, Zhang K. Review of nanofluids for heat transfer applications. Particuology 2009;7:141−50. Available from: https://doi.org/10.1016/j.partic.2009.01.007.

[114] Chand R. Nanofluid technologies and thermal convection techniques. IGI Global; 2017.

[115] Sabiha MA, Saidur R, Hassani S, Said Z, Mekhilef S. Energy performance of an evacuated tube solar collector using single walled carbon nanotubes nanofluids. Energy Convers Manag 2015;105:1377−88. Available from: https://doi.org/10.1016/j.enconman.2015.09.009.

[116] Gan YY, Ong HC, Ling TC, Zulkifli NWM, Wang C-T, Yang Y-C. Thermal conductivity optimization and entropy generation analysis of titanium dioxide nanofluid in evacuated tube solar collector. Appl Therm Eng 2018;145:155−64. Available from: https://doi.org/10.1016/j.applthermaleng.2018.09.012.

[117] Xuan Y, Li Q. Heat transfer enhancement of nanofluids. Int J Heat Fluid Flow 2000;21:58−64. Available from: https://doi.org/10.1016/S0142-727X(99)00067-3.

[118] Siecker J, Kusakana K, Numbi BP. A review of solar photovoltaic systems cooling technologies. Renew Sustain Energy Rev 2017;79:192−203. Available from: https://doi.org/10.1016/j.rser.2017.05.053.

[119] Gorjian S, Sharon H, Ebadi H, Kant K, Scavo FB, Tina GM. Recent technical advancements, economics and environmental impacts of floating photovoltaic solar energy conversion systems. J Clean Prod 2021;278:124285. Available from: https://doi.org/10.1016/j.jclepro.2020.124285.

[120] Good C, Chen J, Dai Y, Hestnes AG. Hybrid photovoltaic-thermal systems in buildings − a review. Energy Procedia 2015;70:683−90. Available from: https://doi.org/10.1016/j.egypro.2015.02.176.

[121] Shakouri M, Ebadi H, Gorjian S. Solar photovoltaic thermal (PVT) module technologies. In: Gorjian S, Shukla ABT-PSEC, editors. Photovoltaic solar energy conversion. 1st (ed.) London: Elsevier; 2020. p. 79−116. Available from: https://doi.org/10.1016/B978-0-12-819610-6.00004-1.

[122] Babu C, Ponnambalam P. The role of thermoelectric generators in the hybrid PV/T systems: a review. Energy Convers Manag 2017;151:368−85. Available from: https://doi.org/10.1016/j.enconman.2017.08.060.

[123] Diwania S, Agrawal S, Siddiqui AS, Singh S. Photovoltaic−thermal (PV/T) technology: a comprehensive review on applications and its advancement. Int J Energy Env Eng 2019;11:33−54. Available from: https://doi.org/10.1007/S40095-019-00327-Y.

[124] Tian Y, Zhao CY. A review of solar collectors and thermal energy storage in solar thermal applications. Appl Energy 2013;104:538−53. Available from: https://doi.org/10.1016/j.apenergy.2012.11.051.

[125] Ibrahim A, Othman MY, Ruslan MH, Mat S, Sopian K. Recent advances in flat plate photovoltaic/thermal (PV/T) solar collectors. Renew Sustain Energy Rev 2011;15:352−65. Available from: https://doi.org/10.1016/j.rser.2010.09.024.

[126] Lamnatou C, Chemisana D. Photovoltaic/thermal (PVT) systems: a review with emphasis on environmental issues. Renew Energy 2017;105:270−87. Available from: https://doi.org/10.1016/j.renene.2016.12.009.

[127] Tonui JK, Tripanagnostopoulos Y. Performance improvement of PV/T solar collectors with natural air flow operation. Sol Energy 2008;82:1−12. Available from: https://doi.org/10.1016/j.solener.2007.06.004.

[128] Mojumder JC, Chong WT, Ong HC, Leong KY. Abdullah-Al-Mamoon. An experimental investigation on performance analysis of air type photovoltaic thermal collector system integrated with cooling fins design. Energy Build 2016;130:272−85. Available from: https://doi.org/10.1016/j.enbuild.2016.08.040.

[129] Kumar R, Rosen MA. Performance evaluation of a double pass PV/T solar air heater with and without fins. Appl Therm Eng 2011;31:1402−10. Available from: https://doi.org/10.1016/j.applthermaleng.2010.12.037.

[130] Hussain F, Othman MYH, Yatim B, Ruslan H, Sopian K, Ibrahim Z. A study of PV/T collector with honeycomb heat exchanger. AIP Conf Proc 2013;1571:10−16. Available from: https://doi.org/10.1063/1.4858622.

[131] Yu C, Li H, Chen J, Qiu S, Yao F, Liu X. Investigation of the thermal performance enhancement of a photovoltaic thermal (PV/T) collector with periodically grooved channels. J Energy Storage 2021;40:102792. Available from: https://doi.org/10.1016/j.est.2021.102792.

[132] Ahmed OK, Mohammed ZA. Influence of porous media on the performance of hybrid PV/thermal collector. Renew Energy 2017;112:378−87. Available from: https://doi.org/10.1016/j.renene.2017.05.061.

[133] Dhiman P, Thakur NS, Kumar A, Singh S. An analytical model to predict the thermal performance of a novel parallel flow packed bed solar air heater. Appl Energy 2011;88:2157−67. Available from: https://doi.org/10.1016/j.apenergy.2010.12.033.

[134] Yousef B, Adam N. Performance analysis for flat plate collector with and without porous media. J Energy South Afr 2008;19:32−42.

[135] Tahmasbi M, Siavashi M, Norouzi AM, Doranehgard MH. Thermal and electrical efficiencies enhancement of a solar photovoltaic-thermal/air system (PVT/air) using metal foams. J Taiwan Inst Chem Eng 2021;124:276−89. Available from: https://doi.org/10.1016/j.jtice.2021.03.045.

[136]   Akbar A, Najafi G, Gorjian S, Kasaeian A, Mazlan M. Performance enhancement of a hybrid photovoltaic-thermal-thermoelectric (PVT-TE) module using nanofluid-based cooling: indoor experimental tests and multi-objective optimization. Sustain Energy Technol Assess 2021;46:101276. Available from: https://doi.org/10.1016/j.seta.2021.101276.

[137]   Dimri N, Tiwari A, Tiwari GN. Thermal modelling of semitransparent photovoltaic thermal (PVT) with thermoelectric cooler (TEC) collector. Energy Convers Manag 2017;146:68−77. Available from: https://doi.org/10.1016/j.enconman.2017.05.017.

[138]   Dimri N, Tiwari A, Tiwari GN. Comparative study of photovoltaic thermal (PVT) integrated thermoelectric cooler (TEC) fluid collectors. Renew Energy 2019;134:343−56. Available from: https://doi.org/10.1016/j.renene.2018.10.105.

[139]   Tiwari S, Agrawal S, Tiwari GN. PVT air collector integrated greenhouse dryers. Renew Sustain Energy Rev 2018;90:142−59. Available from: https://doi.org/10.1016/j.rser.2018.03.043.

[140]   Chamsa-ard W, Brundavanam S, Fung C, Fawcett D, Poinern G. Nanofluid types, their synthesis, properties and incorporation in direct solar thermal collectors: a review. Nanomaterials 2017;7:131. Available from: https://doi.org/10.3390/nano7060131.

[141]   Bianco V, Scarpa F, Tagliafico LA. Numerical analysis of the Al_2O_3-water nanofluid forced laminar convection in an asymmetric heated channel for application in flat plate PV/T collector. Renew Energy 2018;116:9−21. Available from: https://doi.org/10.1016/j.renene.2017.09.067.

[142]   Yazdanifard F, Ameri M, Ebrahimnia-Bajestan E. Performance of nanofluid-based photovoltaic/thermal systems: a review. Renew Sustain Energy Rev 2017;76:323−52. Available from: https://doi.org/10.1016/j.rser.2017.03.025.

[143]   Said Z, Saidur R, Rahim NA. Energy and exergy analysis of a flat plate solar collector using different sizes of aluminium oxide based nanofluid. J Clean Prod 2016;133:518−30. Available from: https://doi.org/10.1016/j.jclepro.2016.05.178.

[144]   Al-Shamani AN, Alghoul MA, Elbreki AM, Ammar AA, Abed AM, Sopian K. Mathematical and experimental evaluation of thermal and electrical efficiency of PV/T collector using different water based nano-fluids. Energy 2018;145:770−92. Available from: https://doi.org/10.1016/j.energy.2017.11.156.

[145]   Sardarabadi M, Passandideh-Fard M. Experimental and numerical study of metal-oxides/water nanofluids as coolant in photovoltaic thermal systems (PVT). Sol Energy Mater Sol Cell 2016;157:533−42. Available from: https://doi.org/10.1016/j.solmat.2016.07.008.

[146]   Ghadiri M, Sardarabadi M, Pasandideh-fard M, Moghadam AJ. Experimental investigation of a PVT system performance using nano ferrofluids. Energy Convers Manag 2015;103:468−76. Available from: https://doi.org/10.1016/j.enconman.2015.06.077.

[147]   Radwan A, Ahmed M, Ookawara S. Performance enhancement of concentrated photovoltaic systems using a microchannel heat sink with nanofluids. Energy Convers Manag 2016;119:289−303. Available from: https://doi.org/10.1016/j.enconman.2016.04.045.

[148]   Al-Waeli AHA, Chaichan MT, Kazem HA, Sopian K. Comparative study to use nano-(Al 2 O 3, CuO, and SiC) with water to enhance photovoltaic thermal PV/T collectors. Energy Convers Manag 2017;148:963−73. Available from: https://doi.org/10.1016/j.enconman.2017.06.072.

[149]   Ajarostaghi SSM, Delavar MA, Dolati A. Numerical investigation of melting process in horizontal shell-and-tube phase change material storage considering different htf channel geometries. Heat Transf Res 2017;48:1515−29. Available from: https://doi.org/10.1615/HeatTransRes.2017015549.

[150] Olfian H, Ajarostaghi SSM, Farhadi M, Ramiar A. Melting and solidification processes of phase change material in evacuated tube solar collector with U-shaped spirally corrugated tube. Appl Therm Eng 2021;182:116149. Available from: https://doi.org/10.1016/j.applthermaleng.2020.116149.

[151] Preet S, Bhushan B, Mahajan T. Experimental investigation of water based photovoltaic/thermal (PV/T) system with and without phase change material (PCM). Sol Energy 2017;155:1104−20. Available from: https://doi.org/10.1016/j.solener.2017.07.040.

[152] Diallo TMO, Yu M, Zhou J, Zhao X, Shittu S, Li G, et al. Energy performance analysis of a novel solar PVT loop heat pipe employing a microchannel heat pipe evaporator and a PCM triple heat exchanger. Energy 2019;167:866−88. Available from: https://doi.org/10.1016/j.energy.2018.10.192.

[153] Liang R, Zhang J, Ma L, Li Y. Performance evaluation of new type hybrid photovoltaic/thermal solar collector by experimental study. Appl Therm Eng 2015;75:487−92. Available from: https://doi.org/10.1016/j.applthermaleng.2014.09.075.

Chapter 3

Advances in solar greenhouse systems for cultivation of agricultural products

Karunesh Kant[1,2], Pascal Biwole[1,4], Ibrahim Shamseddine[1,3], Ghadie Tlaiji[1] and Fabienne Pennec[1]

[1]Université Clermont Auvergne, Clermont Auvergne INP, CNRS, Institut Pascal, Clermont-Ferrand, France, [2]Advanced Materials and Technologies Laboratory, Department of Mechanical Engineering, Virginia Tech, Blacksburg, VA, United States, [3]Université Libanaise, Centre de Modélisation, Ecole Doctorale des Sciences et Technologie, Hadath, Liban, Lebanon, [4]MINES Paris Tech, PSL Research University, PERSEE—Center for Processes, Renewable Energies and Energy Systems, Sophia Antipolis, France

3.1 Introduction

World's population is growing continuously and has almost doubled since the early 1960s. It is predicted to overtake 9.8 billion people by 2050 [1]. The growth of the world's population emphasizes the importance of adapting to the changing food requirements of populations. As a result, investment in the agriculture sector is critical to ensuring food stability and reducing global undernourishment and malnutrition [2]. At present, greenhouse farming is an important industry in several countries as a result of the growing need for agricultural products as a consequence of population growth. In this regard, employing agricultural greenhouses for farming is a supplementary substitute to meet raised food demand year around. The growth of plants depends on various environmental factors such as light, temperature, relative humidity, and carbon dioxide (CO_2) [3]. The microclimate for efficient crop production can be controlled by transparent or semitransparent plastic covering structures as in cultivation greenhouses [4]. The greenhouse structure's shielding not only transmits the useful wavelength of sunlight for photosynthesis but also limits air circulation, which has a more significant effect on plant growth by limiting disease transmission [5]. In greenhouses, environmental factors can be controlled and managed at an optimal level throughout the year, while in open-field farming, there is little control on environmental

Solar Energy Advancements in Agriculture and Food Production Systems.
DOI: https://doi.org/10.1016/B978-0-323-89866-9.00010-9

77

parameters (i.e., light, temperature, water, humidity, and nutrition) that influence plant growth. Therefore, most flowers, vegetables, and other agricultural products need to be transported from one place to another that poses additional costs [6]. Improved environmental control within greenhouses by the provision of essential heating and artificial lighting, as well as intensified production systems, resulted in increased energy consumption [7]. The use of solar energy as a sustainable energy source has recently received considerable attention since employing solar technologies in agricultural greenhouses can assist in reducing production costs, particularly because the solar greenhouse itself is a solar collector [8,9]. Solar energy, being a clean, extensible, and sustainable renewable energy source with a low environmental effect, might be considered as a viable choice for integration with agricultural greenhouses. In this approach, the employment of solar technology can help to reduce greenhouse cultivation systems' reliance on conventional energy sources, resulting in lower green house gas (GHG) emissions.

In this chapter, the authors present the integration of agricultural greenhouses with different solar energy technologies including photovoltaic (PV), photovoltaic-thermal (PVT), and solar thermal collectors. Additionally, the employment of different types of thermal energy storage (TES) units with solar greenhouses is presented and discussed. Further, the modeling thermal behavior of solar-powered greenhouses, economics, and case studies along with future research areas are discussed.

3.2 Solar greenhouse technology

In solar greenhouses, solar energy is collected and stored in a variety of ways and therefore the solar greenhouses differ in their designs. The type of employed solar collection and storage systems depends on many factors including the climate of the location, size of the greenhouse, type of the cultivated plant, the orientation of the greenhouse, and whether a new greenhouse is to be planned or existing is retrofitted [10]. Passive solar greenhouses (PSGs) are designed in a way to collect as much solar energy as possible [11] where the heat is passively collected by the greenhouse. While, in active solar greenhouses, heat is collected by the use of external systems such as PV, PVT, or solar thermal collectors. To intensify the collection of solar heat when thermal collectors are employed, auxiliary energy sources to run external devices for the circulation and/or delivery of the solar energy are required. In both passive and active designs, employing TES units can increase the overall thermal performance of the greenhouse [12]. PSGs have simpler structures and lower capital and operating costs in comparison with active structures. However, the higher thermal efficiency of the active solar greenhouses improves their profitability by partially offsetting the costs [2].

3.2.1 Passive solar greenhouses

PSGs are typically equipped with thermal collection units to maximize the solar gains by capturing incoming solar radiation and storing excess heat in heat storage media. These greenhouses can be grouped according to the heat storage medium and its characteristics. A typical PSG is shown in Fig. 3.1. Several studies on PSGs have been conducted for advanced farming. In this regard, Bazgaou et al. [13] assessed the performance of the passive solar sleeves for heating Canarian greenhouse combined with rock-bed and water as TES. Based on the performance assessment, it was found that tomato production can be improved by 49% compared to the conventional greenhouse. Santamouris et al. [14] designed and fabricated a PSG to reduce heat losses and increase useful solar gains on a daily and seasonal basis. After monitoring the PSG for two years, it was found that the greenhouse can save almost 30% of total heating energy in comparison with an identical conventional greenhouse. Wei et al. [15] investigated the thermal performance of single-span PSGs with removable back walls for the climate in southern Jiangsu province, China (Fig. 3.2). Two types of greenhouse systems, one with a Fully Removable Back Wall (FRG) and the other with a Half-Removable Back Wall (HRG) were investigated. Additionally, the removable wall was integrated with jute fiber. In the sunny days of winter, it was found that the inside temperature of a single-span greenhouse is only 2.9°C. However, the temperature of the greenhouse with removable walls was maintained above 8.2°C. In summer, the maximum difference between the outdoor and the indoor air temperature was 9°C, 6.8°C, and 6.1°C for single-span PSG, HRG, and FRG respectively. Moreover, there was no significant difference in heat release for HRG and FRG. It was found that the removable wall integrated with jute fibreboards is a potential design improvement for the back wall of solar passive greenhouses.

Tong et al. [16] created a new sliding cover and energy-saving solar greenhouse with a circular-type roof and compared it to elliptical-type

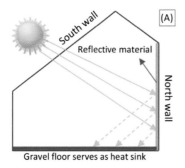

FIGURE 3.1 (A) Schematic view of a typical PSG; (B) photo of a real PSG [1].

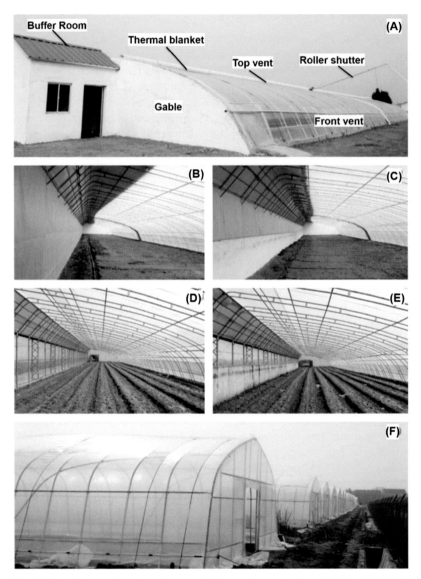

FIGURE 3.2 Photographs of experimental PSGs: (A) External view of FRG and HRG; (B) back wall of FRG in winter; (C) back wall of HRG in winter; (D) back wall of FRG in summer; (E) back wall of HRG in summer; (F) external view of PSGs [15].

greenhouses. The newly built solar greenhouse indicated an improved performance compared to conventional greenhouses. In winter, the effective aperture was increased by 15% and in summer, the cooling load was decreased by 16%. Zhang et al. [17] designed a PSG with heat storage walls for

nonarable lands. To improve the solar energy efficiency and boost "Gobi agriculture," five passive heat-storage north walls were designed for nonarable lands by using nonsoil and locally available inexpensive materials. These included a flange-layered wall (FL), a concrete-layered wall (CL), a gravel-layered wall (GL), an aerocrete brick-layered wall (AL), and a concrete hollow block-layered wall (HL). The impact of these walls on the greenhouse's thermal environment was investigated using unsteady state simulation analysis. The results indicated that FL, CL, and GL can enhance the heat storage/release performance by 5.0%, 38.0%, and 37.3%, and the energy efficiency by 9.2%, 12.9%, and 13.5%, respectively. As a result, the nocturnal interior temperature was averagely increased by 0.7°C, 2.5°C, and 2.5°C, respectively. The results showed a large contribution of FL, CL, and GL toward promoting the thermal environment of solar greenhouses.

Zhang et al. [18] studied the heat transfer performance of the wall with three different external insulation schemes suitable for passive Chinese Solar greenhouses (CSGs). The multilayer wall was not sensitive to the change in the outdoor RH, and the RH of the clay filling in the wall was kept at approximately 99%, which made the wall a stable and efficient heat storage. The main heat storage body was rammed earth in the assembled multilayer wall of the greenhouse, followed by the ordinary reinforced concrete board on the indoor side, while the light energy-saving insulation board on the outdoor side was reported as the smallest heat storage and was mainly used for heat preservation and insulation. The average heat storage ratios were reported as 75.81%, 21.92%, and 2.27%, respectively. They asserted that the 280-mm thick wall on the indoor side has a rapid response to the change in the indoor temperature and can continuously release heat to the interior over 4 days and 21 hours.

3.2.2   Greenhouses integrated with photovoltaic modules

PV systems in greenhouses are a modern technology that reduces their reliance on fossil fuels. Solar PV modules can be integrated into greenhouses in the same way as in buildings, but due to the required transparency of cladding elements to enable sunlight to penetrate inside the greenhouse, different methods should be employed [19,20]. On-grid and off-grid solar PV systems are the common configurations that are combined with agricultural greenhouses. The PV modules' generated power can be directly utilized by the greenhouse. In on-grid PV systems and during the periods of excess production of energy, the surplus is injected into the power grid, while in off-grid PV systems, it is stored in batteries. In periods of deficiency of energy conversion, the batteries provide electricity and in some cases, a complimentary electric generator can be used to supply the electric demand [21]. Using off-grid PV systems is the best choice for installation in locations where there is no access to the local grid due to long distances or complicated topography.

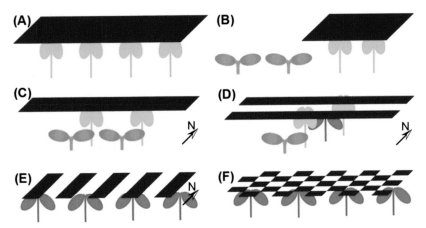

FIGURE 3.3 Effect of the use of opaque PV modules on the crop growth recognized by the color of the crop (yellow, light, and dark green) where PV cover is distributed (A) entirely, (B) partially, (C) longitudinally straight, (D) longitudinally stripped, (E) transversely straight, and (F) checkboard design [23].

Nonetheless, several considerations should be made to optimize the areas covered by PV modules installed on greenhouse enclosures. To achieve optimum growth, cultivated crops need a sufficient amount of light. If the percentage of covering surfaces increases, the amount of received solar radiation inside the greenhouse and consequently the amount of light received by the plants cultivated inside is decreased. For a 1% rise in the PV cover ratio (PVR),[1] the annual decrease in the received solar radiation is estimated as 0.8% [22]. Fig. 3.3 shows the effect of decreasing the quantity of received light below the crucial limit value on the growth trend of the crop [23]. As shown in this figure, the deficiency of light can cause a yellowish color due to a decrease in photosynthesis. As the PVR increases, the internal air temperature decreases, whereas in the case of no ventilation the humidity increases [24].

Cossu et al. [25] inspected the behavior of different horticultural and floricultural crops as a function of PVR that ranged between 25% and 100%. The study consisted of 14 greenhouses located in the southern part of Europe. The coefficient of variation of shadow inside a greenhouse was increased as the PVR increased. As a result, the crop development was not uniquely due to variation in the received light. As a yearly analysis, the coefficient of variation (CV)[2] ranged from 0.31 to 0.60 as the PVR was increased from 25% to 100%. However, as a monthly analysis, the CV ranged from

---

1. PVR is the ratio of projection area of PV modules to surface area of greenhouse.
2. CV is the percentage ratio between the standard deviation and the mean, and quantifies the variability of light distribution on the greenhouse area.

0.51 to 0.92 for 100% PVR and from 0.4 to 0.47 for 25% PVR. Similar outcomes were obtained by Cossu et al. [26] in 2014 in Italy, where the average crop production decreased by 42% (from 10 to 5.2 kg/m^2). Moreover, for the same PVR (50%), Italian PV greenhouses showed some negative impacts on lycopene and sugar content as reported by Bulgari et al. [27]. However, moderate and high demand crops showed an acceptable yield factor for PVR up to 25%. For 10% PVR, tomato crops also showed an acceptable yield factor as reported by Ezzaeri et al. [24]. For tomato and sweet pepper crops, PVR values ranging from 9.8% to 20% resulted in no yield and a decrease in quality [24,28]. The moderate light showed an acceptable yield fraction for PVR up to 50%, except for the yield factor of lettuce. For floricultural crops, petals of roses benefit from moderate shading. Hatamian and Salehi [29] observed protein, carbohydrate, and anthocyanin content quality enhancement of petals in a greenhouse with PVR up to 50%. The development of low light demand crops was not affected by the PVR and remained above the accepted value of 75% even for 100% PVR. Buttaro et al. [30] studied the effect of different covering materials on greenhouses built in Italy, as shown in Fig. 3.4. The use of semitransparent PV modules ensures the passage of a certain percentage of solar radiation. Fig. 3.4A shows the use of traditional opaque PV modules (TPV), Fig. 3.4B shows the use of transparent polycarbonate module (PM) and Fig. 3.4C shows the use of innovative semitransparent PV modules (IPM). From the results, rocket yield growth in TPV was lower compared with IPM and PM.

Luminescent solar concentrators (LSCs) are devices designed to concentrate solar radiation into PV cells. LSCs are composed of colored panels of plastic material that have a special characteristic: they can capture sunlight and concentrate it along their edges, where it is intercepted by small PV cells and converted into electricity [31]. LSCs are semitransparent panels and they can be rationalized by an active slab, made of a highly transparent matrix, generally poly-(methyl methacrylate) (PMMA), containing small amounts of luminescent species. Corrado et al. [32] constructed a LSC greenhouse and an identical control greenhouse with PV cells attached to the roof of both structures as shown in Fig. 3.5. The placement and types of PV cells used in

FIGURE 3.4 Greenhouses integrated with (A) traditional opaque PV modules, (B) transparent polycarbonate, and (C) innovative polycarbonate and PV modules [1].

FIGURE 3.5 Photo of the LSC greenhouse (right) and control greenhouse (left) is shown [32].

the LSC panels were varied for performance comparisons. Solar power generation was monitored continuously for one year, with leading LSC panels exhibiting a 37% increase in power production compared to the reference. The 22.3 m^2 greenhouse was projected to convert a total of 1342 kWh per year. The LSC panels showed no signs of degradation throughout the trial, demonstrating the material's robustness in field conditions. Bashir and Jaghwani [33] prepared red fluorescent polycarbonate (PC) films by solvent-casting technique from PC solution doped with different dye concentrations of perylene dyestuff (KREMER Red 94720). The effect of the dye concentration on the structure and photophysical properties was studied using X-ray diffraction, UV−Vis, and fluorescence spectroscopy. The optimum dye concentration of red fluorescent PC films showed the best emission efficiency for the doping concentration 0.3 wt.%. These results endorsed a new application of this film for active greenhouse LSC dryers (LSCDs) with an enhanced feature of integrated solar-powered convection.

3.2.3 Greenhouses integrated with photovoltaic-thermal modules

Solar PV modules available in the market have an efficiency ranging from 20% to 22% [34]. A substantial portion of the collected solar radiation is dissipated as heat which raises the surface temperature of the PV module and significantly decreases electricity production [35−37]. This heat can be extracted using a hybrid system called the PVT module. A hybrid PVT module has the benefit of both producing electricity and acting as a thermal energy collector. The cooling medium (usually air/water) removes the excess induced heat from the PV module in this hybrid collector, improving overall electric performance. The extracted heat can then be used in low to medium-temperature applications [38]. Fig. 3.6 shows a classification of PVT modules according to the type of the collector, coolant, and material [39]. Concentrating PVT (CPVT) modules use curved reflectors or refractors to focus solar radiation onto multijunction (MJ) or nonsilicon solar cells, resulting in efficiencies of up to 40%. Because of their high optical and thermal

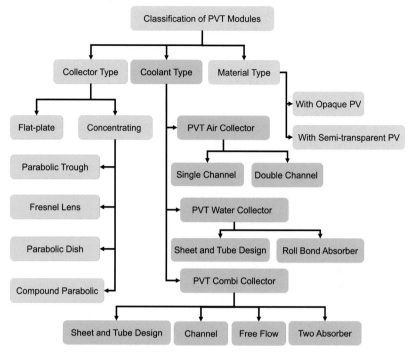

FIGURE 3.6 Different photovoltaic-thermal module technologies according to the type of collector, coolant, and used material [1].

efficiencies, the CPVT modules are among the most efficient hybrid collectors [40,41].

The flat-plate PVT (FP-PVT) collectors are typically installed on greenhouse façades or inclined roofs, where they can also serve as an opaque covering material and a safety barrier that allows the whole structure of the greenhouse to be straighter [42]. Sonneveld et al. [43] built a unique greenhouse with linear Fresnel lenses. The CPVT collector was installed on the greenhouse's roof and mounted using a solar tracking device in their architecture. They reported the generated electricity and the thermal yield as 29 kWh/m^2/year and 518 MJ/m^2 respectively. According to the findings, the CPVT device can be used to monitor the temperature as well as provide illumination inside the greenhouse. Wu et al. [44] performed tests on a CPV collector cover capable of converting surplus light into heat under realistic environmental conditions. They found that when the overall radiation is 1095 W/m^2, the optimal thermal power of 353 W/m^2 can be achieved when the instantaneous efficiency is 32.2%. Furthermore, the findings revealed that the novel covering material would provide uniform indoor brightness, resulting in better internal thermal conditions. Wu et al. [45] conducted an experimental study on a CPVT module in which a gallium arsenide (GaAs)

PV cell was used as a receiver. The main parameters such as temperature distribution of the receiver, electric output and heat, as well as the utilization efficiency of electric and thermal energy were evaluated. The results of evaluations under clear sky conditions showed that the maximum power generation efficiency is about 18% at noon (11:00–13:00), the maximum thermal efficiency of cooling water is about 45%, and the total efficiency of thermal and electrical power is about 55%. Combined with facility agricultural engineering, it comprehensively solves the problem of land use and the utilization efficiency of solar energy systems.

3.2.4 Greenhouse integrated with solar thermal collectors

Because of high energy conversion efficiency, solar thermal collectors have gained a lot of attention. The solar collector and the TES unit are two main components of solar thermal systems [46,47]. Solar thermal collectors are used in greenhouses to capture solar radiation and convert it into heat that can then be transmitted to the indoor space of the greenhouse to provide an optimal thermal atmosphere for the plants grown inside [48,49]. Solar thermal collectors are classified into two categories: nonconcentrating and concentrating [49]. Semple et al. [50] conducted a techno-economic study of a commercial solar greenhouse located in Canada to decrease the energy demand of greenhouse. The energy demand of solar greenhouse with an area of 4000 m^2 was decreased by 64% using flat-plate collectors (FPC) with 861 m^2 heat collection area. The overall coefficient of performance (COP) of the solar greenhouse was 2.9. Lazaar et al. [51] studied a 200 m^2 solar greenhouse integrated with an evacuated tube collector (etc.) with a 200 L TES tank installed under the climate conditions of Tunisia. This system was able to increase the temperature of the inside air by 4°C with 46% energy efficiency of the evacuated tube solar water heater. Xu et al. [49] proposed a water-circulating solar heat collection and release system with an indoor collector constructed of hollow polycarbonate sheets (low-cost, high-strength, and good-durability) for the solar greenhouse. It could thus absorb indoor solar heat during the day and transfer it to the greenhouse for heating at night. The findings of the experiment revealed a daily average heat collection ratio of 72.1%. Awani et al. [52] studied the performance of a greenhouse (100 m^2) in Tunisia with a heating system that included a FPC (2 m^2) connected to a heat pump and a vertical heat exchanger to achieve a comfortable environment inside the greenhouse during the winter. When the outdoor temperature was in the range 15.6°C–20°C, the highest and lowest indoor greenhouse temperatures were recorded as 35°C and 16°C, respectively. Hassanien et al. [53] studied the effect of a coupled system consisting of an etc., a storage tank, and a heat pump on a 32 m^2 solar greenhouse with a plastic cover as shown in Fig. 3.7. The system was able to increase the

FIGURE 3.7 Photos of the heating system: (A) The ETC, (B) the electric heat pump (air to water), and the hot water storage tank, (C) Control greenhouse "without heating system," and (D) the heated greenhouse with heating pipes installed inside the greenhouse [53].

temperature by an average of 2.5°C more than the greenhouse with no heat pump with 45% efficiency and 4.1 years payback period.

The performance of an Active Solar Heating System (ASHS) consisting of two solar water heaters equipped with flat solar collectors, two storage tanks, and heat exchanger pipes was studied by Bazgaou et al. [54]. Experimental results indicated that the ASHS system improves the nocturnal climatic conditions under greenhouse. The thermal comfort created by the ASHS system in the root zone increases the absorption of nutrients, which improves the external quality (color, size, weight, and firmness) and the internal quality (sugar content, acidity, and taste) of tomato fruits. This improvement was also reflected by increasing total tomato yield by 55% in the winter period. In another study, Hussain et al. [55] tested the effectiveness of two Fresnel collectors, one spot and the other linear type, for heating two separate research greenhouses under climate conditions of Chuncheon, South Korea. The results of the experiments revealed that the greenhouse with the spot Fresnel collector performs 7%−12% better than the linear Fresnel collector (Fig. 3.8) with thermal efficiency (extracted thermal energy/incident solar radiation) values ranging from 45% to 70%. The

FIGURE 3.8 Experimental setup of concentrating solar collectors integrated with a greenhouse: (A) Spot Fresnel lens collector system and (B) linear Fresnel lens collector system [55].

techno-economic study showed that the proposed model saves $7344, $8658, and $11,405 over the long term compared to the use of grid electricity, kerosene, and gasoline for heating the greenhouse, respectively.

3.2.5 Solar greenhouses integrated with heat storage

TES systems increase the effectiveness of solar-powered greenhouses by accumulating heat leftover from the daytime operation and using the accumulated energy during nocturnal activities and cloudy days [2,56]. Thermal energy can be stored in the form of sensible, latent, or chemical reactions [56]. Different sensible and latent heat TES materials that can be used for solar greenhouses are given in Table 3.1. Several greenhouses with TES have been developed to improve the environment inside the greenhouses and minimize the need of fossil fuels as an auxiliary heating source.

3.2.5.1 Greenhouses integrated with sensible heat storage

Various types of sensible thermal energy storage (STES) can be integrated into the solar greenhouses to improve their performance. The most common types are rock bed TES (RB-TES), underground TES (UTES), and water TES (WTES) [1]. Guordo et al. [64] integrated RB-TES into a solar greenhouse and found that the crop yield increases by 22% owing to the increase of temperature at night by 3°C and 4.7°C as compared to the traditional greenhouse. Having a higher temperature at night is due to the release of the heat stored in the rock bed during the day. Bazgaou et al. [65] experimented to see how the RB-TES affected the microclimate and crop development in the solar greenhouse. The geometry, material, and date of planting were all kept the same to study the effects of the rock bed on the temperature, humidity, and plant growth within the greenhouse. Fig. 3.9 shows the average development of tomato plants in a greenhouse with (red bars) and without

TABLE 3.1 Sensible and latent heat thermal energy storage materials.

Medium	Temperature range (°C)	Density (kg/m³)	Specific heat (J/kg K)
Rock [57,58]	20	2560	879
Brick [57,58]	20	1800	1880
Reinforced concrete (2%) [57,58]	20	2400	1000
Concrete (high density) [57,58]	20	1900–2400	880
Water [57,58]	0–100	1000	4190
Engine oil [57,58]	Up to 160	888	1880
Ethanol [57,58]	Up to 78	790	2400
Propanol [57,58]	Up to 97	800	2500
Butanol [57,58]	Up to 118	809	2400

Latent heat storage material

Material	Melting temperature (°C)	Latent heat (kJ/kg)	Type of PCM
Polyethylene glycol 600 [59–62]	20–25	146	Organic
D-Lattic acid [60,63]	26	184	Organic
Capric acid [60,63]	32	152	Organic
Docasyl bromide [60]	40	201	Organic
Lauric acid [60,63]	49	178	Organic
Camphene [60]	50	238	Organic
Myristic acid [60,63]	58	199	Organic
Stearic acid [60,63]	69.4	199	Organic
$CaCl_2 \cdot 12H_2O$ [60]	29.8	174	Inorganic
$LiNO_3 \cdot 3H_2O$ [60]	30	189	Inorganic
$KFe(SO_4)_2 \cdot 12H_2O$ [60]	33	173	Inorganic
$LiBr_2 \cdot 2H_2O$ [60]	34	124	Inorganic
$FeCl_3 \cdot 6H_2O$ [60]	37	223	Inorganic
$CoSO_4 \cdot 7H_2O$ [60]	40.7	170	Inorganic
$CaCl_2 \cdot 6H_2O + CaBr_2 \cdot 6H_2O$ (45% + 55%, wt.) [60]	14.7	140	Eutectic
$CaCl_2 + MgCl_2 \cdot 6H_2O$ (50% + 50%, wt.) [60]	25	95	Eutectic

(*Continued*)

TABLE 3.1 (Continued)

Triethylolethane + urea (62.5% + 37.5%, wt.) [60]	29.8	218	Eutectic
$CH_3COONa \cdot 3H_2O$ + NH_2CONH2 (40% + 60%, wt.) [60]	30	200.5	Eutectic
$Mg(NO_3)_3 \cdot 6H_2O$ + NH_4NO_3(61.5% + 38.5%, wt.) [60]	52	125.5	Eutectic

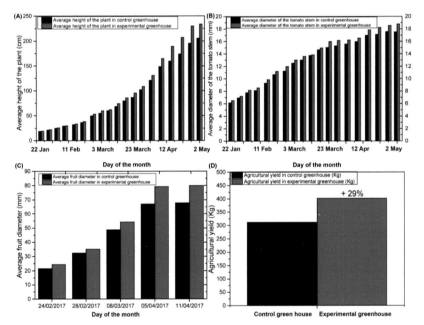

FIGURE 3.9   Average values of the tomato plant (A) height, (B) stem diameter, (C) fruit diameter, and (D) agricultural yield[3] in the greenhouse with (red bars) and without (black bars) rock beds [65].

(black bars) a rock bed. Using rock beds in the solar greenhouse shows significant effects on the plant's growth, plant's length, and stem diameter as well as the size and weight of the product. Further, Bazgaou et al. [66] investigated the effectiveness of a solar heating cooling system (SHCS) in a solar greenhouse microclimate. They concluded that using SHCS to heat and cool

---

3. Agricultural yield is a standard measurement of the amount of agricultural production harvested (yield of a crop) per unit of land area.

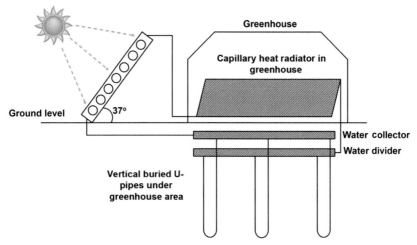

FIGURE 3.10 Schematic drawing of a solar soil heat storage system for use in a greenhouse [67].

the indoor air would increase the greenhouse microclimate and the tomato yield. The findings also showed that the air temperature within the greenhouse integrated with the SHCS is 3°C higher at night and 6°C lower during the day with fewer variations than the conventional greenhouses.

Fig. 3.10 shows a novel low-cost seasonal solar soil heat storage system (SSSHS) for heating applications inside a greenhouse developed by Zhang et al. [67]. During the cold season, the heat energy could be extracted from the UTES and supplied to the greenhouse. They also created a TRNSYS model to simulate the absorption and conservation of solar energy in the soil. Under comparable conditions, the quantities of energy absorbed by the SSSHS and the conventional solar heating system were compared, and it was discovered that the energy consumption of the soil TES is lower than that of the traditional heating process.

Li et al. [68] designed a solar heating system that uses a Fresnel lens to heat greenhouses. Thermal energy was stored using a soil heat storage method. The findings showed that when heating pipes are buried at a depth of 1.65 m, heat transfer to the ground takes around 5 days, resulting in an overall temperature rise of roughly 4°C. Furthermore, it was found that using TES during the coldest season with harsh weather conditions can ensure an interior air temperature over 8°C, guaranteeing the minimum temperature necessary for crop growth. Bezari et al. [69] conducted a comparative study between control and heated greenhouse in a semiarid region. A new design of greenhouse was proposed that consists of an economical rock-bed with the sensible heat technique for heating system in an integrated H-shape channel. The excess diurnal heat captured by the greenhouse is stored in the system and then restored for nocturnal heating. The results obtained indicate that this thermal storage system is efficient and ameliorates the greenhouse climate. The night temperature was improved by 3.2°C and the relative humidity was reduced by 9.6%, compared to a standard greenhouse.

3.2.5.2 Greenhouse integrated with latent heat storage

Latent heat thermal energy storage (LHTES) materials are also known as phase change materials (PCMs) in which thermal energy is stored during phase transition from one state to another. The energy density of LHTES is greater than that of STES. In the transition cycle of the LHTES, the PCM will store a significant amount of thermal energy, and the solid to liquid phase transformation happens at a constant temperature (phase transition temperature). The selection criteria and desirable thermophysical properties of PCM for solar greenhouses were recently discussed by Gorjian et al. [1]. Several researchers have improved the performance of solar greenhouses by integrating PCMs. In this regard, Oztürk et al. [70] integrated paraffin wax as a PCM to the north wall of the solar greenhouse having a $180 \, m^2$ floor area in Turkey. The external heat collection unit consisted of $27 \, m^2$ of south-facing solar air heaters mounted at a 55 degrees tilt angle. During the experimental period, it was found that the average net energy and exergy efficiencies of the system are 40.4% and 4.2%, respectively. Kooli et al. [71] studied the effect of a nocturnal shutter on an insulated greenhouse using a solar air heater with LHTES. The results indicated that the greenhouse with a nocturnal shutter has a 2°C higher indoor temperature than the greenhouse without a shutter. It was noted that the greenhouse internal temperature with the shutter is held at 15°C during the night, while the exterior temperature reached 8°C. Benli et al. [72] investigated the impact of incorporating a PCM storage device into novel solar collectors to meet the greenhouse's heating demands in Turkey. The experimental setup (Fig. 3.11) comprised 10 solar air collectors, each having a surface area of $1.7 \times 0.7 \times 0.025 \, m$ and connected to a storage tank containing $300 \, kg$ of $CaCl_2 0.6H_2O$ PCM. The collected heat was transmitted to the storage tank via hot air and a heat exchanger. In comparison to traditional heating methods, the obtained results showed that proposed collectors combined with the PCM would provide approximately 18%−23% of the total daily thermal energy demand of the greenhouse for 3−4 hours.

Han et al. [73] proposed to introduced an active-passive ventilation wall with latent heat storage (APVW-L) for the effective utilization of solar energy in solar greenhouses during winter as shown in Fig. 3.12. This study demonstrated that the optimized APVW-L could store $5.36 \, MJ/(m^2 \cdot day)$ of solar energy in Beijing. Compared to the identical conventional greenhouses, after midnight, the experimental greenhouse having APVW-L increased the back wall's interior surface temperature by 2.2°C−3.4°C and the average

FIGURE 3.11   View of solar collectors and the experimental greenhouse: (A) Arrangement of FPCs, (B) air circulation lines, (C) solar collectors integrated with the experimental greenhouse, and (D) crop grown inside the solar greenhouse [72].

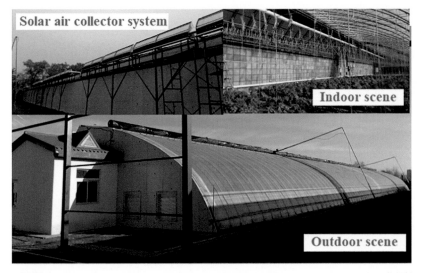

FIGURE 3.12   Appearance of the APVW-L integrated into a solar greenhouse's back wall [73].

indoor air temperature by $0.8°C-1.4°C$. Mirahmad et al. [74] determined the feasibility of a LHTES system for reducing the temperature swing in a greenhouse numerically and the results are compared with experimental values. As a consequence, the diurnal interior temperature can be decreased by $10°C$ by employing a LHTES cooling system.

Table 3.2 presented the findings of studies conducted with different types of solar greenhouses. The most common ways of mounting PV modules on greenhouses are the installation of them on roofs and in very few cases on

TABLE 3.2 Main findings of some studies conducted with different types of solar greenhouses.

Type of solar greenhouse	Main findings	References
Greenhouse + PV	As a yearly analysis, the coefficient of variation (CV) ranged from 0.31 to 0.60 as the PVR was increased from 25% to 100%. However, as a monthly analysis, the CV ranged from 0.51 to 0.92 for 100% PVR and from 0.4 to 0.47 for 25% PVR.	Cossu et al. [25]
Greenhouse + PV	From the results, rocket yield growth in traditional opaque PV modules (TPV) was lower compared with innovative semitransparent PV modules (IPM) and transparent polycarbonate modules (PM).	Buttaro et al. [30]
Greenhouse + PV + LSC	LSC panels exhibiting a 37% increase in power production compared to the reference. The 22.3 m^2 greenhouse was projected to generate a total of 1342 kWh per year, or 57.4 kWh/m^2 if it were composed solely of the leading panel of Criss Cross panel design	Corrado et al. [32]
Greenhouse + PV + LSC	The optimum dye concentration of red fluorescent PC films showed the best emission efficiency for the doping concentration 0.3 wt.%.	Bashir and Jaghwani [33]
Greenhouse + CPVT (Fresnel lens)	The generated electricity and the thermal yield as 29 kWh/m^2/year and 518 MJ/m^2, respectively	Sonneveld et al. [43]
Greenhouse + CPVT	The optimal thermal power was achieved 353 W/m^2 with instantaneous efficiency of 32.2% when the overall radiation is 1095 W/m^2.	Wu et al. [44]

(Continued)

TABLE 3.2 (Continued)

Type of solar greenhouse	Main findings	References
Greenhouse + PVT	The maximum power generation efficiency is about 18% at noon (11:00−13:00), the maximum thermal efficiency of cooling water is about 45%, and the total efficiency of thermal and electrical power is about 55%.	Wu et al. [45]
Greenhouse + solar thermal collectors	The overall COP of the solar greenhouse was obtained at 2.9.	Semple et al. [50]
Greenhouse + ETC	The system was able to increase the temperature of the inside air by 4°C with 46% energy efficiency of the evacuated tube solar water heater	Lazaar et al. [51]
Greenhouse + solar thermal collectors	Findings of the experiment revealed a daily average heat collection ratio of 72.1%	Xu et al. [49]
Greenhouse + FPC	The highest and lowest indoor greenhouse temperatures were recorded as 35°C and 16°C when the outdoor temperature was in the range of 15.6°C and 20°C, respectively	Awani et al. [52]
Greenhouse + ETC	The system was able to increase the temperature by an average of 2.5°C more than the greenhouse with no heat pump with 45% efficiency and 4.1 years payback period	Hassanien et al. [53]
Greenhouse + FPC	ASHS system improves the nocturnal climatic conditions under greenhouse. Total tomato yield was increased by 55% in the winter period.	Bazgaou et al. [54]
Solar greenhouse + TES	Crop yield increases by 22% owing to the increase of temperature at night by 3°C and 4.7°C as compared to the traditional greenhouse.	Guordo et al. [64]
Solar greenhouse + TES	Findings showed that the air temperature within the greenhouse integrated with the SHCS is 3°C higher at night and 6°C lower during the day with fewer variations than the conventional greenhouses	Bazgaou et al. [66]

(Continued)

TABLE 3.2 (Continued)

Type of solar greenhouse	Main findings	References
Solar greenhouse + TES	It was found that using TES during the coldest season with harsh weather conditions can ensure an interior air temperature over 8°C, guaranteeing the minimum temperature necessary for crop growth.	Li et al. [68]
Solar greenhouse + TES	The night temperature was improved by 3.2°C and the relative humidity was reduced by 9.6%, compared to a standard greenhouse.	Bezari et al. [69]
Solar greenhouse + TES	It was found that the average net energy and exergy efficiencies of the system are 40.4% and 4.2%, respectively.	Oztürk et al. [70]
Solar greenhouse + TES	The results indicated that the greenhouse with a nocturnal shutter has a 2°C higher indoor temperature than the greenhouse without a shutter. It was noted that the greenhouse internal temperature with the shutter is held at 15°C during the night, while the exterior temperature reached 8°C.	Kooli et al. [71]
Solar greenhouse + TES	Compared to the identical conventional greenhouses, after midnight, the experimental greenhouse having APVW-L increased the back wall's interior surface temperature by 2.2°C–3.4°C and the average indoor air temperature by 0.8°C–1.4°C.	Han et al. [73]
Solar greenhouse + TES	The diurnal interior temperature was reduced 10°C by employing a LHTES system	Mirahmad et al. [74]

walls of greenhouses. The low efficiency and high prices of PV modules are the most dominant factors that decrease this technology's attractiveness for investment. PVT modules are the most efficient energy conversion systems that can be integrated with greenhouses, providing both thermal and electric requirements of the greenhouse. Solar thermal collectors can be considered as promising alternatives for meeting heating demands in greenhouses, particularly under moderate climatic conditions, because the heating potential of solar thermal collectors is highly dependent on the location of greenhouses in terms of solar radiation availability and local weather conditions. All research studies, including theoretical studies and experimental assessments, have reported improved thermal efficiency and higher interior temperatures with fewer fluctuations in solar greenhouses integrated with TES units, which are especially desired during cold seasons and at night.

3.3   Modeling of solar greenhouses

Numerous studies on the heat and mass transport phenomena in the solar greenhouse system have been performed using mathematical-based models. The majority of these models were developed to evaluate the impact of integrating solar greenhouses with heat storage systems and to examine their energy-efficient design depending on specific weather conditions. There are three major approaches used to explain the thermal behavior of solar greenhouses as knowledge-based models, simulation software, and machine-learning algorithms [1].

3.3.1   Knowledge-based model

The knowledge-based model describes the thermal behavior of solar green-houses from a thermodynamic point of view [75–77]. Jain and Tiwari [78] developed a code in MATLAB® to model the energy balance of a solar greenhouse. The study was conducted on a greenhouse with a surface area of $24 \, m^2$ integrated with a ground air collector (GAC). The predicted plant and room air temperature indicated good agreement with experimental values with an average coefficient of correlation of 0.94 and root mean square (RMS) error 8.4%. An experimental study was conducted by Bargach et al. [79] to compare the thermal efficiency of a $250 \, m^2$ area plastic tunnel solar greenhouse combined with FPCs and a $9 \, m^2$ glass greenhouse associated with selective solar collectors. A first-order differential equation was used to solve the heat balance for both systems. The comparative analysis showed that the first one has better thermal efficiency and is easier to install. The second system contributed to the indoor air conditioning during the day and allowed to reduce its cost by 10%. It was also noted that the solar FPC's system should be used in regions with a moderate climate while the other system can be used in the region where the climate is harsher. An experiment was conducted in Delhi, India by Nayak and Tiwari [80] combining the greenhouse with a PVT collector to assess the thermal performance. Good agreement was observed between the experiment and the simulation output with a RMS error varying between 7.1% and 17.6% and a correlation coefficient ranging between 0.95 and 0.97. Hussain et al. [81] developed a numerical model for solar greenhouse and differential equations were solved using MATLAB software. The experimental measurements and numerical results were compared for two distinct systems. The first system was composed of a

CPVT collector with a glass-reinforced plastic cover and the second without reinforcement enclosure around the absorber structure. As in the previous experiments, energy balance equations were applied for components composing the solar greenhouse that can absorb or exchange heat with the environment. Results revealed an average deviation of about 4.5% between simulated and measured data. The overall outcomes of the CPVT system with glass enclosure were 15% higher than those without covering the collector. Also, by using MATLAB software, a CSGHEAT program was used to examine the heat balance of a Chinese greenhouse model. Gao et al. [82] studied greenhouses integrated with different PV layouts densities. Models included high-density and low-density PV systems that presented, respectively, 1/2 and 1/4 to 1/3 of the greenhouse roof area. Besides, four sun-tracking devices were taken into consideration during the study. A comparison of all cases revealed that the yearly improvement in average irradiance and uniformity of a no-shading sun-tracking high-density panel was around 10.96% and 10.68%, respectively. Zhang et al. [83] developed a mathematical model to evaluate the microlight climate and thermal performance of a Liaoshen-type solar greenhouse (LSG). It was found that the shaded part of the canopy and the radiation intensity of unshaded parts of the north wall drop respectively about 200 and 100 W/m^2 when compared to an empty greenhouse at noon.

3.3.2 Software simulation

A literature survey indicated that the most common greenhouse thermal simulation software are TRNSYS and EnergyPlus. Chen et al. [84] validated an EnergyPlus model with an experiment realized in China. They considered a greenhouse of specific dimensions of about 27 m in length, 5.8 m in span, 2.9 m in ridge height, 2.3 m of north wall height, and 30 degrees inclination of the front roof. The main aim of this research was to determine the optimal orientation for the solar system to maximize the incoming sunlight. Baglivo et al. [85] performed a numerical study on a solar greenhouse considering thermal phenomena with detailed modeling of complex volume discretization, three-dimensional shortwave, and longwave radiative exchange, airflow exchanges. The study was performed by TRNSYS software. Results showed that the cooling and heating energy were respectively 51.4 and 49.1 kWh/m^3. Fluent software has also been used to evaluate energy efficiency by working with CFD models. For example, Tong et al. [86] modeled three different CSGs with 10, 12, and 14 m spans with proportional or same north roof and north wall sizes or the same south roof size. The predictions used two-dimensional CFD simulations of the internal greenhouse environment. Results showed that CSG with all the dimensions varied in the same proportion and with a 14 m span provided the highest indoor air temperature. CSG with the same north roof and wall dimensions and with a 12 m span provided the highest indoor air temperature. From analyzes of the solar heat gains, heat losses, and temperature distributions for each group of dimensions good guidance is given for span selections to the growers. Zhang et al. [87] investigated numerically and experimentally CSGs with different north wall materials. The simulated two-dimensional CFD model was developed with the finite volume-based commercial software, Fluent, to predict the temperature variations. Results demonstrated that polystyrene has an important role in preventing heat loss compared with perforated bricks. Moreover, Liu et al. [88] validated the experimental measurements of a solar greenhouse thermal microclimate by using a CFD model. The results indicated that the north wall of this solar greenhouse could store heat due to an insulation layer of polystyrene.

3.3.3   Machine-learning algorithms

Machine learning algorithms are used to solve complex computational issues. This type of method is efficient for cases that have historically measured data. The machine learning algorithms models makes possible the analysis of thermal behavior of greenhouses and forecast of their energy consumption. In this regards Liu et al. [89] developed a back-propagation neural network (BPNN) to predict the temperature variations of solar greenhouse under the climate conditions of China. The findings concerning the high temperatures showed an agreement index between 0.7 and 0.93 and an RMS error ranging from 1.5°C to 2.2°C. Yu et al. [90] compared the improved particle swarm optimization supported by the least squares support vector machine method (IPSO-LSSVM) with the standard support vector machine (SVM) and BPNN. Results revealed that the IPSO-LSSVM gives more accurate values. In the same way, Taki et al. [91] compared the multiple linear regression (MLR) method with the artificial neural network (ANN) and the multilayer perceptron (MLP) to predict the temperature variation of the indoor air and envelopes of a semisolar greenhouse in Iran. The MLR method was not precise since the Durbin−Watson statistic factor for both the envelopes and the air temperature was about 0.05. Although the ANN showed similar results as the real data, it was concluded that the MLP model is the best model that can predict the real data. The convex bidirectional extreme learning machine (CB-ELM) is a new method that was tested by Zou et al. [92] to visualize the variation of the indoor air temperature and humidity of the proposed greenhouse. This new model was considered to be effective and suitable for predicting values when comparing it with BPNN, SVM, and radial base function (RBF).

3.4   Case studies and economics

Several solar greenhouse projects around the world have been installed or underway for crop production. Some of these projects are discussed in this section. Further, the economics of solar-powered greenhouses are discussed.

3.4.1 Case studies

A CSG with various improvements was installed in Netherland to ensure an appropriate indoor environment for the crops as shown in Fig. 3.13A [93]. In this solar greenhouse, the humidity levels and the temperature are controlled by using ventilation and heaters operated by solar energy. The main improvements in this solar greenhouse are done by using coco peat substrate instead of soil and screening technology to offer thermal insulation during winter and to prevent the entering of heat during summer. A solar green-house in Blacksburg, Virginia USA stores the solar through a subterranean heat sink of soil, rocks, and water beneath its interior planting beds [94]. The heat stored during the day is released at night to maintain a high indoor tem-perature during the winter season. This 52 m^2 greenhouse was constructed with an insulated north roof as opposed to a double-walled polycarbonate roof. Overheating during the day is prevented using venting windows located on the east and west ends of the building as presented in Fig. 3.13B. Ventilation also helps to regulate indoor humidity. The goal of this project is to encourage homeowners, neighborhoods, and local farmers to boost the production of local food. In the same region, Kirkpatrick [95] built a PSG with a solar PV roof. The surface area of the greenhouse is about 46.5 m^2. The north wall is covered by phase change tiles and the concrete foundation is insulated. As a result, the indoor temperature is higher than the outside and achieves a maximum temperature difference of about 15°C.

A solar greenhouse project was installed in Suihua City, Heilongjiang Province, China with an area of 405 ha [96]. It includes 30 solar greenhouses and multispan greenhouses that support the installation of a submersible pump, automatic backwashing filter set, automatic fertilizer machine, and a set of agricultural irrigation monitoring systems. In this project, the indoor air temperature, humidity, soil moisture, and soil PH values are collected in real-time and modified automatically to obtain the best indoor environment for the crops. To maintain food security, a project, presented in Fig. 3.14A,

FIGURE 3.13 (A) A CSG with ventilation and heaters installed in the Netherlands [93], (B) solar greenhouse in Virginia, the USA presenting venting windows on the east and west sides to exhaust the air to the outside and ceiling fans to circulate air inside [94].

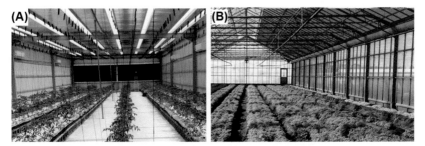

FIGURE 3.14 (A) Inside view of the solar greenhouse project installed in Qatar [97], (B) Commercial solar greenhouse integrated with luminescent solar panels [98].

FIGURE 3.15   The largest solar-roofed greenhouse project installed in Italy [100].

was set in the hot climate of Qatar to sustain about 70% self-sufficiency on solar greenhouse vegetables [97]. The solar agriculture greenhouse is thermally insulated and allows the entry of optimum solar radiation from its roof. The performance of the greenhouse in terms of reducing the cooling load and providing the required solar radiation level was assessed by ray-tracing simulations, cooling load calculations, and field trials. Soliculture, an American based company has carried out extensive crop trials in commercial solar greenhouses integrated with luminescent solar panel Fig. 3.14B to improve crop production as well as solar power generation in the different region such as California, Alberta (Canada), Ontario (Canada), New York [98,99]. The obtained results indicated that the crop yield depends on the type of crop and the climate conditions maintained in the solar greenhouse. The improvement in the various crop production is observed. For example, Herbs, provide up to 50% higher yields, while lettuce takes one week shorter to harvest. These greenhouses that use solar panels lead to capital cost savings of 30% compared to a regular greenhouse. In addition, it improves the plant growth and offsets the electricity bill for 20 years with a 3−7 years payback period.

In Sardinia, Italy, 134 solar-roofed greenhouses are located on 66 acres, as shown in Fig. 3.15, to double the productivity of the land by growing crops and generating electricity [100]. This project can provide 20 MW of electricity and prevent 25,000 tons of CO_2 emissions each year. A semicylindrical greenhouse with solar air collectors was built from galvanized steel supports in Nice, France for the production of mushrooms [101]. The system was well insulated by using polyethylene. Measurements showed that the greenhouse help by saving 27% on the total requirement of 5774 kWh. This greenhouse was not used in July and August since the indoor temperature was more than 20°C.

3.4.2 Economics

The purpose of the optimized designs is to maximize solar energy gains and energy retention inside the greenhouse. The structure, orientation, and envelope characteristics of the whole system are variables that influence the design parameters of the greenhouse and the heating requirements. Besides, the economic performance of solar buildings should be one of the key interests to be taken into account. Bargach et al. [79] compared two types of solar systems installed in Morocco, one using solar FPCs outside the greenhouse and the other based on the selective absorption of solar energy by a heat transfer fluid, using polyethylene alveolar transparent plane collectors inside the greenhouse. The efficiency of the first system was about 49% with a total cost of about $2800 while the second system had an efficiency of about 42% and a total cost of about $2500. The economic and thermal performance of a greenhouse associated with a CPVT system was studied by Imtiaz et al. [81]. The tunnel-type greenhouse situated in South Korea was composed of double layers of polythene cover with a total surface of about 32.1 m². The total cost of the CPVT system was about $5300. It was found that this type of system can save 5400 kWh of electricity consumption, 666 L of kerosene oil consumption, and 567 L of diesel consumption in 5 months. Besides, the heating cost of their system was compared to the cost of other energy sources as shown in Fig. 3.16. It is also clear that the CPVT has the lowest cost during the test period with a maximum cost of about $50 per month. An 18 m² solar greenhouse was built in Sherbrooke, Quebec province, Canada by researchers, using solar energy to heat its indoor environment [102]. The project used six PV modules with power outputs of 165 W. The total cost of this solar greenhouse was about $19,000.

Sethi and Dhiman [103] economically analyzed a solar-cum-biomass hybrid greenhouse. The installation cost of the system is about $6785 and the total cost of the project including the maintenance, the repair, the labor, and the operational costs is about 70,000 $ over 5 years. On the other hand, the total project income including the electricity savings is about $80,000 over 5 years. Benli [104] performed a techno-economic analysis of solar greenhouses. To evaluate greenhouse efficacy, the monthly heat requirement was determined using two types of one-dimensional dynamic calculation

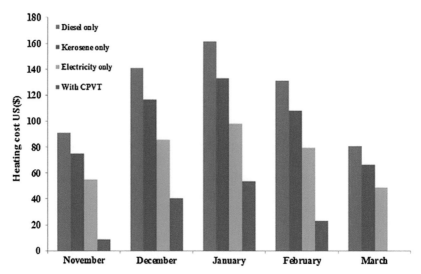

FIGURE 3.16   Monthly cost of different solar greenhouse systems [81].

models in two types of greenhouses at six locations in Turkey. The most important climate factor for greenhouse applications is the sunshine duration to which the greenhouse is exposed; for the Eastern Anatolia Region this is 2664 h/year. In July and August, the air temperature was highest, between 22°C and 30°C. The lowest temperature values were in Bitlis, Van, Erzincan, and Iğdir, and the highest temperature values in Elazig, Malatya. With a simple natural ventilation process, an average daily temperature between 12°C and 24°C can occur. This short-term problem can be solved with effective shading and ventilation. The cost of a 300 m^2 arc-roofed greenhouse was US$3000. The cost of a 240 m^2 gable-roofed greenhouse was US$3000. Mohammadi et al. [105] conducted an exergo-economic analysis and multiobjective optimization of a semisolar greenhouse with experimental validation. The results show that using double-layer glass separated with air-filled space as the greenhouse cover decreases the overall exergy destruction by about 45.36%. The multiobjective optimization results obtained from the Pareto frontier also show total exergy destruction of 8.8623 MJ and mean air unit cost of $15.9164/MJ at the optimum point.

3.5   Summary and prospects

The high energy requirement of agricultural greenhouses is considered a significant obstacle to their widespread implementation around the world. Solar energy, the most abundant alternative energy source can be used to meet the energy demand of greenhouses. This chapter discussed different types of solar greenhouses, as well as the various techniques used to improve their performance.

PSGs are simpler in design and have lower costs, but their profitability is low, while in active solar greenhouses, solar heat collection and PV systems are used to maximize the power received from the sun. Mounting PV modules on greenhouse roofs and walls is an appealing choice for greenhouses, particularly when adequate land is not available. The most noticeable obstacle, however, is the shadowing effect created by PV modules, which may prevent the sufficient entry of solar radiation within the greenhouse, negatively affecting crop production rate. Hybrid PVT collectors are a noteworthy choice for meeting the heat and power requirements of agricultural greenhouses. However, the installation of opaque FP-PVT modules on roofs may cause overheating and a blockage of visible light from entering the interior.

The incorporation of TES units into solar greenhouses can raise their thermal yield by enhancing both the inside temperature of the greenhouse and the temperature of the cultivated crops inside. The integration of solar greenhouses with TES benefits the long-term operation of these sustainable facilities, especially in areas with high fuel costs and harsh winters. Despite the advantages of using TES units in solar greenhouses, additional research in terms of technical and economic evaluations is still required to make this technology a commercial solution for use in solar greenhouses.

Mathematical models can explain and predict the thermal behavior of solar greenhouses. The knowledge-based models are linked to physical parameters and use energy balance equations to explain the thermal activity of solar greenhouses. Machine learning algorithms can help to explain the dynamics of greenhouses, but they are insufficiently accurate since they can only produce the most reliable results from the testing data collection, which is not necessarily available for greenhouses.

Despite the many advantages that can be attributed to solar energy systems and their implementation into agricultural greenhouses, they have also been linked with environmental effects that are directly dependent on the scale of the projects and the characteristics of the location where the solar greenhouse is implemented. In this regard, environmental impact assessments (EIAs) and life cycle assessments (LCAs) of solar greenhouses are highly recommended for future research. With more technological and economic advancements, as well as the implementation of encouraging policies and appealing mechanisms, new solar greenhouses are expected to provide a great global opportunity to facilitate sustainable growth in the agriculture sector in the future.

Acknowledgments

The authors want to acknowledge the Université Clermont Auvergne postdoctoral grant 2020−21.

References

[1] Gorjian S, Calise F, Kant K, Ahamed MS, Copertaro B, Najafi G, et al. A review on opportunities for implementation of solar energy technologies in agricultural greenhouses. J Clean Prod 2021;. Available from: https://doi.org/10.1016/j.jclepro.2020.124807.

[2] Shukla A, Sharma A, Kant K. Solar greenhouse with thermal energy storage: a review. Curr Sustain Energy Rep 2016;3:58–66. Available from: https://doi.org/10.1007/s40518-016-0056-y.

[3] Martzopoulou A, Vafiadis D, Fragos V. Energy gain in passive solar greenhouses due to CO2 enrichment. Energies 2020;13:1242. Available from: https://doi.org/10.3390/en13051242.

[4] Sapounas A, Katsoulas N, Slager B, Bezemer R, Lelieveld C. Design, control, and performance aspects of semi-closed greenhouses. Agronomy 2020;10:1739. Available from: https://doi.org/10.3390/agronomy10111739.

[5] Cooper PI, Fuller RJ. A transient model of the interaction between crop, environment and greenhouse structure for predicting crop yield and energy consumption. J Agric Eng Res 1983;28:401–17. Available from: https://doi.org/10.1016/0021-8634(83)90133-6.

[6] Brinckmann JA. Geographical indications for medicinal plants:globalization, climate change, quality and market implications for geo-authentic botanicals. World J Tradit Chin Med 2015;1:16–23. Available from: https://doi.org/10.15806/j.issn.2311-8571.2014.0020.

[7] Naghibi Z, Carriveau R, Ting DS-K. Improving clean energy greenhouse heating with solar thermal energy storage and phase change materials. Energy Storage 2020;2. Available from: https://doi.org/10.1002/est2.116.

[8] Abdel-Ghany AM, Al-Helal IM. Solar energy utilization by a greenhouse: general relations. Renew Energy 2011;36:189–96. Available from: https://doi.org/10.1016/j.renene.2010.06.020.

[9] Sethi VP, Sharma SK. Survey and evaluation of heating technologies for worldwide agricultural greenhouse applications. Sol Energy 2008;82:832–59. Available from: https://doi.org/10.1016/j.solener.2008.02.010.

[10] Gorjian S, Ebadi H, Najafi G, Singh Chandel S, Yildizhan H. Recent advances in net-zero energy greenhouses and adapted thermal energy storage systems. Sustain Energy Technol Assess 2021;43:100940. Available from: https://doi.org/10.1016/j.seta.2020.100940.

[11] Panwar NL, Kaushik SC, Kothari S. Solar greenhouse an option for renewable and sustainable farming. Renew Sustain Energy Rev 2011;15:3934–45. Available from: https://doi.org/10.1016/j.rser.2011.07.030.

[12] Du X, Li P, Zhao C, Sang G. The effect of PCM with different thermos-physical parameters on indoor temperature of Xi'an solar greenhouse. IOP Conf Ser Earth Environ Sci 2021;631:012014. Available from: https://doi.org/10.1088/1755-1315/631/1/012014.

[13] Bazgaou A, Fatnassi H, Bouharroud R, Elame F, Ezzaeri K, Gourdo L, et al. Performance assessment of combining rock-bed thermal energy storage and water filled passive solar sleeves for heating Canarian greenhouse. Sol Energy 2020;198:8–24. Available from: https://doi.org/10.1016/j.solener.2020.01.041.

[14] Santamouris M, Argiriou A, Vallindras M. Design and operation of a low energy consumption passive solar agricultural greenhouse. Sol Energy 1994;52:371–8.

[15] Wei B, Guo S, Wang J, Li J, Wang J, Zhang J, et al. Thermal performance of single span greenhouses with removable back walls. Biosyst Eng 2016;141:48–57. Available from: https://doi.org/10.1016/j.biosystemseng.2015.11.008.

[16] Tong X, Sun Z, Sigrimis N, Li T. Energy sustainability performance of a sliding cover solar greenhouse: solar energy capture aspects. Biosyst Eng 2018;176:88–102. Available from: https://doi.org/10.1016/j.biosystemseng.2018.10.008.

[17] Zhang X, Lv J, Dawuda MM, Xie J, Yu J, Gan Y, et al. Innovative passive heat-storage walls improve thermal performance and energy efficiency in Chinese solar greenhouses for non-arable lands. Sol Energy 2019;190:561−75. Available from: https://doi.org/10.1016/j.solener.2019.08.056.

[18] Zhang G, Shi Y, Liu H, Fei Z, Liu X, Wei M, et al. Heat transfer performance of an assembled multilayer wall in a Chinese solar greenhouse considering humidity. J Energy Storage 2021;33:102046. Available from: https://doi.org/10.1016/j.est.2020.102046.

[19] Allardyce CS, Fankhauser C, Zakeeruddin SM, Grätzel M, Dyson PJ. The influence of greenhouse-integrated photovoltaics on crop production. Sol Energy 2017;155:517−22. Available from: https://doi.org/10.1016/j.solener.2017.06.044.

[20] Harjunowibowo D, Ding Y, Omer S, Riffat S. Recent active technologies of greenhouse systems−a comprehensive review. Bulg J Agric Sci 2018;24:158−70. Available from: http://www.agrojournal.org/24/01-22.pdf.

[21] Gorjian S, Zadeh BN, Eltrop L, Shamshiri RR, Amanlou Y. Solar photovoltaic power generation in Iran: development, policies, and barriers. Renew Sustain Energy Rev 2019;106:110−23. Available from: https://doi.org/10.1016/j.rser.2019.02.025.

[22] Cossu M, Cossu A, Deligios PA, Ledda L, Li Z, Fatnassi H, et al. Assessment and comparison of the solar radiation distribution inside the main commercial photovoltaic greenhouse types in Europe. Renew Sustain Energy Rev 2018;94:822−34. Available from: https://doi.org/10.1016/j.rser.2018.06.001.

[23] Yano A, Cossu M. Energy sustainable greenhouse crop cultivation using photovoltaic technologies. Renew Sustain Energy Rev 2019;109:116−37. Available from: https://doi.org/10.1016/j.rser.2019.04.026.

[24] Ezzaeri K, Fatnassi H, Bouharroud R, Gourdo L, Bazgaou A, Wifaya A, et al. The effect of photovoltaic panels on the microclimate and on the tomato production under photovoltaic canarian greenhouses. Sol Energy 2018;173:1126−34. Available from: https://doi.org/10.1016/j.solener.2018.08.043.

[25] Cossu M, Yano A, Solinas S, Deligios PA, Tiloca MT, Cossu A, et al. Agricultural sustainability estimation of the European photovoltaic greenhouses. Eur J Agron 2020;118:126074. Available from: https://doi.org/10.1016/j.eja.2020.126074.

[26] Cossu M, Murgia L, Ledda L, Deligios PA, Sirigu A, Chessa F, et al. Solar radiation distribution inside a greenhouse with south-oriented photovoltaic roofs and effects on crop productivity. Appl Energy 2014;133:89−100. Available from: https://doi.org/10.1016/j.apenergy.2014.07.070.

[27] Bulgari R, Cola G, Ferrante A, Franzoni G, Mariani L, Martinetti L. Micrometeorological environment in traditional and photovoltaic greenhouses and effects on growth and quality of tomato (*Solanum lycopersicum* L.). Ital J Agrometeorol 2015;20:27−38.

[28] Trypanagnostopoulos G, Kavga A, Souliotis Y. Tripanagnostopoulos, greenhouse performance results for roof installed photovoltaics. Renew Energy 2017;111:724−31. Available from: https://doi.org/10.1016/j.renene.2017.04.066.

[29] Hatamian M, Salehi H. Physiological characteristics of two rose cultivars (*Rosa hybrida* L.) under different levels of shading in greenhouse conditions. J Ornam Plants 2017;7:147−55.

[30] Buttaro D, Renna M, Gerardi C, Blando F, Serio F, Santamaria P. Soilless production of wild rocket as affected by greenhouse coverage with photovoltaic modules. Acta Sci Pol Hortorum Cultus 2016;15:129.

[31] Kant K, Nithyanandam K, Pitchumani R. Analysis and optimization of a novel hexagonal waveguide concentrator for solar thermal applications. Energies 2021;14:2146. Available from: https://doi.org/10.3390/en14082146.

[32] Corrado C, Leow SW, Osborn M, Carbone I, Hellier K, Short M, et al. Power generation study of luminescent solar concentrator greenhouse. J Renew Sustain Energy 2016; 8:043502. Available from: https://doi.org/10.1063/1.4958735.

[33] El-Bashir SM, Al-Jaghwani AA. Perylene-doped polycarbonate coatings for acrylic active greenhouse luminescent solar concentrator dryers. Results Phys 2020;16:102920. Available from: https://doi.org/10.1016/j.rinp.2019.102920.

[34] Most Efficient Solar Panels: Solar panel efficiency explained | energysage https://news. energysage.com/what-are-the-most-efficient-solar-panels-on-the-market/; 2021 [accessed 21.06.21].

[35] Kant K, Shukla A, Sharma A, Biwole PH. Heat transfer studies of photovoltaic panel coupled with phase change material. Sol Energy 2016;140:151−61. Available from: https://doi.org/10.1016/j.solener.2016.11.006.

[36] Kant K, Shukla A, Sharma A, Biwole PH. Thermal response of poly-crystalline silicon photovoltaic panels: Numerical simulation and experimental study. Sol Energy 2016;134:147−55. Available from: https://doi.org/10.1016/j.solener.2016.05.002.

[37] Idoko L, Anaya-Lara O, McDonald A. Enhancing PV modules efficiency and power output using multi-concept cooling technique. Energy Rep 2018;4:357−69. Available from: https://doi.org/10.1016/j.egyr.2018.05.004.

[38] Kant K, Pitchumani R, Shukla A, Sharma A. Analysis and design of air ventilated building integrated photovoltaic (BIPV) system incorporating phase change materials. Energy Convers Manag 2019;196:149−64. Available from: https://doi.org/10.1016/j.enconman.2019.05.073.

[39] Diwania S, Agrawal S, Siddiqui AS, Singh S. Photovoltaic−thermal (PV/T) technology: a comprehensive review on applications and its advancement. Int J Energy Environ Eng 2020;11.

[40] Kurtz S, Geisz J. Multijunction solar cells for conversion of concentrated sunlight to electricity. Opt Express 2010;. Available from: https://doi.org/10.1364/oe.18.000a73.

[41] Dimroth F, Grave M, Beutel P, Fiedeler U, Karcher C, Tibbits TND, et al. Wafer bonded four-junction GaInP/GaAs//GaInAsP/GaInAs concentrator solar cells with 44.7% efficiency. Prog Photovolt Res Appl 2014;22:277−82. Available from: https://doi.org/10.1002/pip.2475.

[42] Ghani S, Bakochristou F, ElBialy EMAA, Gamaledin SMA, Rashwan MM, Abdelhalim AM, et al. Design challenges of agricultural greenhouses in hot and arid environments − a review. Eng Agric Environ Food 2019;12:48−70. Available from: https://doi.org/10.1016/j.eaef.2018.09.004.

[43] Sonneveld PJ, Swinkels GLAM, van Tuijl BAJ, Janssen HJJ, Campen J, Bot GPA. Performance of a concentrated photovoltaic energy system with static linear Fresnel lenses. Sol Energy 2011;85:432−42. Available from: https://doi.org/10.1016/j.solener.2010.12.001.

[44] Wu G, Yang Q, Fang H, Zhang Y, Zheng H, Zhu Z, et al. Photothermal/day lighting performance analysis of a multifunctional solid compound parabolic concentrator for an active solar greenhouse roof. Sol Energy 2019;180:92−103. Available from: https://doi.org/10.1016/j.solener.2019.01.007.

[45] Wu G, Yang Q, Zhang Y, Fang H, Feng C, Zheng H. Energy and optical analysis of photovoltaic thermal integrated with rotary linear curved Fresnel lens inside a Chinese solar greenhouse. Energy 2020;197:117215. Available from: https://doi.org/10.1016/j.energy.2020.117215.

[46] Gorjian S, Tavakkoli Hashjin T, Ghobadian B, Banakar A. A thermal performance evaluation of a medium-temperature point-focus solar collector using local weather data and

artificial neural networks. Int J Green Energy 2015;12:493−505. Available from: https://doi.org/10.1080/15435075.2013.848405.

[47] Ketabchi F, Gorjian S, Sabzehparvar S, Shadram Z, Ghoreishi MS, Rahimzadeh H. Experimental performance evaluation of a modified solar still integrated with a cooling system and external flat-plate reflectors. Sol Energy 2019;187:137−46. Available from: https://doi.org/10.1016/j.solener.2019.05.032.

[48] Gorjian S, Ebadi H, Calise F, Shukla A, Ingrao C. A review on recent advancements in performance enhancement techniques for low-temperature solar collectors. Energy Convers Manag 2020;222:113246. Available from: https://doi.org/10.1016/j.enconman.2020.113246.

[49] Xu W, Song W, Ma C. Performance of a water-circulating solar heat collection and release system for greenhouse heating using an indoor collector constructed of hollow polycarbonate sheets. J Clean Prod 2020;253:119918. Available from: https://doi.org/10.1016/j.jclepro.2019.119918.

[50] Semple L, Carriveau R, Ting DS-K. A techno-economic analysis of seasonal thermal energy storage for greenhouse applications. Energy Build 2017;154:175−87. Available from: https://doi.org/10.1016/j.enbuild.2017.08.065.

[51] Lazaar M, Bouadila S, Kooli S, Farhat A. Comparative study of conventional and solar heating systems under tunnel Tunisian greenhouses: thermal performance and economic analysis. Sol Energy 2015;120:620−35. Available from: https://doi.org/10.1016/j.solener.2015.08.014.

[52] Awani S, Chargui R, Kooli S, Farhat A, Guizani A. Performance of the coupling of the flat plate collector and a heat pump system associated with a vertical heat exchanger for heating of the two types of greenhouses system. Energy Convers Manag 2015;103:266−75. Available from: https://doi.org/10.1016/j.enconman.2015.06.032.

[53] Hassanien RHE, Li M, Tang Y. The evacuated tube solar collector assisted heat pump for heating greenhouses. Energy Build 2018;169:305−18. Available from: https://doi.org/10.1016/j.enbuild.2018.03.072.

[54] Bazgaou A, Fatnassi H, Bouharroud R, Ezzaeri K, Gourdo L, Wifaya A, et al. Effect of active solar heating system on microclimate, development, yield and fruit quality in greenhouse tomato production. Renew Energy 2021;165:237−50. Available from: https://doi.org/10.1016/j.renene.2020.11.007.

[55] Imtiaz Hussain M, Ali A, Lee GH. Performance and economic analyses of linear and spot Fresnel lens solar collectors used for greenhouse heating in South Korea. Energy 2015;90:1522−31. Available from: https://doi.org/10.1016/j.energy.2015.06.115.

[56] Kant K, Shukla A, Sharma A. Advancement in phase change materials for thermal energy storage applications. Sol Energy Mater Sol Cell 2017;172:82−92. Available from: https://doi.org/10.1016/j.solmat.2017.07.023.

[57] ISO-10456, Building materials and products—Hygrothermal properties—Tabulated design values and procedures for determining declared and design thermal values, 2007.

[58] Tatsidjodoung P, Le Pierrès N, Luo L. A review of potential materials for thermal energy storage in building applications. Renew Sustain Energy Rev 2013;18:327−49. Available from: https://doi.org/10.1016/j.rser.2012.10.025.

[59] Baetens R, Jelle BPBP, Gustavsen A. Phase change materials for building applications: a state-of-the-art review. Energy Build 2010;42:1361−8. Available from: https://doi.org/10.1016/j.enbuild.2010.03.026.

[60] Sharma A, Tyagi VV, Chen CR, Buddhi D. Review on thermal energy storage with phase change materials and applications. Renew Sustain Energy Rev 2009;13:318−45. Available from: https://doi.org/10.1016/j.rser.2007.10.005.

[61] Su W, Darkwa J, Kokogiannakis G. Review of solid−liquid phase change materials and their encapsulation technologies. Renew Sustain Energy Rev 2015;48:373−91. Available from: https://doi.org/10.1016/j.rser.2015.04.044.

[62] Zhu N, Ma Z, Wang S. Dynamic characteristics and energy performance of buildings using phase change materials: a review. Energy Convers Manag 2009;50:3169−81. Available from: https://doi.org/10.1016/j.enconman.2009.08.019.

[63] Kant K, Shukla A, Sharma A. Performance evaluation of fatty acids as phase change material for thermal energy storage. J Energy Storage 2016;6:153−62. Available from: https://doi.org/10.1016/j.est.2016.04.002.

[64] Gourdo L, Fatnassi H, Tiskatine R, Wifaya A, Demrati H, Aharoune A, et al. Solar energy storing rock-bed to heat an agricultural greenhouse. Energy 2019;169:206−12. Available from: https://doi.org/10.1016/j.energy.2018.12.036.

[65] Bazgaou A, Fatnassi H, Bouhroud R, Gourdo L, Ezzaeri K, Tiskatine R, et al. An experimental study on the effect of a rock-bed heating system on the microclimate and the crop development under canarian greenhouse. Sol Energy 2018;176:42−50. Available from: https://doi.org/10.1016/j.solener.2018.10.027.

[66] Bazgaou A, Fatnassi H, Bouharroud R, Ezzaeri K, Gourdo L, Wifaya A, et al. Efficiency assessment of a solar heating cooling system applied to the greenhouse microclimate. Mater Today Proc 2020;. Available from: https://doi.org/10.1016/j.matpr.2019.10.101.

[67] Zhang L, Xu P, Mao J, Tang X, Li Z, Shi J. A low cost seasonal solar soil heat storage system for greenhouse heating: design and pilot study. Appl Energy 2015;156:213−22. Available from: https://doi.org/10.1016/j.apenergy.2015.07.036.

[68] Li Z, Ma X, Zhao Y, Zheng H. Study on the performance of a curved fresnel solar concentrated system with seasonal underground heat storage for the greenhouse application. J Sol Energy Eng Trans ASME 2018;141:1−9. Available from: https://doi.org/10.1115/1.4040839.

[69] Bezari S, Bekkouche S, Benchatti A. Investigation and improvement for a solar greenhouse using sensible heat storage material. FME Trans 2021;49:154−62. Available from: https://doi.org/10.5937/fme2101154B.

[70] Öztürk HH. Experimental evaluation of energy and exergy efficiency of a seasonal latent heat storage system for greenhouse heating. Energy Convers Manag 2005;46:1523−42. Available from: https://doi.org/10.1016/j.enconman.2004.07.001.

[71] Kooli S, Bouadila S, Lazaar M, Farhat A. The effect of nocturnal shutter on insulated greenhouse using a solar air heater with latent storage energy. Sol Energy 2015;115:217−28. Available from: https://doi.org/10.1016/j.solener.2015.02.041.

[72] Benli H, Durmuş A. Performance analysis of a latent heat storage system with phase change material for new designed solar collectors in greenhouse heating. Sol Energy 2009;83:2109−19. Available from: https://doi.org/10.1016/j.solener.2009.07.005.

[73] Han F, Chen C, Hu Q, He Y, Wei S, Li C. Modeling method of an active−passive ventilation wall with latent heat storage for evaluating its thermal properties in the solar greenhouse. Energy Build 2021;238:110840. Available from: https://doi.org/10.1016/j.enbuild.2021.110840.

[74] Mirahmad A, Sadrameli SM. A comparative study on the modeling of a latent heat energy storage system and evaluating its thermal performance in a greenhouse. Heat Mass Transf Und Stoffuebertragung 2018;54:2871−84. Available from: https://doi.org/10.1007/s00231-018-2316-4.

[75] Taki M, Ajabshirchi Y, Ranjbar SF, Rohani A, Matloobi M. Modeling and experimental validation of heat transfer and energy consumption in an innovative greenhouse structure.

Inf Process Agric 2016;3:157−74. Available from: https://doi.org/10.1016/j.inpa.2016. 06.002.

[76] Effat MB, Shafey HM, Nassib AM. Solar greenhouses can be promising candidate for CO2 capture and utilization: mathematical modeling. Int J Energy Environ Eng 2015; 6:295−308. Available from: https://doi.org/10.1007/s40095-015-0175-z.

[77] Esmaeli H, Roshandel R. Optimal design for solar greenhouses based on climate conditions. Renew Energy 2020;145:1255−65. Available from: https://doi.org/10.1016/j.renene. 2019.06.090.

[78] Jain D, Tiwari GN. Modeling and optimal design of ground air collector for heating in controlled environment greenhouse. Energy Convers Manag 2003;44:1357−72. Available from: https://doi.org/10.1016/S0196-8904(02)00118-8.

[79] Bargach M, Tadili R, Dahman A, Boukallouch M. Comparison of the performance of two solar heating systems used to improve the microclimate of agricultural greenhouses in Morocco. Renew Energy 2004;29:1073−83. Available from: https://doi.org/10.1016/ S0960-1481(03)00101-0.

[80] Nayak S, Tiwari GN. Energy and exergy analysis of photovoltaic/thermal integrated with a solar greenhouse. Energy Build 2008;40:2015−21. Available from: https://doi.org/ 10.1016/j.enbuild.2008.05.007.

[81] Imtiaz Hussain M, Ali A, Lee GH. Multi-module concentrated photovoltaic thermal system feasibility for greenhouse heating: model validation and techno-economic analysis. Sol Energy 2016;135:719−30. Available from: https://doi.org/10.1016/j.solener.2016.06.053.

[82] Gao Y, Dong J, Isabella O, Santbergen R, Tan H, Zeman M, et al. Modeling and analyses of energy performances of photovoltaic greenhouses with sun-tracking functionality. Appl Energy 2019;233−234:424−42. Available from: https://doi.org/10.1016/j.apenergy.2018. 10.019.

[83] Zhang Y, Henke M, Li Y, Yue X, Xu D, Liu X, et al. High resolution 3D simulation of light climate and thermal performance of a solar greenhouse model under tomato canopy structure. Renew Energy 2020;160:730−45. Available from: https://doi.org/10.1016/j. renene.2020.06.144.

[84] Chen C, Li Y, Li N, Wei S, Yang F, Ling H, et al. A computational model to determine the optimal orientation for solar greenhouses located at different latitudes in China. Sol Energy 2018;165:19−26. Available from: https://doi.org/10.1016/j.solener.2018.02.022.

[85] Baglivo C, Mazzeo D, Panico S, Bonuso S, Matera N, Congedo PM, et al. Data from a dynamic simulation in a free-floating and continuous regime of a solar greenhouse modelled in TRNSYS 17 considering simultaneously different thermal phenomena. Data Br 2020;33:106339. Available from: https://doi.org/10.1016/j.dib.2020.106339.

[86] Tong G, Christopher DM, Zhang G. New insights on span selection for Chinese solar greenhouses using CFD analyses. Comput Electron Agric 2018;149:3−15. Available from: https://doi.org/10.1016/j.compag.2017.09.031.

[87] Zhang X, Wang H, Zou Z, Wang S. CFD and weighted entropy based simulation and optimisation of Chinese solar greenhouse temperature distribution. Biosyst Eng 2016; 142:12−26. Available from: https://doi.org/10.1016/j.biosystemseng.2015.11.006.

[88] Liu X, Li H, Li Y, Yue X, Tian S, Li T. Effect of internal surface structure of the north wall on Chinese solar greenhouse thermal microclimate based on computational fluid dynamics. PLoS One 2020;15:e0231316. Available from: https://doi.org/10.1371/journal.pone.0231316.

[89] Liu S, Xue Q, Li Z, Li C, Gong Z, Li N, et al. An air temperature predict model based on BP neural networks for solar greenhouse in North China. J China Agric Univ 2015; 20:176−84.

[90] Yu H, Chen Y, Hassan SG, Li D. Prediction of the temperature in a Chinese solar green-house based on LSSVM optimized by improved PSO. Comput Electron Agric 2016;122:94−102. Available from: https://doi.org/10.1016/j.compag.2016.01.019.

[91] Taki M, Ajabshirchi Y, Ranjbar SF, Rohani A, Matloobi M. Heat transfer and MLP neural network models to predict inside environment variables and energy lost in a semi-solar greenhouse. Energy Build 2016;110:314−29. Available from: https://doi.org/10.1016/j.enbuild.2015.11.010.

[92] Zou W, Yao F, Zhang B, He C, Guan Z. Verification and predicting temperature and humidity in a solar greenhouse based on convex bidirectional extreme learning machine algorithm. Neurocomputing 2017;249:72−85. Available from: https://doi.org/10.1016/j.neucom.2017.03.023.

[93] High-tech Chinese solar greenhouse project gets underway in Netherlands - Produce Business UK; 2021. https://www.producebusinessuk.com/high-tech-chinese-solar-green-house-project-gets-underway-in-netherlands/ [Date accessed 30-March-2021].

[94] Solaripedia, Green Architecture & Building, Projects in Green Architecture & Building; 2021. https://www.solaripedia.com/13/projects.html [Date accessed 30-March-2021].

[95] Building a Solar PV Greenhouse, LinkedIn; 2021. Available from: https://www.linkedin.com/pulse/building-solar-pv-greenhouse-dave-kirkpatrick/ [Date accessed 30-March-2021].

[96] Chinese agricultural water S&T; 2021. http://toronto.china-consulate.org/eng/st/40/ [Date accessed 30 March-2021].

[97] Qatar University to develop Qatar's first solar agriculture greenhouse; 2021. https://the-peninsulaqatar.com/article/21/09/2020/QU-to-develop-Qatar%E2%80%99s-first-solar-agriculture-greenhouse [Date accessed 30-March-2021].

[98] Soliculture, Soliculture/Value; 2021. http://www.soliculture.com/value/ [Date accessed 30-March-2021].

[99] Soliculture, plant response; 2021. http://www.soliculture.com/plant-response/ [Date accessed 30-March-2021].

[100] Largest solar-roofed greenhouse project in the world now operational in Italy- Greenhouse Canada; 2021. https://www.greenhousecanada.com/largest-solar-roofed-greenhouse-project-in-the-world-now-operational-in-italy-20201/ [Date accessed 30-March - 2021].

[101] Solar Greenhouse for the production of "Pleurote" mushrooms, Project, ENG, CORDIS, European Commission; 2021, https://cordis.europa.eu/project/id/SE.-00608-83 [Date accessed 30-March-2021].

[102] Solar greenhouse for crops − PV Magazine International; 2021, https://www.pv-magazine.com/2020/12/08/solar-greenhouse-for-crops/ [Date accessed 30-March-2021].

[103] Sethi VP, Dhiman M. Design, space optimization and modelling of solar-cum-biomass hybrid greenhouse crop dryer using flue gas heat transfer pipe network. Sol Energy 2020;206:120−35. Available from: https://doi.org/10.1016/j.solener.2020.06.006.

[104] Benli H. Techno-economic analysis of solar greenhouses. Int J Veg Sci 2020; 26:249−61. Available from: https://doi.org/10.1080/19315260.2019.1664698.

[105] Mohammadi B, Ranjbar F, Ajabshirchi Y. Exergoeconomic analysis and multi-objective optimization of a semi-solar greenhouse with experimental validation. Appl Therm Eng 2020;164:114563. Available from: https://doi.org/10.1016/j.applthermaleng.2019.114563.

Chapter 4

Photovoltaic water pumping systems for irrigation: principles and advances

Pietro Elia Campana[1], Iva Papic[2], Simson Jakobsson[2] and Jinyue Yan[1]
[1]*Mälardalen University, Future Energy Center, Västerås, Sweden,* [2]*KTH — Royal Institute of Technology, School of Industrial Engineering and Management, Stockholm, Sweden*

4.1 Introduction

Irrigation plays a fundamental role in sustaining our world's food supply because it increases yields and reduces crop yields' vulnerability to weather phenomena such as drought and climate changes. Although harvested land has not shown significant growth since the 1960s, irrigation, together with the mechanization of the agricultural sector and the use of agro-chemicals, has led to a linear increase in crop yield from the beginning of the Green Revolution until today, as Fig. 4.1 shows. According to the Food and Agriculture Organization (FAO) of the United Nations, 40% of global crop production comes from irrigated areas, which cover 20% of the global harvested area [1,2]. The area equipped with irrigation systems more than doubled from 1961 to 2009 and it is projected to reach 337 million ha by 2050 [3]. Recently, by using a remote-sensing approach based on analysis of the

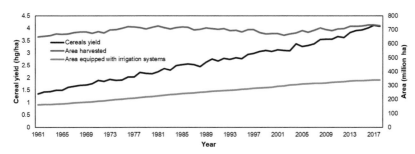

FIGURE 4.1 Cereal yields, and harvested and irrigated areas, 1961–2018 [1].

Solar Energy Advancements in Agriculture and Food Production Systems.
DOI: https://doi.org/10.1016/B978-0-323-89866-9.00007-9
113

normalized difference vegetation index (NDVI), Meier et al. [4] have shown that global irrigated areas retrieved from statistical databases may have been underestimated by as much as 18%. Globally, the agricultural sector consumes 70% of total freshwater withdrawals, and most of this is used in the irrigation sector. Irrigation is responsible for most of the groundwater depletion worldwide [5]. Megatrends such as population growth, dietary shifts, and climate change are likely to lead to global food demand increasing further (and thus to concomitant increases in irrigated areas and freshwater consumption to help meet that demand), as it is projected to increase by between 59% and 98% by 2050 [6].

Nevertheless, projections on the irrigation sector are still uncertain. Wada et al. [7] projected an increase in groundwater extraction of 39% by 2050 and a growth in irrigated areas of about 20 million ha by 2050 as compared to the period 2005−07. A study conducted by the International Institute for Applied Systems Analysis (IIASA) showed that although irrigated areas are going to increase under three different Shared Socioeconomic Pathways, the expected increase in irrigation efficiency could offset an increase in irrigated land in the order of 10%−15% by 2050, thus leading to a nonincrease in total water demand in the irrigation sector [8]. Rosa et al. [9] identified two different sustainable paths for converting rain-fed agricultural land to irrigated land to support food production under a changing climate for 300 million and 1.4 billion more people worldwide. The first one was based on the exploitation of small-scale water storage solutions and deficit irrigation, while the second one was based on the expansion of large-scale water storage systems. On the other hand, according to FAO [3], although during the period 1961−2009 the global irrigated area increased at a rate of 1.6% per annum since 2009 the trend has been significantly slowing and it is projected to be 0.1% per annum until 2050. This is mostly due to issues related to water scarcity. For instance, in areas like the Middle East, North Africa, and Central Asia—and in India and China, where more than 80% of the freshwater withdrawals are used in the agricultural sector—serious water scarcity issues are already a threat (i.e., unsustainable depletion of groundwater resources) [3]. Puy [10] analyzed three datasets concerning irrigation areas, including the FAO areas equipped for irrigation [1,2], and identified a superlinear growth rate between the expansion of irrigated areas and population growth. This superlinear trend was most accentuated in the Americas, Asia, and Europe and was justified by Puy [10] with the concept of "*aggregate effect*".

Mekonnen and Gerbens-Leenes [11] estimated that the global consumptive water footprint for food production is going to increase by 22% by 2090. The projected increase is mostly related to climate and land-use changes. Mekonnen and Gerbens-Leenes [11] identified that increasing water productivity and shifting diets towards crops distinguished by low water input are some of the pathways towards sustainable water management. Several pathways towards reducing water consumption and maximizing water use efficiency were listed by Cole et al. [12], including combining weather prediction with hydrological modeling, satellite observations, and distributed Internet of Things (IoT) sensors. According to Mancosu et al. [13], improving water productivity through crop management strategies, improvements in irrigation efficiencies, and the development and wide use of decision support tools are key strategies for sustainable management of water in agriculture to meet the growing demand for food. For instance, Davis et al. [14] have shown that increasing the production of millets and sorghum in India can lead to an improved nutritional supply and a reduced irrigation volume in the order of 5% to 21%.

From a water–food–energy nexus perspective, the irrigation sector shows not only close relationships between water consumption and food production but also a close relationship with the energy sector. The irrigation sector is one of the major electricity consumers worldwide [15]. The conventional energy sources for irrigation are the electric grid and fossil fuels, such as diesel and petrol. Considering only pumps powered by the electric grid, the estimated global electricity consumption is 62 TWh/year [16]. This statistic does not take into consideration the energy consumption of diesel-powered pumps, which represent a common irrigation energy source in rural and remote areas without access to mains electricity, especially in developing countries.

Alternative and renewable options for meeting the energy requirements for irrigation include solar, wind, biogas, and hydropower-driven water pumping systems, or hybrid solutions that combine the electricity grid or fossil-fuel-powered generators with renewable energy technologies. Traditionally, renewables-based solutions have primarily been implemented for areas without access to the grid and for which grid extension was not a technically and economically feasible solution. Among the renewable solutions, photovoltaic water pumping systems (PVWPSs) have dominated the market for irrigation due to their several advantages over both renewable and nonrenewable solutions. One of their major advantages as compared to other renewable solutions is the extremely low operation and maintenance costs that are key to their success, especially in rural and remote areas where the supply of spare parts is difficult and the availability of skilled technicians is limited. Another technical advantage of PVWPSs as compared to, for instance, wind-driven water pumping systems for irrigation is that their capacity to supply water better matches the crop water requirements. This is because crop water requirements depend mostly on solar radiation patterns. From an economic perspective, the drop in PV module prices has made other renewable solutions noncompetitive with PVWPSs. As compared to nonrenewable solutions, PVWPSs have the economic and technical advantage of not relying on fuels whose supply to rural and remote areas can be difficult and expensive. Moreover, from an environmental perspective, PVWPSs do not create any kind of pollution (i.e., greenhouse gas [GHG] emissions, soil contamination) during their operation.

Although PVWPSs have dominated the off-grid market for renewables-based water pumping systems, in recent years, the application areas have been greatly expanded by the drastic drop in PV module prices, which allowed the technology to reach grid/water parity. The increased economic competitiveness of PVWPSs has opened up the market for grid-connected configurations and hybrid grid—PVWPSs solutions.

4.2 Photovoltaic water pumping systems

The main components of a PVWPS are the PV array, a power control unit that matches power production with the power requirements, an electric motor, and a water pump. The operating principle of PVWPSs is to transform solar energy into electricity through the PV modules, and then to convert the electricity into mechanical energy via an electric motor that drives a water pump to lift water. The PV modules supply the electricity in the form of direct current (DC) either to a DC pump through a DC/DC converter for optimal operation, or by going through a DC/alternating current (AC) inverter that converts the direct current into alternating current to supply an AC motor-pump. The lifted water can then be supplied directly to the specific application or can be stored in a water tank or basin. The storage system is important when water must be available during the night or in cloudy conditions. As a storage system, there is a possibility of including batteries in the system, but this is generally not common due to the higher cost, maintenance, and complexity of the system. A generalized PVWPS with a submersible water motor pump and a water tank as a storage system is depicted in Fig. 4.2. Several system configurations are available, with differences ranging from the choice of the PV array technology to the type of water source, which can shape the selection of water pump and storage system. An overview of system choices and configurations is provided in Fig. 4.3.

To meet different power, voltage, and current demands, several PV modules are combined in an array. They can be connected in series to increase the output voltage, or in parallel to increase the output current. The array can be mounted on a structure with a fixed tilt or one- or two-axis solar trackers. Manual adjustments to the fixed tilted system can be performed to increase seasonal performances in terms of power/water output. For instance, a three-times-a-day manual adjustment was reported in Sontake and Kalamkar [17]. The tracking option allows for a higher harvest of solar radiation and thus more power and water output. The two-axis tracking system also allows more stable production throughout the day, which also has positive effects on the pressure stabilization of the water pumping and distribution system [18]. This is a positive aspect for irrigation systems directly connected to a PVWPS since it stabilizes the water pressure at the emitters. Another way to increase system performance is to cool and clean the panels since high

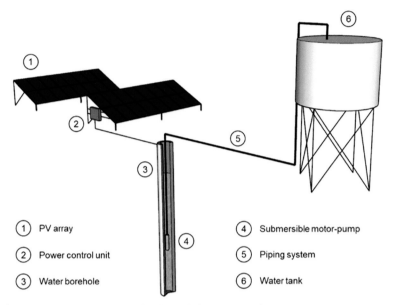

FIGURE 4.2 Main components of a photovoltaic water pumping system.

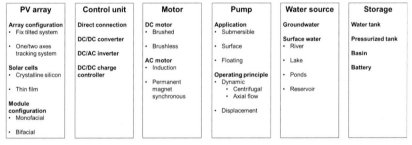

FIGURE 4.3 Overview of conventional photovoltaic water pumping system components and configurations.

temperatures and dust decrease their efficiency. There are three main types of PV modules technology utilized on the market: monocrystalline, polycrystalline, and thin film. Monocrystalline and polycrystalline are made from crystalline silicon and are the most common on the market, accounting for 95% of worldwide electricity production [19]. One of the latest trends in PV modules is the adoption of bifacial PV modules that can produce electricity from the front and rear sides of the modules. According to the latest projections from the German Mechanical Engineering Industry Association [20], bifacial solar panels will represent more than 70% of the market by 2030.

For small applications, the PV array can be directly connected to a DC motor-pump without any type of electrical controller in one of the oldest and simplest system configurations studied since the 1960s [21]. In this configuration, the terminal voltage and armature current of the motor correspond to the PV system voltage and current, and this leads to a close relationship between pump speed and solar irradiance [22]. Such a configuration is less efficient than configurations where the PVWPS is equipped with electrical controllers such as maximum power point trackers (MPPT), DC/DC converters, or DC/AC inverters with variable frequency drives if AC motor-pumps are used. Shabaan et al. [23] compared a directly coupled DC PVWPS with an MPPT-coupled DC PVWPS. The PVWPS equipped with MPPT and DC/DC converter registered an increase in PV power output between 20.1% and 82.7% over the system without MPPT. A review of algorithms used to extract the MPP is provided in Periasamy et al. [24]. If the storage system is a battery, a DC/DC charge controller is included in the electrical system to control the charging and discharging processes of the battery. In this configuration, a further DC/DC converter or DC/AC inverter is integrated into the system to drive the DC or AC motor-pump unit. Typically, for optimal and safe operation, other electrical controllers and equipment are integrated into the PVWPS that, for example, avoid dry running of the motor-pump, define minimum and maximum operating powers, or safely disconnect the PV array from the rest of the electric circuit (i.e., circuit breakers).

One of the major classifications of PVWPSs is related to the choice of the electric motor: DC or AC. DC motors can be either brushed or brushless. Brushless DC motors offer higher performances and lower operation and maintenance costs as compared to brushed DC motors [25]. Sontake and Kalamkar [17] reported that one of the most common choices for PVWPSs is to use permanent magnet brushless DC motors. Although DC motors are typically marked by high efficiency, DC motors are more expensive than AC motors in terms of investment and maintenance cost. This is one of the reasons why AC motors are generally preferred over DC motors. Other reasons for which AC motors are preferred over DC motors include availability, longer lifespan, rugged design, reliability, and higher flexibility in terms of control strategies [24,26]. DC motors are mostly used in low-power solar water pumping systems of below 5 kW [27]. Santra [28] compared 1-hp DC and AC pumps used for the operation of different types of microirrigation systems in India. The comparison showed that the DC pumping system had a longer operating time as compared to the AC system, but the AC system could provide a higher operating pressure. From an economic perspective, the DC system was slightly more competitive than the AC system.

The performance of a solar pump is often described with a pump efficiency curve, which shows variations in water flow versus power inputs for a given total head [18]. AC induction motors are used in PVWPSs in two main configurations: multi and single stage. In the multistage configuration, a DC/DC converter equipped with MPPT extracts the maximum power from the PV array, and then a DC/AC inverter transforms the DC voltage into variable voltage and frequency. In the single-stage configuration, the DC/DC converter is removed and the MPPT is embedded into the DC/AC inverter. Permanent magnet synchronous motors (PMSM) are becoming more popular in solar water pumping applications because they have several advantages, including high power-to-weight ratio, high torque-to-inertia ratio, and quick acceleration and deceleration capability [29]. One of the disadvantages of PMSM is that they require encoders for rotor position and speed measurements and those encoders are costly and not reliable when used in harsh environments such as in submersible water pumping systems. To increase the reliability of PMSM in solar water pumping applications, Murshid and Singh [29] proposed a solution to encoders by using a voltage sensor and an algorithm for computing rotor speed without relying on sensors. Ibrahim et al. [30] highlighted that PMSM has the disadvantage of using rare Earth materials, with consequent economic and availability issues. The authors proposed a ferrite magnet synchronous reluctance motor to tackle this disadvantage. Yadav et al. [31] compared a two-stage sine-wave pump controller with MPPT (SPCM) with a single-stage variable frequency drive (VFD). The SPCM showed an 18% increased efficiency and enhanced power quality as compared to the VFD.

Based on the pump installation, solar pumps are classified into submersible, surface, and floating water pumps. Submersible solar pumps lift water from deep wells, whereas surface pumps are installed close to water bodies, typically in shallow wells, rivers, basins, or ponds. Floating solar pumps are typically installed in rivers, ponds, and reservoirs and have the ability to adjust the hydraulic height given their installation on floating platforms. A further classification of solar water pumps is based upon the functioning principle: dynamic and positive displacement pumps. Dynamic pumps, such as centrifugal and axial flow pumps, increase the speed of the water through a rotating impeller and then transform kinetic energy into pressure. Positive displacement pumps use the reciprocating motion of pistons or diaphragms to first move water into a cavity and then to discharge the water. The main disadvantage of dynamic pumps as compared to positive displacement pumps is the lower efficiency. This disadvantage is offset by the lower maintenance and related costs as compared to positive displacement pumps [27].

Depending on the water pumping application, PVWPSs can be equipped with different storage solutions. The most common solution is to install an elevated water tank that stores water and releases it when water is required, using gravity. For large applications, basins or ponds can be used to store water. If the hydraulic head at the outlet of the water pumping system is not adequate, a secondary water pump that can be installed to boost the pressure. A pressure water tank is a solution that can avoid the secondary water pump. The optimal size of the water tank has been the object of several studies because it is a key component that guarantees the reliability of the system, especially for drinking purposes. Muhsen et al. [32] used the concept of loss-of-load probability (LLP) for the optimal design of PVWPSs for drinking and irrigation applications with special consideration of the water tank. Other storage solutions include batteries or hybrid solutions (i.e., multienergy storage systems). Soenen et al. [33] compared battery storage and a water tank as a storage solution for PVWPSs for an off-grid community. They concluded that the battery storage solution had a 22% lower life-cycle cost than the storage tank solution. Yahyaoui et al. [34] investigated the optimal design of PVWPSs for an irrigation system equipped with multienergy storage, as shown in Fig. 4.4, to guarantee the reliability of the system. By performing a sensitivity analysis considering different system configurations and locations, the authors concluded that the PVWPS with battery was the optimal system configuration from a cost perspective.

The main off-grid applications of PVWPSs are the provision of water for drinking purposes or general domestic uses in remote off-grid communities. Livestock watering has been another traditional application of PVWPSs to allow grazing in areas with no surface water resources and electric grid. Another conventional application of PVWPSs is irrigation for crop production to guarantee food security in remote areas without access to the electric grid. Other applications include: (1) seawater desalination and general water purification [35], (2) reforestation or afforestation, (3) halting desertification, or (4) ecological restoration [36], and the provision of ecological waterscapes or urban water systems [37]. Knowledge gaps still exist both at the micro-level, such as in the design of electrical and mechanical equipment to increase system efficiency, and at macro levels concerning system optimization and system integration with the environment with special consideration of water−food−energy nexus aspects. For instance, an accurate assessment of water requirements is essential for the optimal design of solar water pumping systems, and this can differ greatly depending on the application or combination of selected applications. For

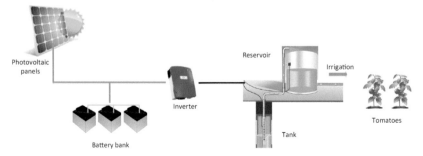

FIGURE 4.4 Multienergy storage photovoltaic water pumping system for irrigation [34].

instance, livestock water requirements vary with the livestock age and size, activity, and lactation, while irrigation for crops or trees depends on the season, crop or tree type, and method of delivery, among other things. Domestic use of water is affected to a great extent by behavioral, societal, and climatic conditions.

Aliyu et al. [38] reviewed the approaches involved in the designing of PVWPSs. Most of those approaches are based on the dynamic simulation of the PVWPS output water, the correct design of the system components, and costs. As compared to conventional water pumping systems working at constant water flow, PVWPSs work at variable water flows as a function of incoming solar radiation and thus of PV power production. This makes more challenging to estimate the water flow of PVWPSs as compared to conven- tional systems. This barrier is typically overcome by providing the character- istic curve of the solar pump in terms of water flow versus power input versus hydraulic head. This makes it possible to accurately estimate the water flow given the system configuration (i.e., hydraulic head) and the dynamic electricity output from the PV system. Hamidat and Benyoucef [39] and Ould-Amrouche et al. [40] conducted experimental research activities to estimate the performance of solar pumps in terms of power input to the motor-pump unit versus output water flow as a function of the hydraulic head. The experiments were performed at the Algerian Center of Development of Renewable Energies using a pumping test facility, as shown in Fig. 4.5, composed of a water well, a PV system, several solar pumps, and an air/water-regulated pressure vessel to guarantee a stable pressure during the experiments. A typical characteristic curve for a 400-W nominal power centrifugal pump is provided in Fig. 4.6.

To avoid performing pumping tests in order to predict the performances of solar water pumping systems, Alonso Abella et al. [41] developed a methodology to estimate the characteristic curve of PVWPSs equipped with variable frequency drives and AC pumps by using manufacturers' typical performance curves. To estimate the dynamic performances of PVWPSs in terms of water flow as a function of power input and variation of total hydraulic head, Meunier et al. [42] developed a fourth-order surface to approximate the characteristic curves provided by a manufacturer of solar water pumping systems. To accurately depict the variation of the operating hydraulic head during the pumping, a detailed model of the well drawdown was also integrated. A review of the factors affecting the performances of PVWPSs and the related characteristic curve is provided in Hadwan and Alkholidi [43].

Caton [44] compared two simple design strategies for PVWPSs with the simulation results of a PVWPS model. The simple design approaches were the peak sun hour procedure based on the monthly average daily solar irradiation to estimate the PV power peak and the hydraulic-energy rule-of- thumb approach that relates the hydraulic energy to the PV power peak.

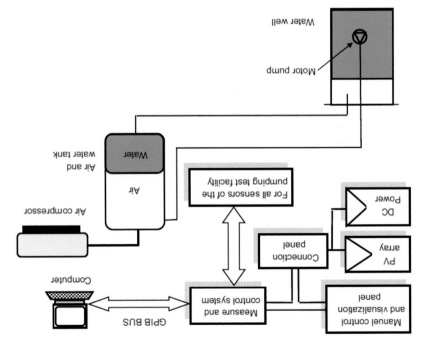

FIGURE 4.5 Photovoltaic water pumping test facility at the Algerian Center of Development of Renewable Energies [39].

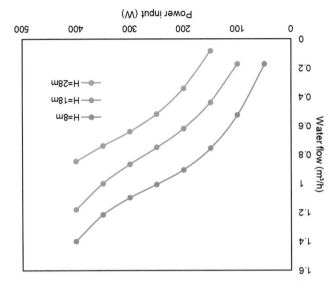

Power input (W)

FIGURE 4.6 Characteristic curves of a 400-W solar pump at different hydraulic heads ($H = 8$, 18, and 28 m). Adapted from Ould-Amrouche et al. [40].

By comparing the results of the simulation models applying the two simple design procedures, the author concluded that those could lead to an undersized system due to the difficulties in depicting the dynamic efficiencies of the subsystems composing the PVWPS. The author also proposed several tracking options that have the scope to increase the PVWPS water output, as well as proposing the promotion of community engagement. Ebaid et al. [45] provided a detailed design approach for each component of a PVWPS to serve a specific well within the Disi-Mudawara to Amman Water Conveyance Project in Jordan. The PV system was designed using and comparing two approaches as in Caton [44]: Worst Month Method Analysis and Peak Sun Hours Analysis. The authors also provided a detailed design of the inverter, charge controller, battery capacity, cables, and pump. To accurately design the system, the authors also performed a detailed analysis of the system conversion losses including, shading, dust accumulation, mismatching losses, and cabling losses. Bouzidi [46] developed a novel sizing method based on the loss of power supply probability (LPSP) concept defined as the ratio of total water deficit volume to total required water volume. The optimal system design was attained through a sensitivity analysis consisting in varying the values of the decisional variables (i.e., the number of PV modules and storage factor) and calculating the corresponding life-cycle cost. Gualteros and Rousse [47] developed a sizing tool for PVWPS in Python language. The package facilitates component selection, optimal system design, dynamic simulations for a typical meteorological year, and financial analysis. A schematic diagram of the tool and its main components is provided in Fig. 4.7. The optimization is performed by calculating all possible combinations of PV system capacity, tilt angles, and reservoir capacity, and

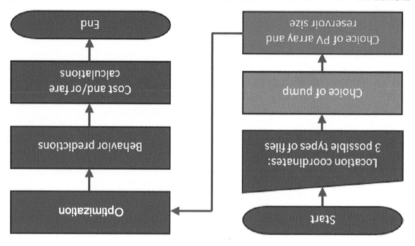

FIGURE 4.7 Schematic diagram of the photovoltaic water pumping system tool developed by Gualteros and Rousse [47].

the resulting water shortage probability (WSP) indicator, and the cost of the system. The optimal solution is the one that guarantees the highest reliability at the lowest cost. The effectiveness of the model was tested by showing the improvements in system design and related costs using two case studies in previously published research. Olcan [48] classified the optimal sizing methods as "intuitive" (i.e., based on simple approaches like the "worst month"), "numerical" (i.e., based on dynamic system simulations), "analytical" (i.e., based on indicators such the system reliability), "heuristic" (i.e., based on genetic algorithm multiobjective optimization), and "hybrid".

It is worth mentioning that solar water pumping systems can also be driven by solar thermal systems that convert solar radiation into mechanical work through Rankine, Brayton, or Stirling thermodynamic cycles. The mechanical power generated by the cycles can be used directly or indirectly through the production of electricity for water pumping. Shahverdi et al. [49] investigated the performances of a parabolic-trough-collector-driven Organic Rankine Cycle to produce electricity for water pumping. The solar field had an area of 22.6 m^2 and was able to produce 2.2 kW of electrical power to drive the water pumping system with an overall system efficiency at the optimal point of 12.19%. A schematic diagram of the investigated solar thermal water pumping system is depicted in Fig. 4.8. Ali [50] compared three solar water pumping technologies for irrigation in Sudan, including PVWPS, parabolic trough water pumping systems, and concentrating dish water pumping systems. PVWPSs showed the lowest energy efficiency among the investigated solutions but at the same time showed the lowest levelized cost of energy equal to US$0.033/kWh. The levelized cost of energy for the

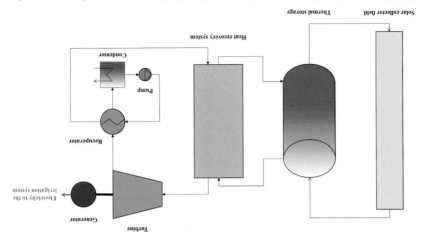

FIGURE 4.8 Schematic diagram of a solar thermal driven water pumping system using Organic Rankine Cycle to supply electric power for an irrigation system [49]. (Figure redrawn from the original.)

parabolic trough water pumping systems and for the concentrating dish water pumping systems were 0.075 US$/kWh and 0.062 US$/kWh, respectively. The author performed a sensitivity analysis to study the effect of energy conversion efficiency on the levelized cost of energy of thermal water pumping solutions. Nevertheless, even at the highest energy conversion efficiencies, the thermal water pumping solutions for irrigation could not compete with PVWPSs in terms of levelized cost of energy.

4.3   Photovoltaic water pumping systems for irrigation

PVWPSs for irrigation can have multiple system layouts depending on water volumes, pressure requirements, and irrigation scheduling. If irrigation can be applied during the daytime, one of the simplest configurations is to directly connect the PVWPS to the irrigation system. Such configuration is preferable for large-scale systems where the daily water volumes, especially during the peak of the irrigation seasons, would not economically justify any type of storage system. Such configuration requires an accurate design of both the PVWPS and the irrigation system so that water pressure and flow variations due to the intermittency of solar radiation match the pressure and flow requirements of the irrigation systems. For small applications or applications where irrigation must be applied during the night, a storage system complements the PVWPS. The proper design of PVWPS for irrigation depends on the accurate assessment of energy needs and energy conversion. The energy needs for irrigation are related to several factors, the most important being crop water requirements, the irrigation system, and the distance—both vertical and horizontal—of the water source from the point of application. The crop water requirements can be estimated with the FAO Penman–Monteith method [51] and the resulting monthly energy consumption for irrigation E_{irr} (kWh) can be estimated with the following equation [52]:

$$E_{irr} = cf \frac{\frac{(ET_c - P_e) \times A \times TDH}{\eta_{irr} \times (1 - LR)}}{\eta_m \eta_p}, \tag{4.1}$$

where, cf is a conversion factor that takes into account the density of water, gravity acceleration, and the conversion between Joule and kWh, ET_c is the evapotranspiration in cultural conditions (m), P_e is the effective precipitation (m), η_{irr} is the efficiency of the irrigation system (%), LR is the leaching requirement (%), A is the irrigated area (m^2), TDH is the total dynamic head (m), η_m is the motor efficiency (%), and η_p is the pump efficiency (%). An overview of the factors affecting the energy consumption for irrigation is provided in Fig. 4.9.

Different types of irrigation systems require different water volumes for the same crop since part of the water required by the plants is lost through evaporation due to irrigation system inefficiencies. Irrigation is typically applied

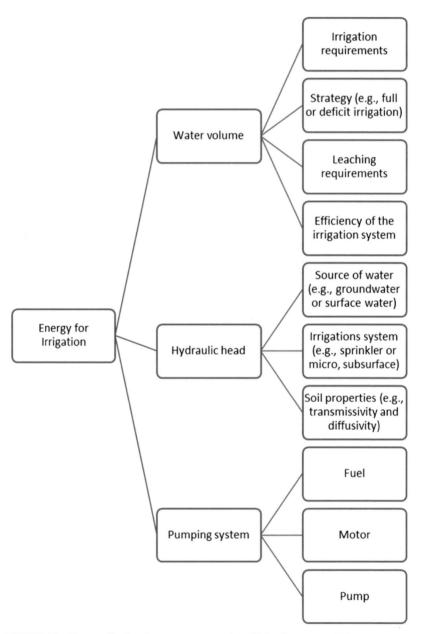

FIGURE 4.9   Factors affecting the energy consumption of irrigation systems [52].

through four different types of irrigation systems: surface irrigation, sprinkler irrigation, microirrigation, and subsurface irrigation. The range of irrigation efficiency for these types of irrigation systems is summarized in Fig. 4.10.

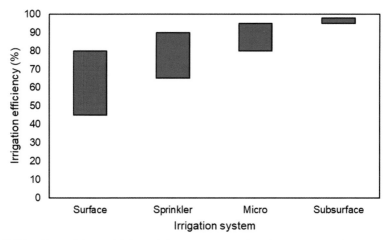

FIGURE 4.10   Efficiency of different irrigation systems. *Adapted from Irmak et al. [53].*

As it shows, irrigation efficiency increases from surface irrigation to subsurface irrigation. Subsurface irrigation systems supply water to crops through buried pipes and have the highest efficiency because they eliminate the loss of water through superficial evaporation.

The total dynamic head is the total equivalent height that the fluid has to be pumped up to guarantee the pressure required to properly operate the irrigation system. It equals the sum of the static head, drawdown, the head required to operate the irrigation system, and the continuous and concentrated water head losses. The pressure requirements depend on where the water is available, but also on the pressure required to properly run the irrigation system. At parity of crop and water sources, surface and subsurface irrigation systems do not require high pressure. A surface irrigation system works basically at 0 bar [54,55], while subsurface irrigation systems typically require operating pressures of less than 1 bar. Medium pressure levels are required for microirrigation systems. Daccache et al. [54] and Espinosa-Tasón et al. [55] reported 1 bar as the operating pressure for microirrigation systems. Higher pressures in the order of 3 bar are required to drive sprinkler-based irrigation systems [54,55].

Typically, most of the PV off-grid energy systems are designed for the month of the year with the lowest irradiation, to guarantee the highest system reliability. In some cases, it is more suitable to design the system for the month of the year when energy requirements are highest. In the case of PVWPSs for irrigation, to match the monthly water demand and supply, the design considers the month with the highest design ratio, defined as the ratio between monthly average daily irrigation water requirements and monthly average solar irradiation [56]. An example of identifying the highest design ratio for irrigation of grassland is depicted in Fig. 4.11, where, based on the

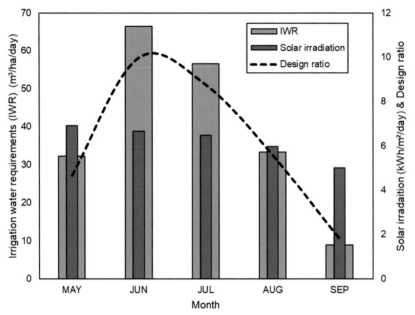

FIGURE 4.11   Design procedure for photovoltaic water pumping systems for irrigation based on design ratio. *Adapted from Campana et al. [56].*

monthly averaged daily irrigation water requirements and solar irradiation, June is the month for which PVWPSs need to be designed to guarantee the maximum reliability.

The adoption of the design ratio is particularly helpful for the design of PVWPSs for irrigation that supply water for multiple crops grown during the year, as in Bouzidi and Campana [57] and Zhang et al. [58].

By following this design approach, Eq. (4.2) determines the PVWPS peak capacity to meet the irrigation water requirements [59,60]:

$$P_{p,PVWP} = \frac{cf}{mf[1 - \alpha_c(T_{cell} - T_{STC})]\eta_m\eta_p} \max_m \frac{TDH_m^{clim} IWR_{t,m}^{clim}}{E_{s,m}^{clim}}, \quad (4.2)$$

where, *mf* is the matching factor that can be adjusted to consider power losses during the lifetime of the PV generator and other derating factors, such as soiling or shading losses, α_C is the temperature coefficient of the PV modules (%/°C), T_{cell} is the PV cell temperature (°C) that for design purposes can be assumed as the highest monthly daily operating cell temperature, T_{STC} is the temperature under standard test conditions (25°C), $IWR_{t,m}$ represents the total monthly average daily irrigation water requirement (m³/day), and $E_{s,m}$ is the monthly average daily solar irradiation hitting the PV array (kWh/m²/day). The function *max* indicates that the design of the PVWPSs must be conducted for month *m* that has the highest design ratio. The $IWR_{t,m}$

is given by the sum of the *IWR* of the *j*th crop with *N* equal to the total number of irrigated crops, as follows:

$$IWR_{t,m}^{clim} = \sum_{j=1}^{N} IWR_{j,m}^{clim} \quad m = \{Jan, Feb, ...Dec\} \quad (4.3)$$

The superscript *clim* emphasizes that an accurate design of PVWPSs should also take into consideration how climate change scenarios might affect *TDH*, *IWR*, and E_s during the lifetime of the system. The *TDH* can be affected by water tables falling due to climate changes, such as different precipitation patterns that accordingly affect the replenishment of the groundwater resources, and anthropogenic factors such as overpumping. The *IWR* can also be affected by climate changes due to changes in the precipitation patterns that can affect the effective precipitation and thus soil water balances but also changes in other climatological parameters patterns such as changes in ambient temperature, solar radiation, wind speed, and relative humidity. Those climate patterns might also affect the crop growing cycle with a consequent impact on *IWR* volumes and distribution during the irrigation season. Salem et al. [61] assessed the impact of climate change on the groundwater level in an irrigated region of Bangladesh by combining a climate model and a hydrological model. The aim of the study was to quantify the impact that climate change under different representative concentration pathways (RCPs) [62], had on the groundwater level and accordingly on irrigation costs. In all the investigated scenarios, the groundwater level declined, increasing irrigation costs. Wu et al. [63] investigated climate-change and anthropogenic effects on seven of the world's largest mid-latitude aquifers. The authors showed that most of the impact on groundwater resources under the RCP marked by a radiative forcing of 8.5 W/m^2 (RCP 8.5) was associated with a reduction in snowmelt and increased evapotranspiration rather than with a reduction in precipitation. Anthropogenic pumping was the main contributor to the depletion of groundwater resources in most of the investigated aquifers. Bouras et al. [64] assessed the impact of climate changes on wheat yield and wheat irrigation requirements under scenarios RCP 4.5 and RCP 8.5 from the MEDiterranean COordinated Regional Climate Downscaling EXperiment (Med-CORDEX) [65]. The authors showed that the projected climate changes are likely to affect the growing cycle of wheat by shortening it by up to 50 days. The shortened crop growing cycle could significantly decrease the crop water requirements and lead to temporal peaks in crop water requirements being different as compared to today.

Several studies have focused on the design of solar irrigation systems using the approach defined in Eq. (4.2). Deveci et al. [66] designed a solar-powered micro drip irrigation system for 100 acacia trees and 30 oak trees by comparing a directly coupled PVWPS with a PVWPS equipped with battery. The authors highlighted that directly connected irrigation systems can decrease irrigation performances and increase system costs, especially if irrigation must be applied during periods marked by low levels of solar radiation because it leads to oversizing. The PVWPS equipped with battery was 63% more economically competitive than the directly coupled system in terms of investment costs. The PVWPS equipped with battery also showed better performances because of the constant flow rate. Eltawil et al. [67] designed a PVWPS to supply water to more than 230 date palm trees divided into two irrigation fields. The system used variable frequency drives for the AC pump and bubblers as emitters. While designing the system the authors considered a mismatching factor equal to 0.85 to consider the lifetime performance of the PV system in supplying the required energy for water pumping.

The drawback of using a design approach simply based on Eq. (4.2) is that a lot of details in the performances of the subsystems are lost. This drawback can lead to an oversized system and thus economic losses, or to an undersized system and thus to a likely system failure. In the context of PVWPSs for irrigation, Campana et al. [18] suggested combining a simple design approach to a more detailed dynamic simulation of the system to verify the design process and conduct further system optimization, as shown in Fig. 4.12. The authors

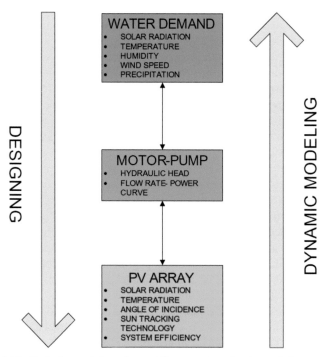

FIGURE 4.12 Designing and dynamic modeling procedure for photovoltaic water pumping systems [18].

combined detailed hourly simulations of the PV system and motor-pump system with hourly simulations of the crop water requirements to analyze the match between water supply and water demand.

The optimization of PVWPSs for irrigation is a complex task because the system itself is an integrated system with close relationships among at least three macro-areas, including energy, water, and food. Moreover, despite being a renewable solution, the exploitation of water resources might lead the technology to be a nonsustainable solution. Thus, the optimal design of the system should take into careful consideration the availability of water resources and the dynamic effect of groundwater response to pumping. Glasnovic and Margeta [59] developed an integrated hybrid simulation and optimization model, PVPS-Irrigation implemented in MATLAB®, for optimally sizing PVWPSs for irrigation. Concerning the submodels, the model included the PVWPS, the borehole, the soil, the crop, and the irrigation method. The optimization algorithm was dynamic programming and the objective function was to minimize the PVWPS capacity while at the same time meeting the crop water requirements. The model constraints were addressed in the parallel simulation model. Campana et al. [68] proposed an integrated optimization approach based on system dynamic simulations that include solar energy conversion into electricity, water resource response to pumping, crop water requirements, and crop yield response to water, as shown in Fig. 4.13.

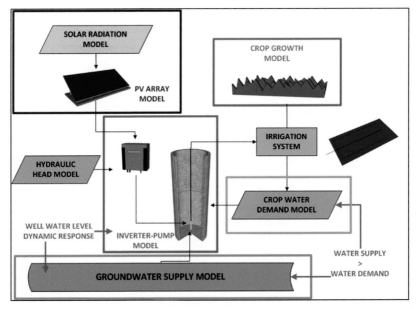

FIGURE 4.13 Overview of the modeling blocks involved in the optimization of integrated photovoltaic water pumping systems for irrigation [68].

The optimization problem was to maximize the annualized profit related to the implementation of PVWPSs for irrigation of forage and maximize the system reliability using a genetic algorithm as a solver. The optimization problem was constrained by the hourly decline in the groundwater level and the daily pumped water limited by the available water resources. Part of this work was related to the development of a novel control system for PVWPSs for irrigation presented in Campana et al. [69]. The control system had the twofold objective of matching the daily crop water demand and water supply and interacting with the groundwater resources by sensing the decline in groundwater level and adjusting the pump speed to avoid an excessive decline in the groundwater table. López-Luque et al. [70] proposed an optimal design for directly coupled PVWPSs for irrigation of olive orchards based on the maximization of economic profitability, quantified using the net present value as indicator. The optimal PVWPS capacity was selected by varying the PVWPS capacity and performing dynamic simulations of the system also taking into account the crop yield response to water and accounting costs and revenues generated by the PVWPS implementation. The highest net present values were attained with PVWP capacities varying between 150 and 300 W_p/ha.

Reca et al. [71] developed an optimization model for PVWPSs for greenhouse irrigation. In particular, the authors investigated the optimal design and operation as a function of the irrigation sectors per hectare taking into consideration the pressure requirements of the noncompensating emitters and relative discharge performances. Two operational strategies were implemented and compared, the first being based on the irrigation of a single sector at a time, and the second being based on the irrigation of a variable number of sectors depending on the dynamic performances of the solar pumps in terms of power, water flow, and pressure. From a net present value perspective, the first operational strategy had out-performed the second and the PVWPS was able to become profitable if at least four sectors were irrigated. Yahyaoui et al. [72] developed an optimal design algorithm for a PVWPS for tomato irrigation. The objectives of the optimal sizing tool were system reliability in terms of crop water requirements and protection of the battery against overdischarge and overcharge. The decisional variables of the optimization algorithm were PV system capacity, battery capacity, and reservoir capacity. The tool developed by the authors was validated against the commercial software PVsyst, showing good agreement. The developed algorithm was applied in three countries (i.e., Tunisia, Spain, and Qatar) to support a cost analysis. Rezk et al. [73] used the HOMER package to optimally design an off-grid PVWPS for irrigation equipped with battery as energy storage and reverse osmosis for brackish water desalination, as shown in Fig. 4.14. The location of the study was marked by underground brackish water with a salinity of 2500 mg/L. To meet the wheat irrigation water requirements standards in terms of salinity, the groundwater is pumped through a reverse osmosis unit to produce water with a salinity of 800 mg/L. The authors compared this

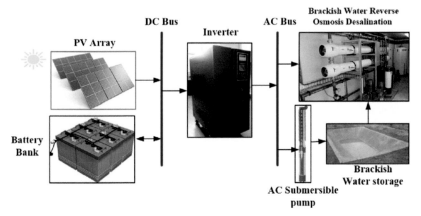

FIGURE 4.14 Schematic diagram of photovoltaic water pumping systems used for removing salt from brackish water to meet water requirements for irrigation [73].

option against other several options including the installation of diesel generators or the extension of the electric grid. The investigated PVWPS option showed the lowest cost of energy as compared to the alternatives.

Zavala et al. [74] developed a novel analytical method for the optimal management of a multisector irrigation system served by PVWPS for an olive farm near Almeria, Spain. The optimization method finds the number of irrigation sectors that can be irrigated simultaneously based on the available power/water pumped by the PVWPS. This is attained by two primary objective functions—minimize the irrigation deficit and maximize the soil moisture level at the end of the irrigation seasons. The authors mainly compared two scenarios—in the first, the sectors were irrigated successively, while in the second the sectors were operated simultaneously. The results of the optimization model were that the first scenario required a 3.3-kW$_p$ PVWPS while the second scenario required a 2.75-kW$_p$ PVWPS. Using the LPSP concept, Bouzidi and Campana [57] developed an optimal design algorithm for the irrigation of multiple crops in the Saharan regions of Algeria. The optimal PVWPS design was performed for date palm, though further optimization was focused on reducing the excess water produced by the PVWPS. By combining tomato, wheat, and sweet peppers with date palm, the usage of the excess water produced by the PVWPS was improved from 56% to 86%. García et al. [75] developed an algorithm called the Smart Photovoltaic Irrigation Manager (SPIM) to optimize water production from PVWPSs and daily irrigation scheduling. The algorithm considers all the main operational parameters of a solar irrigation system, including soil properties and the hydraulic characteristic of the irrigation sectors for the optimal operation of the system. The model was satisfactorily implemented in a real case study for the irrigation of an olive orchard in the south of Spain. García et al. [76] developed a model for optimal sizing of PVWPSs and irrigation networks called Model for Optimal Photovoltaic Irrigation System Sizing (MOPISS). The model was developed in MATLAB and used a genetic algorithm for the optimization. The optimization problem was defined as multiobjective and consisted of one objective function to minimize the cost of the irrigation system, including the pump unit, and a further objective function to minimize the cost of the PV system. To show the effectiveness of the developed model, the authors applied it to a previously designed and operating PVWPS for irrigation serving a 13.4-ha olive orchard at the University of Córdoba. The results of the optimization model showed that the entire system cost of the operating PVWPS could be reduced by about 23%−38%.

4.4 Recent advances in photovoltaic water pumping systems for irrigation

Recent advances in PVWPSs for irrigation have been attained at different levels. At the components or subcomponents level, recent advances have consisted in improving efficiencies, such as improving the conversion efficiency of controllers and inverters or improving the efficiency of the controller coupled with the motor unit through newly developed algorithms. Novel control strategies have been developed also relating to the adoption of artificial intelligence algorithms for forecasting and control. At the system level, the major advances in PVWPSs for irrigation have focused on hybridization to guarantee stable water flow and hydraulic conditions in the irrigation network. Moreover, interfacing PVWPSs for irrigation with the main electric grid and related operational strategies have also brought other recent advances. The drastic drop in PV module prices has also extended the applications of PVWPSs for irrigation—for instance, to on-grid solutions. This economic aspect has also favored system integration with other novel concepts, such as agrivoltaics. Other recent advances in PVWPSs for irrigation have focused on using Internet of Things (IoT) technologies and cyber-physical systems for monitoring and management.

Elkholy and Fathy [77] developed an optimal control strategy for PVWPSs based on a three-level inverter and induction motors. The extraction of maximum power at any irradiance and temperature levels was supported by an Artificial Neural Network. The optimal operational strategy consisted in using the metaheuristic Teaching Learning Based Optimization algorithm to control inverter voltage and frequency to support the extraction of maximum power from the PV array and minimize motor losses. To better control and optimize the operation of PVWPSs with induction motors, Errouha et al. [78] proposed a new method based on Fuzzy Logic Control to improve the conventional Direct Torque Control. The simulation results showed that the proposed controller could lead to increased pumped water volumes at parity of PV system capacity during three seasons (i.e., hot, moderate, and cold). During the entire considered simulations period, the PVWPS with the newly developed control could pump about 1132 m^3, against 1071 m^3 for the conventional controller. Lakshmiprabha and Govindaraju [79] proposed an isolated DC/DC converter integrated with a Pulse Width Modulated (PWM) Inverter to drive a Brushless DC motor for irrigation applications. The MPPT algorithm was the drift-free Optimum Power Point Tracking. The developed inverter had the characteristics of higher voltage gain and galvanic isolation, and better performances under partial shading conditions as compared to existing inverters. Cordeiro et al. [80] proposed a novel three-level quadratic Boost DC–DC converter coupled with a switched reluctance motor for PVWPSs applications. The authors also proposed a novel MPPT algorithm based on temporal variation in the power and voltage. The laboratory experimental results showed good accuracy with the simulation results, leading to the conclusion that the proposed technical improvements in PVWPSs controller can increase the solar energy conversion efficiency and guarantee a more continuous water flow output. A review of further novel control and energy management strategies for PVWPSs is provided in Poompavai and Kowsalya [81]. Concerning the use of artificial intelligence techniques, Ammar et al. [82] investigated the effects of PV power forecasting on the performance of PVWPSs for irrigation equipped with both battery and water tank as storage systems. Three forecasting techniques were employed: empirical models, Feed Forward Neural Network, and Adaptive Neuro-Fuzzy Inference System. These last techniques showed the best accuracy for both sunny and cloudy days. The Adaptive Neuro-Fuzzy Inference System was embedded in the energy management system of the PVWPS to meet the load and avoid overcharging and discharging of the battery.

Typically, the selection of a pump in a conventional water pumping system is based on the hydraulic system curve, the hydraulic head (H) versus water flow (Q) curve of the pump, and the best efficiency point of the pump, as shown in Fig. 4.15. This selection process is based on the fact that conventional water pumping systems usually work at the same duty point since the water flow is constant, given that the frequency is also constant. This does not apply to PVWPSs because water flow and frequency show a linear relationship with solar radiation. This implies that the duty point varies with the level of solar radiation. To overcome this design shortcoming, Almeida et al. [83] proposed a new selection method for pumps in large-scale PVWPSs. The novel method is based on the consideration of how the efficiency varies in a wider range of frequencies. The proposed method was compared with the traditional method and the authors showed that the proposed one could lead to higher system efficiencies and higher pumped water volumes per kW$_p$ installed of the PV system (i.e., 1779 m^3/kW$_p$ for the PVWPS with the pump selected with the proposed method, versus 1476 m^3/kW$_p$).

Similarly, dos Santos et al. [84] developed a new methodology to identify the best efficiency point of centrifugal pumps driven by PVWPSs. The new

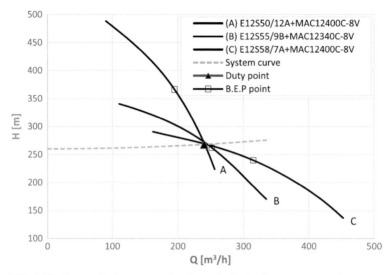

FIGURE 4.15   Pump selection process based on: the hydraulic system curve (system curve) and hydraulic head (H) versus water flow (Q) curve of three pumps, the system duty point, and the best efficiency points of the selected pumps [83].

operating point, called Solar Best Efficiency Point, was compared against the conventional best efficiency point used in the selection of centrifugal pumps in conventional water pumping applications. The developed methodology consisted in calculating the PV power profile given the profiles of tilted solar radiation and ambient temperature, calculating the power profile at the pump's shaft, calculating the corresponding water flows and system efficiency, and finally deriving the water head corresponding to the maximum efficiency. As compared to using the best efficiency point, the use of the solar best efficiency point could lead to higher system efficiencies and higher daily pumped water volumes.

To better match water demand and supply on a seasonal, daily, or even a subhourly basis, hybrid configurations can represent a technical and economical solution to other configurations that rely, for instance, on storage systems. The hybridization with other renewable energies such as wind power, as shown in Fig. 4.16, relies on the seasonal and daily complementarity between solar and wind resources that can be experienced in several locations around the Earth.

Stoyanov et al. [85] evaluated the integration of several PV and wind technologies into a hybrid system for the irrigation of wine grapes and cherries at eight locations in Bulgaria. The system optimization was carried out by setting three objective functions: minimizing the system investment costs, minimizing the surplus of electrical energy converted, and maximizing the fulfillment of crop water requirements. The variables of the optimization

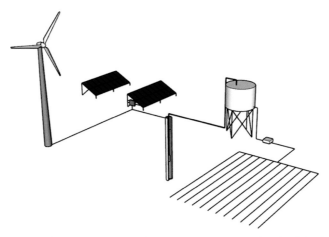

FIGURE 4.16 Hybrid photovoltaic and wind water pumping system for microirrigation.

problem were PV, wind turbine, water tank, and battery capacities. The authors highlighted that those sites marked by a significant complementarity between solar and wind resources show better performances in terms of values attained by the objective functions. Xiang et al. [86] investigated a hybrid solution for irrigation of 22.6 ha of Cassava in Guangxi Province, China. The solution consisted of a wind turbine of 44 kW$_r$ and a PV system of 22 kW$_p$. Nevertheless, due to the limited wind potential of the area (i.e., low average wind speed), the hybrid system was not considered to be the most viable solution due to its cost-effectiveness. The author concluded that oversizing the PVWPS by 50% was a more suitable solution to meet the irrigation water requirements.

Li et al. [87] focused on the optimization of a hybrid PV–wind–battery system for drip irrigation of tomatoes in a greenhouse in Weinan, China. The authors set the objective function as minimization of the annual system cost, while the optimization model constraints were battery state of charge and system reliability. The results of the optimization model were validated through an experimental setup used to provide energy for water pumping for fifteen days. During this period the battery state of charge was kept always above 50%. Campana et al. [56] investigated solar and wind water pumping systems for irrigation of grassland in Hails, Inner Mongolia, China. Solar water pumping systems showed better performances than wind water pumping systems for irrigation due to the better match between water supply and crop water demand. From a cost perspective, assuming the average costs for PVWPSs and wind water pumping systems, the authors concluded that the best solution is location-dependent since it is related to the specific solar and wind conditions during the irrigation season. Nevertheless, hybrid configurations were not studied. A comprehensive review of PV and wind hybrid systems for water pumping systems with special consideration of electrical coupling, power conditioning systems, control strategies, and challenges is provided in Angadi et al. [88].

To guarantee constant operating pressure in a large-scale olive drip irrigation system in Portugal, Almeida et al. [89] retrofitted an existing diesel-powered irrigation system with a 140-kW$_p$ PVWPS with a north–south horizontal axis tracker. The solar tracker aimed to stabilize the power production, water flow, and pressures during the day, and at the same reduce the system power peak by increasing energy conversion. The original system configuration included a 250-kVA diesel generator, two 45-kW centrifugal pumps, a soft-starter, and a 55-kW frequency converter. The schematic diagram of the retrofitted system is shown in Fig. 4.17. To preserve the current irrigation system and its operation and to maximize the PV power penetration, three operational strategies were developed and implemented in the hybrid water pumping system. The operational strategies were based on the availability of power from the PV system and were as follows: (1) "only PV", in which pumps 1 and 2 are fed by the PV system; (2) "hybrid mode", in which pump 1 is fed by the diesel generator and

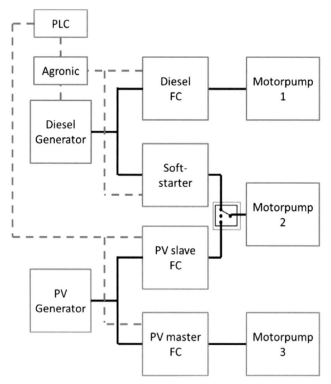

FIGURE 4.17 Schematic diagram of hybrid photovoltaic–diesel water pumping system for irrigation [90].

pump 3 is fed by the PV generator; and (3) "Only Diesel", in which pumps 2 and 3 are fed by the PV generator. In 2018, the PV share during the irrigation season was 36% and it reached its peak in August at 53%.

Both in hybrid and nonhybrid PVWPSs solutions for irrigation, several authors have highlighted the surplus of energy converted by the PV system as compared to the irrigation needs, especially for irrigation systems designed to serve a single crop. An example of energy balance is provided in Fig. 4.18.

This has been identified as an area for improving the economic competitiveness of off-grid PVWPSs for irrigation by, for instance, powering other off-grid loads during the nonirrigation season (e.g., other farming loads), or combining irrigation of other crops that shows irrigation requirements complementary to the main crop, or combining solar irrigation with DC grids. Another recent solution, driven mostly by the reduction in PV module prices over the last decade, is to interconnect the PVWPSs to the main grid through dual-mode inverters. This makes it possible to hybridize the irrigation system, guaranteeing stable hydraulic performances as in PV−diesel configurations, but also to exchange the surplus of electricity with the grid. Barrueto Guzmán et al. [92] investigated the implementation of grid-connected PVWPSs for irrigation facilities in Chile. One of the main motivations of this study was that since 2013 more than 1500 off-grid PVWPSs for fruit orchard irrigation were installed with capacities ranging

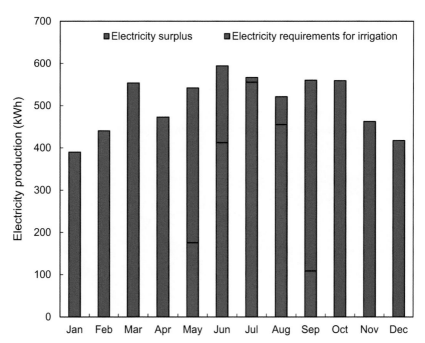

FIGURE 4.18 Photovoltaic water pumping potential electricity production, electricity requirements for irrigation, and electricity surplus. Adapted from Campana et al. [91].

between 1 and 3 kW$_p$. At the same time, 99% of the population in Chile have access to the electric grid. The authors compared off-grid PVWPSs and on-grid PVWPSs, and the latter showed significantly better performances in terms of payback period as compared to the former. This was due to two main aspects: the possibility to exchange the surplus electricity with the grid, and a design approach that for on-grid PVWPSs significantly reduced the peak power. Hassan and Kamran [93] studied the possibility of hybridizing PVWPSs with the electric grid. This was motivated by the fact that diesel water pumping systems and off-grid PVWPSs are very costly for farmers in poor areas of Pakistan. The proposed hybrid PVWPS was connected to the grid through a DC bus that powers the variable frequency drive connected to a three-phase AC induction motor. The advantage of this solution is that it provides flexibility to the PV system, since PV modules can be added whenever the farmer has the economic capacity to further invest in the PVWPS side of the system. The authors conducted a life-cycle cost analysis for a reference 10-hp AC water pump and compared the hybrid system to a fully off-grid PVWPS. The payback period of the hybrid solution was 0.8 years against 3.1 years for the off-grid solution.

From a system integration perspective, combining agrivoltaic systems and PVWPSs, as shown in Fig. 4.19, is a natural win-win solution that also increases the PV system self-consumption rate. A synergistic combination of PVWPSs with an agrivoltaic system was suggested by Proctor et al. [94] while analyzing the benefits of agrivoltaic systems and their alignment with the Green New Deal resolution [95]. The authors identified this integrated solution as an opportunity for the decarbonization of the agricultural sector. Al-Agele et al. [96] studied the water productivity under an agrivoltaic system in Oregon, USA by combining tomato growth with irrigation using a small-scale DC pump. Two irrigation strategies were performed: full and deficit irrigation. In the first treatment, water was applied up to field capacity when the soil water content reached 75%, while in the second treatment irrigation was applied when the soil water content

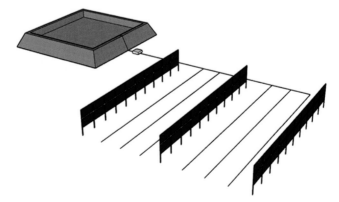

FIGURE 4.19 Photovoltaic water pumping system for microirrigation integrated into an agrivoltaic system.

reached 40%. The highest water productivity values were registered between adjacent rows of PV modules under deficit irrigation treatment. The water productivity was 53.98 kg/m^3 higher than the control plot with the same irrigation strategy. A further novel system integration was studied by Ye et al. [97] for large-scale water transfer projects such as the China's "South-to-North water diversion" project, as shown in Fig. 4.20. Water transfer projects are of vital importance for supplying water in areas with high water scarcity for energy conversion, as well as for irrigation, food production in general, and domestic consumption. Given the scale of water transfer projects, they require a huge amount of electricity for water pumping that can be offset by using the channel space to build PV systems that can both produce electricity for water pumping and act as water-saving techniques by reducing evaporation losses.

IoT technologies have evolved enormously in recent years and have been implemented in several fields to remotely monitor and control systems. IoT technologies thus represent a cost-effective opportunity for the remote operation of PVWPSs, reducing human intervention and thus associated costs. Al-Ali et al. [98] developed an IoT platform for smart solar irrigation. The system was based upon the National Instruments controller, myRIO, that interfaces the soil moisture, humidity, temperature, and flow rate sensors, with a relay board to switch the PVWPS on and off, as shown in Fig. 4.21.

FIGURE 4.20   Integration of photovoltaic water pumping system for water transfer projects [97].

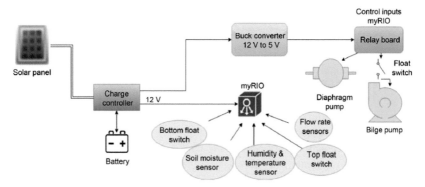

FIGURE 4.21   Internet-of-things based solar irrigation system [98].

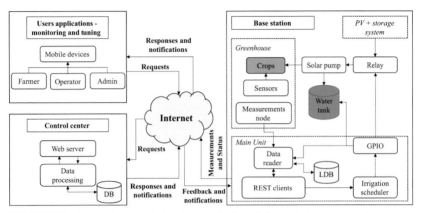

FIGURE 4.22 Proposed system design for cyberization of photovoltaic water pumping systems for drip irrigation in a greenhouse [100]. (Figure redrawn from the original.)

The controller is also equipped with Wi-Fi communication that enables the transfer of operational conditions monitored by the sensors to a remote web server as well as allowing the end-user to remotely control the system. Three operation models were implemented by the authors and consisted in "Local control mode", which allows the end-user to start and stop the system, "Mobile monitoring and control mode", which allows remote monitoring and control, and "Fuzzy logic-based control mode", which allows the automatic operation of the system based on sensor readings and control algorithm. The developed small-scale prototype can be scaled up and allows the remote operation of PVWPSs. Gimeno-Sales et al. [99] developed an open-source monitoring system for a PVWPS equipped with a lithium-ion battery as a storage device. Rasberry Pi was used as a low-cost device for gathering data from all the sensors and for wireless communication. The authors also developed a graphical user interface in the open-source software Grafana for visualizing the monitored parameters.

Selmani et al. [100] proposed a new approach for the cyberization of PVWPSs for drip irrigation in greenhouses that included a base station, a control center, and a remote monitoring and tuning unit, as shown in Fig. 4.22. The base station aggregates all the sensors and control units for the monitoring and operation of the crops and different components forming the PVWPS unit through low-cost devices such as Arduino microcontrollers. The control center receives and analyses all the data received by the base stations and allows the operation of the base station active components to be controlled both automatically through mathematical algorithms and manually through feedbacks and actions performed by end-users through the interface via HTTP requests.

4.5 Case studies

Several PVWPSs for irrigation projects have been implemented and are under implementation worldwide. This section briefly describes a few development and demonstration projects. Despite being implemented in different countries, the main scope of these projects was to provide a renewable solution to water pumping for irrigation. Moreover, the common objective of those few projects presented in this section was to prove the technical and economic feasibility of PVWPSs for irrigation and at the same time to guarantee that the technology did not negatively interfere with the available water resources.

In 2009, the Asian Development Bank supported a development and demonstration project focused on grassland conservation in the Qinghai Province that combined solar irrigation with water and ecological management [101]. The process of grassland desertification is mainly caused by human activities such as overgrazing, as shown in Fig. 4.23. Climate change, different distributions of precipitation, and other factors that drive desertification are presenting a significant threat to valuable grasslands used for raising livestock. A 2-kW$_p$ PVWPS for irrigation was installed to halt grassland desertification. The small-scale PVWPS for irrigation showed that the increased grassland productivity could lead to better incomes for rural communities and at the same time reduce overgrazing and halt desertification. A follow-up project using the same pilot PVWPS showed that irrigation of a desertified grassland could increase its productivity up to 20−30 times without imposing any threat to locally available water resources [102].

Afterward, a larger project entitled "Demonstration and Scale Up of Photovoltaic (PV) Solar Water Pumping Technology and System for the Conservation of Grassland and Farmland in China" founded by the Swedish International Development Cooperation Agency was implemented at three pilot sites in Qinghai, Inner Mongolia, and Xinjiang to further study the integration of PVWPSs with water-saving irrigation techniques. The overall objectives of the project were to evaluate the technical and economic feasibility of PVWPSs for grassland irrigation, to evaluate the ecological impacts

FIGURE 4.23 Difference in grassland productivity between fenced and overgrazed grassland in Inner Mongolia, China.

(e.g., carbon sequestration due to improved grassland restoration [103]), to evaluate the effect of solar pumping on water resources [104,105], and to develop novel business models for scaling up and commercialization [106,107]. The PVWPSs installed in Inner Mongolia were part of the Experiment and Demonstration Station for Water and Soil Conservation, Science and Technology, belonging to the Ministry of Water Resource (MWR). One of the pilot PVWPSs was located next to the village of Xilamu Rensumu in Wanlanchabu desert grassland area (41.32°N, 111.22°E, and 1590 m above mean sea level) and supplied water for a 1-ha drip irrigation system. The schematic diagram of the pilot PVWPS is provided in Fig. 4.24. The pilot PVWPS was used to validate simulation and optimization models and assess crop water requirements and the effect of water pumping on available water resources [107].

Recently, the European Commission has founded the project "MArket uptake of an innovative irrigation Solution based on LOW WATer-ENergy consumption" (MASLOWATEN) [108]. The main objective of the project was to show the technical and economic viability of PVWPSs for large-scale irrigation combined with water-saving irrigation technologies for a 30% water reduction in the irrigation sector. Within the project framework [90], it is investigated the economic viability of five large-scale PVWPSs with capacities between 40 and 360 kW_p in Spain, Portugal, Italy, and Morocco,

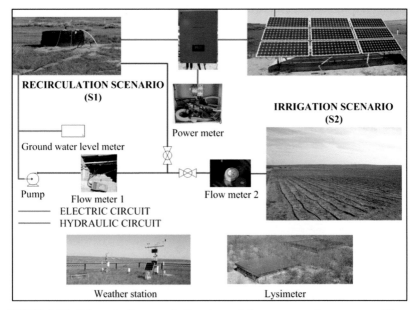

FIGURE 4.24 Schematic diagram of pilot photovoltaic water pumping system at Xilamu Rensumu in Wanlanchabu desert grassland area, Inner Mongolia, China [68].

FIGURE 4.25   Hybrid photovoltaic water pumping system for large-scale olive orchard irrigation installed in Portugal [109].

covering different system configurations available on the market. The experimental facility in Portugal is depicted in Fig. 4.25. All the projects showed positive economic viability with an internal rate of return varying between 10 and 28%. A further economic analysis was extended to the Economic Community of West African States region [110]. Large-scale PVWPSs were shown to be profitable in all the investigated locations, with payback periods varying between 2.1 and 10 years. The authors suggested implementing PVWPSs that substitute both current diesel-powered, and electric-grid-powered irrigation systems. PVWPSs with water storage tanks showed better economics as compared to those working at constant water pressure.

"Solar Irrigation for Agricultural Resilience in South Asia" (SoLAR-SA) is a project started by the Swiss Agency for Development and Cooperation (SDC) and the International Water Management Institute (IWMI) to provide climate-smart PVWPSs solutions for irrigation in South Asia [111]. South Asia (i.e., Bangladesh, India, Nepal, and Pakistan) is one of the regions on Earth that most relies on groundwater irrigation. Annually, approximately 250 km³ of groundwater is used in the irrigation sector, which counts more than 25 million installed water pumps for irrigation with a market value of US$3.78 billion of electricity use [112]. There are currently more than 250,000 solar irrigation pumps installed and the market is growing exponentially, with several issues concerning groundwater depletion. Through research and pilot sites spread among the participating countries, the project aims to develop PVWPSs solutions for irrigation that are climate-resilient, gender-equitable, socially-inclusive, and groundwater responsive [111]. The expected outcomes are technical, financial, and institutional innovations that can guarantee improved living conditions and lower the impact on groundwater resources.

The "Solar PV Powered Pumping Systems Project" is funded by the African Development Fund for the spread of PVWPSs for irrigation in Sudan [113]. The project aims to reduce farmers' dependency on fossils fuels, improve crop productivities, and promote better living conditions through the implementation of solar irrigation systems for 1170 farmers. A review of other programs and projects concerning PVWPSs for irrigation is provided in Nagpal [15] and in Hartung and Pluschke [114]. Other implemented projects can be found at the websites of the major solar pump manufacturers.

4.6 Economics, challenges, and prospects

The PV module price is the main price component in the initial investment cost of PVWPSs, followed by the inverter for AC solar water pumping applications, and by the pump unit, as shown in Fig. 4.26 [68]. During the last decade, PV module prices have significantly dropped, reaching levels of US\$220−320/kW$_p$ at the end of 2020, a price reduction in the order of 50%−70% as compared to 2013 [115]. The decline in PV module price has removed previous financial barriers to the widespread adoption of the technology in terms of both numbers and capacities. Currently, large-scale PVWPSs in the order of 40 kW$_p$ to 1 MW$_p$ are becoming more common in the market [116].

Traditionally, PVWPSs have competed with diesel-powered water pumping systems for irrigation in off-grid areas. In most of the studies conducted on this research topic that include economic and financial analyses, PVWPSs showed higher investment costs but lower life-cycle costs as compared to diesel-powered systems [92]. The few exceptions include case studies where national or regional regulations subsidize diesel prices—the main item costs in the life-cycle cost analysis for diesel-powered irrigation systems. An example of this market distortion was presented in Bouzidi and Campana [57], where the country in question heavily subsidizes the diesel price, meaning that PVWPSs for irrigation cannot compete with diesel systems even in terms of life-cycle costs. Rizi et al. [117] compared different alternatives for irrigation water pumping in Iran through a life-cycle cost analysis. Those alternatives included off-grid PVWPSs with and without battery, grid extension, and diesel systems. Due to the high subsidies on diesel and electricity, PVWPSs could compete with electric pumps connected to the grid, only if the capacity of the PVWPS was lower than 3 kW$_p$ and the grid extension was higher than 2 km.

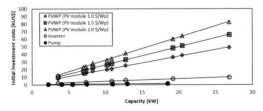

FIGURE 4.26 Photovoltaic water pumping systems initial investment costs as a function of capacity and components [68].

García et al. [118] compared the economic and environmental impacts of diesel water pumping systems, on- and off-grid PVWPSs, and electric-grid-powered pumping systems for irrigation in Spain. The economic results showed that off-grid PVWPSs had the highest investment costs, but the lowest life-cycle costs as compared to the investigated alternatives. From a life-cycle assessment viewpoint, diesel water pumping systems for irrigation showed the highest global warming potential. Irrigation systems driven by off-grid PVWPSs, and on-grid PVWPSs with and without an existing electric grid extension showed global warming potentials equal to 121, 19, and 29 g CO_2-eq/kWh, respectively. Todde et al. [109] investigated the environmental impacts of two large-scale PVWPSs—one located in Morocco and one in Portugal. The system located in Morocco was a hybrid PVWPS connected to the grid, while the one in Portugal was a hybrid PVWPS coupled with a diesel generator. Both systems were serving large-scale olive orchards. The energy payback times were 1.98 and 4.58 years for the systems in Morocco and Portugal, respectively. The carbon payback times were 1.86 and 9.16 years for Morocco and Portugal, respectively. The global warming potentials were 48 and 103.5 g CO_2-eq/kWh, respectively. The authors pointed out that the environmental performances of the investigated large-scale PVWPSs for irrigation were not only due to the significant difference in solar energy potentials between the two considered countries/locations but were also connected to other factors including the operational strategies of the solar irrigation systems. Hammad and Ebaid [119] also showed that, for underground water pumping in Jordan, large-scale PVWPSs were the most viable solution from a life-cycle perspective among different alternatives including diesel generators, electric grids, and hybrid solutions. The cost of energy for off-grid PVWPSs was the lowest and equal to US$0.136/kWh. The grid-connected solution had a cost of energy equal to US$0.144/kWh, while the diesel generator showed the highest cost of energy equal to US$0.239/kWh.

A summary of the irrigation system costs from the FAO database is provided in Fig. 4.27 [120]. Except for the outliers among the surface irrigation systems, sprinkler systems showed the highest average cost as compared to surface and localized irrigation systems. Chukalla et al. [121] highlighted that the higher efficiencies of microirrigation and subsurface irrigation systems as compared to surface and sprinkler irrigation systems are connected to higher investment costs, as shown in Fig. 4.28. Considering only the investment and operational costs, the specific investment costs (i.e., US$/ha) increase from surface to subsurface irrigation, but on the other hand the operational costs related to water and energy consumption decrease from sprinkler to subsurface irrigation systems, since lower water volumes and operating pressure are required. The optimal choice of the irrigation system consists thus in finding the best trade-off between investment and operational costs, and corresponding benefits.

One of the major remaining challenges of PVWPSs is associated with the water−food−energy nexus aspects of technology integration. While, as

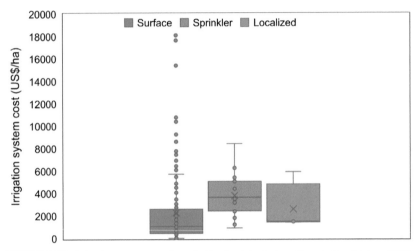

FIGURE 4.27 Irrigation system costs from the Food and Agriculture Organization database [120].

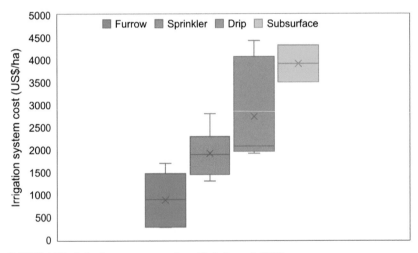

FIGURE 4.28 Irrigation system costs from Chukalla et al. [121].

mentioned in the previous sections, the technology is based on a renewable technology with low global warming potential, the extraction of groundwater or the general exploitation of water resources might threaten the sustainability of the technology. Shah et al. [122] analyzed PVWPSs from a water-–energy perspective with special consideration of groundwater irrigation in South Asia. The authors highlighted that PVWPSs can represent a threat to the already scarce groundwater resources because the falling prices of PV modules make PVWPSs more accessible to farmers, even without subsidies. This represents a challenge for central and local governments to know the actual number of solar irrigation systems installed and to control the effects on groundwater resources. The authors highlighted that policies are needed to balance both increasing incomes in poor communities through PVWPSs and preserving groundwater resources. Closas and Rap [123] highlighted that most of the current PVWPSs programs have focused on the spread of the technology without taking into consideration its limitations in the framework of the water–food–energy nexus, especially on aspects related to overexploitation of water resources. Access to water can lead to the phenomenon called "*repressed demand*" [124] that can lead to an increased water extraction exceeding the actual demand. Moreover, in irrigation projects, overexploitation of water resources can also be connected to a lack of knowledge of the actual crop water requirements. Gupta [125] analyzed the water––food–energy nexus of solar pumps in the state of Rajasthan, where more than 25% (about 37,000) of the pumps installed are PVWPSs. By analyzing data from 414 rural farmers, one of the main findings was that the installation of PVWPSs has led to an increase in average water consumption by farmers, especially those who did not have access to water pumping before (i.e., "repressed demand"). One policy recommendation given by the author was to promote the connection of PVWPSs to the grid. By doing so and guaranteeing generous feed-in-tariffs, the government can indirectly monitor the amount of power used for irrigation and thus quantify groundwater extraction, and at the same time can preserve groundwater resources because farmers prefer to benefit from selling the electricity surplus rather than using the power for irrigation [122]. Rathore et al. [126] provided a comprehensive overview of the key barriers for PVWPSs for irrigation in India, including: lack of knowledge of the technology; lack of technicians, project managers, and engineers in the field of solar irrigation systems; lack of infrastructures; and policy barriers. The author made a series of policy recommendations to overcome the identified barriers. When it comes to groundwater exploitation, the authors suggested a series of regulations that include the connection of PVWPSs for irrigation to the main grid, limiting the capacity of solar pumps, promoting community-based PVWPSs for irrigation, and promoting water-saving irrigation techniques at the same level as solar pumps.

4.7   Conclusions

Irrigation has a crucial role in sustaining the current and future world population by increasing and stabilizing crop yields. Irrigation can be energy-intensive, depending on the water source and irrigation system, and can cause serious threats to the available water resources. It is thus important that a water–food–energy nexus perspective is adopted to analyze the effects of irrigation. Solar-powered irrigation systems (in particular solar PV) integrated with water-saving irrigation techniques represent a viable solution to decarbonize the irrigation sector, especially in those areas that heavily rely on diesel-powered water pumping systems, and to reduce pressure on water resources. The drastic drop in PV module prices that has occurred in the last decade has enormously contributed to the spread of the technology in both on- and off-grid areas, as well as increasing the number and capacities of installations. This chapter has aimed to review mainly solar PVWPSs with a special focus on systems for irrigation purposes, highlighting recent advances on both the micro and macro level, and analyzing current challenges. From a components perspective, recent advances are focusing on improving components and system performances, in terms of both energy conversion efficiencies and operational conditions (i.e., the operating conditions of irrigation systems in solar PVWPSs without water tanks). Artificial intelligence algorithms and IoT devices help to achieve better performances, monitoring, and control. From a system perspective, one of the main trends is the hybridization of PVWPSs with other renewable energies, diesel generators, and the electric grid. Further integration into novel PV applications, such as agrivoltaics, is also an interesting trend that further allows synergies to be harnessed between water, energy and food sectors, and sustainable development targets to be reached. The evident water—food—energy nexus interdependencies in PVWPSs require integrated governance to guarantee higher crop yields and better living conditions without compromising the current and future availability of water resources.

Acknowledgments

Pietro Elia Campana acknowledges the financial support received from the Swedish Energy Agency for the project "Evaluation of the first agrivoltaic system in Sweden".

References

[1] FAO—Food and Agriculture Organization of the United Nations, AQUASTAT. Available from: http://www.fao.org/aquastat/en/; 2012 [accessed 15.04.21].

[2] FAO—Food and Agriculture Organization of the United Nations, FAOSTAT. Available from: http://www.fao.org/faostat/en/; 2012 [accessed 15.04.21].

[3] FAO—Food and Agriculture Organization of the United Nations, The future of food and agriculture — Trends and challenges. Rome; 2017.

[4] Meier J, Zabel F, Mauser W. A global approach to estimate irrigated areas—a comparison between different data and statistics. Hydrol Earth Syst Sci 2018;22(2):1119—33.

[5] Boretti A, Rosa L. Reassessing the projections of the world water development report. NPJ Clean Water 2019;2(1):1—6.

[6] Valin H, Sands RD, Van der Mensbrugghe D, Nelson GC, Ahammad H, Blanc E, et al. The future of food demand: understanding differences in global economic models. Agric Econ 2014;45(1):51—67.

[7] Wada Y, Flörke M, Hanasaki N, Eisner S, Fischer G, Tramberend S, et al. Modeling global water use for the 21st century: the water futures and solutions (WFaS) initiative and its approaches. Geoscientific Model Dev 2016;9(1):175—222.

[8]   IIASA—International Institute for Applied Systems Analysis. Available from: https://ar16. iiasa.ac.at/agricultural-water-demand/; 2021 [accessed 14.05.21].

[9]   Rosa L, Chiarelli DD, Sangiorgio M, Beltran-Peña AA, Rulli MC, D'Odorico P, et al. Potential for sustainable irrigation expansion in a 3 C warmer climate. Proc Natl Acad Sci 2020;117(47):29526−34.

[10] Puy A. Irrigated areas grow faster than the population. Ecol Appl 2018;28(6):1413−19.

[11] Mekonnen MM, Gerbens-Leenes W. The water footprint of global food production. Water 2020;12(10):2696.

[12] Cole MB, Augustin MA, Robertson MJ, Manners JM. The science of food security. NPJ Sci Food 2018;2(1):1−8.

[13] Mancosu N, Snyder RL, Kyriakakis G, Spano D. Water scarcity and future challenges for food production. Water 2015;7(3):975−92.

[14] Davis KF, Chhatre A, Rao ND, Singh D, Ghosh-Jerath S, Mridul A, et al. Assessing the sustainability of post-Green Revolution cereals in India. Proc Natl Acad Sci 2019; 116(50):25034−41.

[15] Nagpal D. Solar pumping for irrigation: Improving livelihoods and sustainability. Abu Dhabi: IRENA; 2016.

[16] UNESCO (United Nations Educational, Scientific and Cultural Organization), The United Nations World Water Development Report 2014. Water and Energy Volume 1, UNESCO, France. Available from: http://unesdoc.unesco.org/images/0022/002257/225741E.pdf; 2014.

[17] Sontake VC, Kalamkar VR. Solar photovoltaic water pumping system-A comprehensive review. Renew Sustain Energy Rev 2016;59:1038−67.

[18] Campana PE, Li H, Yan J. Dynamic modelling of a PV pumping system with special consideration on water demand. Appl energy 2013;112:635−45.

[19] Fraunhofer ISE. Photovoltaics report, Fraunhofer ISE and PSE Conferences & Consulting GmbH, Freiburg; 14 March 2019.

[20] VDMA. International Technology Roadmap for Photovoltaic (ITRPV), 12th Ed.; 2021.

[21] Shepovalov OV, Belenov AT, Chirkov SV. Review of photovoltaic water pumping system research. Energy Rep 2020;6:306−24.

[22] Chandel SS, Naik MN, Chandel R. Review of performance studies of direct coupled photovoltaic water pumping systems and case study. Renew Sustain Energy Rev 2017; 76:163−75.

[23] Shabaan S, El-Sebah MIA, Bekhit P. Maximum power point tracking for photovoltaic solar pump based on ANFIS tuning system. J Electr Syst Inf Technol 2018;5(1):11−22.

[24] Periasamy P, Jain NK, Singh IP. A review on development of photovoltaic water pumping system. Renew Sustain Energy Rev 2015;43:918−25.

[25] Zahab EEA, Zaki AM, El-sotouhy MM. Design and control of a standalone PV water pumping system. J Electr Syst Inf Technol 2017;4(2):322−37.

[26] Abouda S, Nollet F, Chaari A, Essounbouli N, Koubaa Y. Direct torque control-DTC of induction motor used for piloting a centrifugal pump supplied by a photovoltaic generator. Int J Electrical, Computer Commun Eng 2013;7(8):1110−15.

[27] Chandel SS, Naik MN, Chandel R. Review of solar photovoltaic water pumping system technology for irrigation and community drinking water supplies. Renew Sustain Energy Rev 2015;49:1084−99.

[28] Santra P. Performance evaluation of solar PV pumping system for providing irrigation through micro-irrigation techniques using surface water resources in hot arid region of India. Agric Water Manag 2021;245:106554.

[29] Murshid S, Singh B. Reduced sensor-based PMSM driven autonomous solar water pumping system. IEEE Trans Sustain Energy 2019;11(3):1323−31.

[30] Ibrahim MN, Rezk H, Al-Dhaifallah M, Sergeant P. Modelling and design methodology of an improved performance photovoltaic pumping system employing ferrite magnet synchronous reluctance motors. Mathematics 2020;8(9):1429.

[31] Yadav K, Sastry OS, Wandhare R, Sheth N, Kumar M, Bora B, et al. Performance comparison of controllers for solar PV water pumping applications. Sol Energy 2015; 119:195−202.

[32] Muhsen DH, Ghazali AB, Khatib T. Multiobjective differential evolution algorithm-based sizing of a standalone photovoltaic water pumping system. Energy Convers Manag 2016;118:32−43.

[33] Soenen C, Reinbold V, Meunier S, Cherni JA, Darga A, Dessante P, et al. Comparison of tank and battery storages for photovoltaic water pumping. Energies 2021;14(9):2483.

[34] Yahyaoui I, Atieh A, Serna A, Tadeo F. Sensitivity analysis for photovoltaic water pumping systems: energetic and economic studies. Energy Convers Manag 2017;135:402−15.

[35] Jones LE, Olsson G. Solar photovoltaic and wind energy providing water. Glob Chall 2017;1(5):1600022.

[36] Olsson A, Campana PE, Lind M, Yan J. Potential for carbon sequestration and mitigation of climate change by irrigation of grasslands. Appl Energy 2014;136:1145−54.

[37] Shao W, Zhu M, Liu J, Weng B, Xiang C, Gong J, et al. Photovoltaic water lifting and ecological water supplement for Xiang'an water system in Xiamen City. Energy Procedia 2017;142:230−5.

[38] Aliyu M, Hassan G, Said SA, Siddiqui MU, Alawami AT, Elamin IM. A review of solar-powered water pumping systems. Renew Sustain Energy Rev 2018;87:61−76.

[39] Hamidat A, Benyoucef B. Systematic procedures for sizing photovoltaic pumping system, using water tank storage. Energy Policy 2009;37(4):1489−501.

[40] Ould-Amrouche S, Rekioua D, Hamidat A. Modelling photovoltaic water pumping systems and evaluation of their CO_2 emissions mitigation potential. Appl Energy 2010; 87(11):3451−9.

[41] Alonso Abella M, Lorenzo E, Chenlo F. PV water pumping systems based on standard frequency converters. Prog Photovoltaics: Res Appl 2003;11(3):179−91.

[42] Meunier S, Heinrich M, Quéval L, Cherni JA, Vido L, Darga A, et al. A validated model of a photovoltaic water pumping system for off-grid rural communities. Appl Energy 2019;241:580−91.

[43] Hadwan M, Alkholidi A. Assessment of factors influencing the sustainable performance of photovoltaic water pumping systems. Renew Sustain Energy Rev 2018;92:307−18.

[44] Caton P. Design of rural photovoltaic water pumping systems and the potential of manual array tracking for a West-African village. Sol Energy 2014;103:288−302.

[45] Ebaid MS, Qandil H, Hammad M. A unified approach for designing a photovoltaic solar system for the underground water pumping well-34 at Disi aquifer. Energy Convers Manag 2013;75:780−95.

[46] Bouzidi B. New sizing method of PV water pumping systems. Sustain Energy Technol Assess 2013;4:1−10.

[47] Gualteros S, Rousse DR. Solar water pumping systems: A tool to assist in sizing and optimization. Sol Energy 2021;225:382−98.

[48] Olcan C. Multi-objective analytical model for optimal sizing of stand-alone photovoltaic water pumping systems. Energy Convers Manag 2015;100:358−69.

[49] Shahverdi K, Bellos E, Loni R, Najafi G, Said Z. Solar-driven water pump with organic Rankine cycle for pressurized irrigation systems: a case study. Therm Sci Eng Prog 2021;25:100960.

[50] Ali B. Comparative assessment of the feasibility for solar irrigation pumps in Sudan. Renew Sustain Energy Rev 2018;81:413—20.

[51] Allen RG, Pereira LS, Raes D, Smith M. Crop evapotranspiration-Guidelines for computing crop water requirements-FAO Irrigation and drainage paper 56. FAO, Rome, 300(9), D05109; 1998.

[52] Campana PE, Lastanao P, Zainali S, Zhang J, Landelius T, Melton F. Towards an operational irrigation management system for Sweden with a water-food-energy nexus perspective; 2021. EarthArXiv preprint EarthArXiv: 10.31223/X5SW41.

[53] Irmak S, Odhiambo LO, Kranz WL, Eisenhauer DE. Irrigation efficiency and uniformity, and crop water use efficiency; 2011.

[54] Daccache A, Ciurana JS, Diaz JR, Knox JW. Water and energy footprint of irrigated agriculture in the Mediterranean region. Environ Res Lett 2014;9(12):124014.

[55] Espinosa-Tasón J, Berbel J, Gutiérrez-Martín C. Energized water: Evolution of water-energy nexus in the Spanish irrigated agriculture, 1950—2017. Agric Water Manag 2020; 233:106073.

[56] Campana PE, Li H, Yan J. Techno-economic feasibility of the irrigation system for the grassland and farmland conservation in China: photovoltaic vs. wind power water pumping. Energy Convers Manag 2015;103:311—20.

[57] Bouzidi B, Campana PE. Optimization of photovoltaic water pumping systems for date palm irrigation in the Saharan regions of Algeria: increasing economic viability with multiple-crop irrigation. Energy Ecol Environ 2020;1—28.

[58] Zhang C, Campana PE, Yang J, Yu C, Yan J. Economic assessment of photovoltaic water pumping integration with dairy milk production. Energy Convers Manag 2018;177:750—64.

[59] Glasnovic Z, Margeta J. A model for optimal sizing of photovoltaic irrigation water pumping systems. Sol energy 2007;81(7):904—16.

[60] Campana PE. PV water pumping systems for agricultural applications (Doctoral dissertation, Mälardalen University); 2015.

[61] Salem GSA, Kazama S, Shahid S, Dey NC. Impacts of climate change on groundwater level and irrigation cost in a groundwater dependent irrigated region. Agric Water Manag 2018;208:33—42.

[62] The Intergovernmental Panel on Climate Change. Available from: https://www.ipcc.ch/; 2021 [accessed 20.07.21].

[63] Wu WY, Lo MH, Wada Y, Famiglietti JS, Reager JT, Yeh PJF, et al. Divergent effects of climate change on future groundwater availability in key mid-latitude aquifers. Nat Commun 2020;11(1):1—9.

[64] Bouras E, Jarlan L, Khabba S, Er-Raki S, Dezetter A, Sghir F, et al. Assessing the impact of global climate changes on irrigated wheat yields and water requirements in a semi-arid environment of Morocco. Sci Rep 2019;9(1):1—14.

[65] Ruti PM, Somot S, Giorgi F, Dubois C, Flaounas E, Obermann A, et al. MED-CORDEX initiative for Mediterranean climate studies. Bull Am Meteorological Soc 2016;97(7):1187—208.

[66] Deveci O, Onkol M, Unver HO, Ozturk Z. Design and development of a low-cost solar powered drip irrigation system using systems modeling language. J Clean Prod 2015; 102:529—44.

[67] Eltawil MA, Alhashem HA, Alghannam AO. Design of a solar PV powered variable frequency drive for a bubbler irrigation system in palm trees fields. Process Saf Environ Prot 2021;.

[68] Campana PE, Li H, Zhang J, Zhang R, Liu J, Yan J. Economic optimization of photovoltaic water pumping systems for irrigation. Energy Convers Manag 2015;95:32—41.

[69] Campana PE, Zhu Y, Brugiati E, Li H, Yan J. PV water pumping for irrigation equipped with a novel control system for water savings. Energy Procedia 2014;61:949−52.

[70] López-Luque R, Reca J, Martínez J. Optimal design of a standalone direct pumping photovoltaic system for deficit irrigation of olive orchards. Appl Energy 2015;149:13−23.

[71] Reca J, Torrente C, López-Luque R, Martínez J. Feasibility analysis of a standalone direct pumping photovoltaic system for irrigation in Mediterranean greenhouses. Renew Energy 2016;85:1143−54.

[72] Yahyaoui I, Atieh A, Tadeo F, Tina GM. Energetic and economic sensitivity analysis for photovoltaic water pumping systems. Sol Energy 2017;144:376−91.

[73] Rezk H, Abdelkareem MA, Ghenai C. Performance evaluation and optimal design of stand-alone solar PV-battery system for irrigation in isolated regions: a case study in Al Minya (Egypt). Sustain Energy Technol Assess 2019;36:100556.

[74] Zavala V, López-Luque R, Reca J, Martínez J, Lao MT. Optimal management of a multisector standalone direct pumping photovoltaic irrigation system. Appl Energy 2020; 260:114261.

[75] García AM, García IF, Poyato EC, Barrios PM, Díaz JR. Coupling irrigation scheduling with solar energy production in a smart irrigation management system. J Clean Prod 2018;175:670−82.

[76] García AM, Perea RG, Poyato EC, Barrios PM, Díaz JR. Comprehensive sizing methodology of smart photovoltaic irrigation systems. Agric Water Manag 2020;229:105888.

[77] Elkholy MM, Fathy A. Optimization of a PV fed water pumping system without storage based on teaching-learning-based optimization algorithm and artificial neural network. Sol Energy 2016;139:199−212.

[78] Errouha M, Derouich A, Motahhir S, Zamzoum O, El Ouanjli N, El Ghzizal A. Optimization and control of water pumping PV systems using fuzzy logic controller. Energy Rep 2019;5:853−65.

[79] Lakshmiprabha KE, Govindaraju C. An integrated isolated inverter fed bldc motor for photovoltaic agric pumping systems. Microprocessors Microsyst 2020;79:103276.

[80] Cordeiro A, Pires VF, Foito D, Pires AJ, Martins JF. Three-level quadratic boost DC-DC converter associated to a SRM drive for water pumping photovoltaic powered systems. Sol Energy 2020;209:42−56.

[81] Poompavai T, Kowsalya M. Control and energy management strategies applied for solar photovoltaic and wind energy fed water pumping system: a review. Renew Sustain Energy Rev 2019;107:108−22.

[82] Ammar RB, Ammar MB, Oualha A. Photovoltaic power forecast using empirical models and artificial intelligence approaches for water pumping systems. Renew Energy 2020;153:1016−28.

[83] Almeida RH, Ledesma JR, Carrêlo IB, Narvarte L, Ferrara G, Antipodi L. A new pump selection method for large-power PV irrigation systems at a variable frequency. Energy Convers Manag 2018;174:874−85.

[84] dos Santos WS, Torres PF, Brito AU, Manito ARA, Pinto Filho GF, Monteiro WL, et al. A novel method to determine the optimal operating point for centrifugal pumps applied in photovoltaic pumping systems. Sol Energy 2021;221:46−59.

[85] Stoyanov L, Bachev I, Zarkov Z, Lazarov V, Notton G. Multivariate analysis of a wind−PV-based water pumping hybrid system for irrigation purposes. Energies 2021; 14(11):3231.

[86] Xiang C, Liu J, Yu Y, Shao W, Mei C, Xia L. Feasibility assessment of renewable energies for cassava irrigation in China. Energy Procedia 2017;142:17−22.

[87] Li D, Zhu D, Wang R, Ge M, Wu S, Cai Y. Sizing optimization and experimental verification of a hybrid generation water pumping system in a greenhouse. Math Probl Eng 2020;2020.

[88] Angadi S, Yaragatti UR, Suresh Y, Raju AB. Comprehensive review on solar, wind and hybrid wind-PV water pumping systems-an electrical engineering perspective. CPSS Trans Power ElectrAppl 2021;6(1):1−19.

[89] Almeida RH, Carrêlo IB, Martínez-Moreno F, Carrasco LM, Narvarte L. A 140 kW hybrid PV-diesel pumping system for constant-pressure irrigation. In 33rd European Photovotaic Solar Energy Conference and Exhibition; 2017.

[90] Carrêlo IB, Almeida RH, Narvarte L, Martinez-Moreno F, Carrasco LM. Comparative analysis of the economic feasibility of five large-power photovoltaic irrigation systems in the Mediterranean region. Renew Energy 2020;145:2671−82.

[91] Campana PE, Olsson A, Li H, Yan J. An economic analysis of photovoltaic water pumping irrigation systems. Int J Green Energy 2016;13(8):831−9.

[92] Barrueto Guzmán A, Barraza Vicencio R, Ardila-Rey JA, Núñez Ahumada E, González Araya A, Arancibia Moreno G. A cost-effective methodology for sizing solar PV systems for existing irrigation facilities in Chile. Energies 2018;11(7):1853.

[93] Hassan W, Kamran F. A hybrid PV/utility powered irrigation water pumping system for rural agricultural areas. Cogent Eng 2018;5(1):1466383.

[94] Proctor KW, Murthy GS, Higgins CW. Agrivoltaics Align with green new deal goals while supporting investment in the US' rural economy. Sustainability 2021;13(1):137.

[95] United States of America House of Representative. H. Res. 109: Recognizing the Duty of the Federal Government to Create a Green New Deal; United States of America House of Representative: Washington, DC, USA, 2019.

[96] AL-agele HA, Proctor K, Murthy G, Higgins C. A case study of tomato (*Solanum lycopersicon* var. Legend) production and water productivity in agrivoltaic systems. Sustainability 2021;13(5):2850.

[97] Ye B, Jiang J, Liu J. Feasibility of coupling pv system with long-distance water transfer: a case study of China's "south-to-north water diversion.". Resources Conserv Recycling 2021;164:105194.

[98] Al-Ali AR, Al Nabulsi A, Mukhopadhyay S, Awal MS, Fernandes S, Ailabouni K. IoT-solar energy powered smart farm irrigation system. J Electron Sci Technol 2019; 17(4):100017.

[99] Gimeno-Sales FJ, Orts-Grau S, Escribá-Aparisi A, González-Altozano P, Balbastre-Peralta I, Martínez-Márquez CI, et al. PV monitoring system for a water pumping scheme with a lithium-ion battery using free open-source software and iot technologies. Sustainability 2020;12(24):10651.

[100] Selmani A, Oubehar H, Outanoute M, Ed-Dahhak A, Guerbaoui M, Lachhab A, et al. Agricultural cyber-physical system enabled for remote management of solar-powered precision irrigation. Biosyst Eng 2019;177:18−30.

[101] Radstake F, Yeager C. Greener pastures from the Sun: Solar photovoltaic-driven irrigation in Qinghai Province; 2012.

[102] Yan J, Wang H, Gao Z. Qinghai pasture conservation using solar photovoltaic (PV)-driven irrigation. Asian Development Bank, Final Report ADB RSC-C91300 (PRC); 2010.

[103] Olsson A, Campana PE, Lind M, Yan J. PV water pumping for carbon sequestration in dry land agriculture. Energy Convers Manag 2015;102:169−79.

[104] Yu Y, Liu J, Wang H, Liu M. Assess the potential of solar irrigation systems for sustaining pasture lands in arid regions−acase study in Northwestern China. Appl Energy 2011;88(9):3176−82.

[105] Xu H, Liu J, Qin D, Gao X, Yan J. Feasibility analysis of solar irrigation system for pastures conservation in a demonstration area in Inner Mongolia. Appl energy 2013; 112:697−702.
[106] Zhang C, Campana PE, Yang J, Yan J. Economic performance of photovoltaic water pumping systems with business model innovation in China. Energy Convers Manag 2017;133:498−510.
[107] Zhang J, Liu J, Campana PE, Zhang R, Yan J, Gao X. Model of evapotranspiration and groundwater level based on photovoltaic water pumping system. Appl energy 2014; 136:1132−7.
[108] MASLOWATEN. Available from: https://maslowaten.eu/; 2021 [accessed 21.07.21].
[109] Todde G, Murgia L, Deligios PA, Hogan R, Carrelo I, Moreira M, et al. Energy and environmental performances of hybrid photovoltaic irrigation systems in Mediterranean intensive and super-intensive olive orchards. Sci Total Environ 2019;651:2514−23.
[110] Lorenzo C, Almeida RH, Martínez-Núñez M, Narvarte L, Carrasco LM. Economic assessment of large power photovoltaic irrigation systems in the ECOWAS region. Energy 2018;155:992−1003.
[111] SoLAR. Available from: https://solar.iwmi.org/; 2021 [accessed 21.07.21].
[112] Verma S, Kashyap D, Shah T, Crettaz M, Sikka A. SoLAR: Solar Irrigation for Agriculture Resilience; 2018.
[113] African Development Fund. Available from: https://projectsportal.afdb.org/dataportal/VProject/show/P-SD-FF0-001; 2021 [accessed 21.07.21].
[114] Hartung H, Pluschke L. The benefits and risks of solar-powered irrigation—A global overview. United Nations Food and Agriculture Organization; 2018.
[115] IRENA. Renewable Power Generation Costs in 2020. Abu Dhabi: International Renewable Energy Agency; 2021.
[116] Narvarte L, Almeida RH, Carrêlo IB, Rodríguez L, Carrasco LM, Martinez-Moreno F. On the number of PV modules in series for large-power irrigation systems. Energy Convers Manag 2019;186:516−25.
[117] Rizi AP, Ashrafzadeh A, Ramezani A. A financial comparative study of solar and regular irrigation pumps: case studies in eastern and southern Iran. Renew Energy 2019;138:1096−103.
[118] García AM, Gallagher J, McNabola A, Poyato EC, Barrios PM, Díaz JR. Comparing the environmental and economic impacts of on-or off-grid solar photovoltaics with traditional energy sources for rural irrigation systems. Renew Energy 2019; 140:895−904.
[119] Hammad M, Ebaid MS. Comparative economic viability and environmental impact of PV, diesel and grid systems for large underground water pumping application (55 wells) in Jordan. Renewables: Wind, Water, Sol 2015;2(1):1−22.
[120] FAO − Food and Agriculture Organization of the United Nations. AQUASTAT. Available from: http://www.fao.org/aquastat/en/overview/archive/investment-costs; 2012 [accessed 15.04.21].
[121] Chukalla AD, Krol MS, Hoekstra AY. Marginal cost curves for water footprint reduction in irrigated agriculture: guiding a cost-effective reduction of crop water consumption to a permit or benchmark level. Hydrol Earth Syst Sci 2017;21(7):3507−24.
[122] Shah T, Verma S, Durga N. Karnataka's smart, new solar pump policy for irrigation. Economic Political Wkly 2014;10−14.
[123] Closas A, Rap E. Solar-based groundwater pumping for irrigation: Sustainability, policies, and limitations. Energy Policy 2017;104:33−7.

[124] Fedrizzi MC, Ribeiro FS, Zilles R. Lessons from field experiences with photovoltaic pumping systems in traditional communities. Energy Sustain Dev 2009;13(1):64–70.

[125] Gupta E. The impact of solar water pumps on energy-water-food nexus: Evidence from Rajasthan, India. Energy Policy 2019;129:598–609.

[126] Rathore PKS, Das SS, Chauhan DS. Perspectives of solar photovoltaic water pumping for irrigation in India. Energy Strategy Rev 2018;22:385–95.

Chapter 5

Agrivoltaics: solar power generation and food production

Max Trommsdorff[1,2], Ipsa Sweta Dhal[1], Özal Emre Özdemir[1],
Daniel Ketzer[3], Nora Weinberger[3] and Christine Rösch[3]

[1]*Fraunhofer Institute for Solar Energy Systems ISE, Freiburg im Breisgau, Germany,*
[2]*Wilfried Guth Chair, Department of Economics, University of Freiburg, Wilhelmstr, Freiburg im, Breisgau, Germany,* [3]*Karlsruhe Institute of Technology (KIT), Karlsruhe, Germany*

5.1 Introduction

As a part of the global clean energy transition, the increased deployment of ground-mounted PV (GM-PV) systems depends on the availability of land. In some regions, scarce land resources can lead to competition between agriculture and PV land use, threatening both food and energy security. Agrivoltaics is a method to combine agricultural and electricity production on the same unit of land, which significantly increases land-use efficiency and has the potential to contribute towards mitigation of related land-use conflicts. Additionally, agrivoltaics is expected to stabilize agricultural yields in regions that are vulnerable to the effects of climate change by providing weather protection and shading and might contribute to strengthen and vitalize rural economies and livelihoods.

Adolf Goetzberger and his colleague Armin Zastrow were the first to propose the concept of agrivoltaics in the early 1980s [1]. However, it was only about 10 years ago that agrivoltaics gained traction. In Japan, pioneer Akira Nagashima analyzed crop growth below PV modules within the first research pilot systems in 2004 and promoted the technology under the heading of "solar sharing" which led to the first governmental support scheme implemented in 2012 [2]. In 2014, China installed the first large-scale agrivoltaic systems and, still, today remains the country with the largest installed capacity in the world [3]. France was the first European country to systematically support agrivoltaics with regular tenders starting in 2017. This development was largely driven by the research of Christian Dupraz at the French Institut National de la Recherche Agronomique (INRAé) and the Sun'Agri R&D program [4]. Other countries that implemented or plan to implement

Solar Energy Advancements in Agriculture and Food Production Systems.
DOI: https://doi.org/10.1016/B978-0-323-89866-9.00012-2

159

governmental supporting schemes are the United States of America in the state of Massachusetts, South Korea, India, Israel, Germany, and Italy. An overview of the policies of those countries can be found in Section 5.7.

In 2021, agrivoltaics emerged as a market-ready technology with a globally installed capacity of more than 14 GWp. In most subtropical and semiarid regions, however, agrivoltaics remains widely unknown even though the technical potential appears to be very high especially in these regions.

5.1.1 Definition and classification of agrivoltaics

The title of the first scientific publication on agrivoltaics "Potatoes under the collector" indicates that the original idea of dual land use referred to a high elevation of PV modules to harvest electricity and to cultivate food crops on the ground below [5]. This could be regarded as the classical agrivoltaics design also known as overhead agrivoltaics, horizontal agrivoltaics, or stilted agrivoltaics. Similarly, the term "solar sharing" used in Japan illustrates the understanding of a higher and a lower layer to harvest twice.

Since 2015, however, new approaches of ground-mounted vertical PV in parallel to agricultural areas were developed. This interspace agrivoltaics or vertical agrivoltaics raised the need for a definition and classification of agrivoltaics.

In 2018, Lasta and Konrad [6] were the first to propose a classification, distinguishing between arable farming, PV greenhouses, and buildings. However, the authors did not yet address highly elevated and ground-mounted agrivoltaics. Brecht et al. [7] suggested another classification defining crop production and livestock as the two main applications of agrivoltaic systems. While the authors also included highly elevated and ground-mounted systems, a newer suggestion by Gorjian et al. [8] assigns typical tilt and tracking technologies of PV modules and agricultural application to both

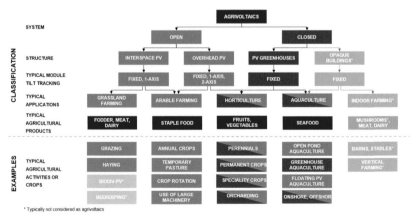

FIGURE 5.1 Classification of PV and agricultural colocation approaches [9].

classes. Fig. 5.1 illustrates a revised version of the classification as used in Gorjian et al. [9]. In this chapter, we only discuss open agrivoltaic systems as PV greenhouses are described in Chapter 5.

5.1.2 Current state of research and technology

Agrivoltaic research includes a wide range of fields such as agronomical, microclimatic, and PV design studies and relies on inter- and transdisciplinary research. Weselek et al. [10] adopted an approach of qualitative, selective literature review with the main focus on agronomical questions including an overview of selected crop shading impact studies not directly connected to agrivoltaics. Aroca-Delgado et al. [11] provide a review of various shading systems, investigating the general compatibility of crops and PV modules. Further qualitative review studies include Jain et al. [12] and, with a wider focus on techno-ecological synergies of PV technologies, Hernandez et al. [13].

As the number of published papers significantly increased from 2016 onwards, protocol-driven, keyword-based literature reviews have been conducted for instance by Toledo and Scognamiglio [14] who also included PV greenhouse-related research, for a total count of 215 papers published before the end of 2020. From the end of 2020 to mid-2021 at least 20 more papers on agrivoltaics were published, not including the conference proceedings of the first international conference on agrivoltaics, the AgriVoltaic2020. Fig. 5.2 shows the steep increase of research in the first wave after 2011 and then again after 2015.

Besides published papers, technical reports are a major source of scientific work, for instance by the US-based National Renewable Energy Laboratory (NREL) [15−17] or the German Fraunhofer Institute for Solar Energy Systems ISE [18,19].

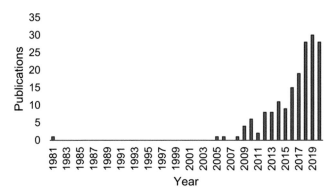

FIGURE 5.2 Scientific papers published, directly related to agrivoltaics and photovoltaic greenhouses [14].

TABLE 5.1 Selected research and development teams around the world.

Country	Research program/main organization	Unique research focus
China	Fudan University [20].	Agrivoltaic system mathematical modeling and simulations accompanying the development of agrivoltaics in China since 2017.
France	Sun'Agri, Institut national de la recherche agronomique (INRAé) and industry partner itk [4,21].	Extensive experience since 2010, among others, with a single tracking system in combination with horticulture, also on utility-scale systems. Extensive research on water balance-related benefits and agronomic decision support tools.
Japan	University of Chiba, Chiba Eco Energy co. Ltd. [22].	Extensive experience and research from 2013/14 onwards as well as operation of several agrivoltaic "solar sharing" systems and policy research. One of the leading initiatives in Japan.
Germany	Fraunhofer ISE [23], University of Hohenheim [24]. Karlsruhe Institute of Technology (KIT), Institute for Technology Assessment and Systems Analysis (ITAS) [25].	Experience in transdisciplinary research, agronomic considerations, international technology transfer, artificial intelligence and integrated modeling approaches since 2013/14.
Belgium	Katholieke Universiteit Leuven (KU Leuven) [26].	Focus on orchard applications and designs, researching horticulture PV since 2017/18.
USA	University of Massachusetts Clean Energy Extension and Centre for Agricultural, Food, and Environment [27].	Reviewing design applications for agrivoltaics promotion scheme. Research on cranberry and horticulture since 2015/16.
USA	University of Arizona [28].	Research for semiarid climate contexts. Exemplary groundwork since 2016/17 on water balance and irrigation benefits for crops in nonrainfed conditions. Educational approach with agrivoltaics as learning lab.

(Continued)

TABLE 5.1 (Continued)

Country	Research program/main organization	Unique research focus
USA	InSPIRE, U.S. Department of Energy (DOE), National Renewable Energy Laboratory (NREL) [29].	Research coordination since 2018 and consolidation of low impact PV strategies, PV greenhouses, agrivoltaics, pollinator habitats, and more.
Italy	Università Cattolica del Sacro Cuore, Rem Tec S.r.l [30].	Special focus on dual-axis tracking systems, simulation, life-cycle assessment, and system evaluation since 2015/16.
India	ICAR—Central Arid Zone Research Institute (CAZRI) [31].	Special focus on crop and water management in arid as well as semiarid regions (Rajasthan) since 2016/17.

Table 5.1 represents a list of selected research teams around the world. Along with increasing stakeholder interest in the field, research is expanding in many countries with new teams forming every year.

5.2 Agricultural aspects

To address the challenges of climate change, efficient water use, and the desire for higher yields, crop protection measures in agriculture are steadily rising—among them the use of greenhouses, foil tunnels, and hail protection nets. At the same time, the agricultural sector is moving more into focus as one of the largest emitters of greenhouse gases, especially methane and nitrous oxide. Due to more frequent weather extremes such as heavy rainfall and hail and the resulting yield fluctuations, farms are more and more exposed to economic challenges.

The dual-use of farmland for food production and PV power generation represents an opportunity to address these challenges simultaneously. In horticulture and berry production, agrivoltaics could reduce the use of or replace plastic foils and/or hail nets providing shelter against hail or frost damage as well as sunburn on crops. In arid regions, shading offers benefits by reducing evapotranspiration and water demand. While synergies in horticulture applications seem to be the highest, applications in arable farming have a much larger area potential in most countries. If the produced electricity is directly used on site, agrivoltaics could also contribute to reducing the carbon footprint of the farming unit. The upcoming use of electrified land machines and electricity storage technologies might facilitate this development.

Thus, agrivoltaics can increase the productivity of the farmland and aid in enhancing the overall income of the farmers. It may also contribute towards diversifying the income of the farmers by facilitating the growth of various crops under the installed PV modules and the revenue generated from electricity sales or land lease rents from the owner of the agrivoltaic system. However, due to different cultivation systems, variety-specific traits, and machine employment, the agricultural application affects the design of the agrivoltaic system [18]. For the farmer, the installation of an agrivoltaic system also represents a long-term decision with consequences on the use of agricultural machinery and the range of crops suitable for cultivation over the lifetime of the system of 20 years or more.

One challenge of agrivoltaics is to determine a reasonable allocation of solar radiation between energy generation and crop production. Shading caused by PV modules is probably the most crucial factor when considering agricultural aspects, but also associated microclimatic changes can affect crop growth and development [32].

5.2.1 Light availability

The major and most obvious impact of an agrivoltaic system is that the PV modules reduces the average available light for the crops at ground level. Field trials and experiments show that plants react differently to changing light conditions. The theoretical relationship between light and photosynthesis is illustrated by the so-called light response curve.

Fig. 5.3 shows how a sun-loving and a more shade-tolerant crop can respond to increasing light intensity, assuming no other factors are growth limiting. Generally, at low light levels, the rate of photosynthesis increases linearly with increasing light intensity. Because plants cannot reach their full growth potential, low light intensities mean stress [34]. As the light intensity rises, the growth of assimilation rate starts leveling off until a species-specific light saturation point is reached. Further increases in light intensity do not cause a change in the rate of photosynthesis as the capacity of light-harvesting reactions is finite [35]. The exact shape of the light response curve at canopy level under field conditions differs from plant to plant.

The light saturation point is a crucial criterion for defining the shading ratio of an agrivoltaic system or, once the system is installed, for determining the suitability of crops to be cultivated in the system. The lower the light saturation point the more shade can be given to a crop without experiencing yield losses. Though, if the shading rates are adjusted accordingly, it is theoretically possible to grow all crops in an agrivoltaic system. It must be emphasized that when other factors (e.g., water) are already limiting crop

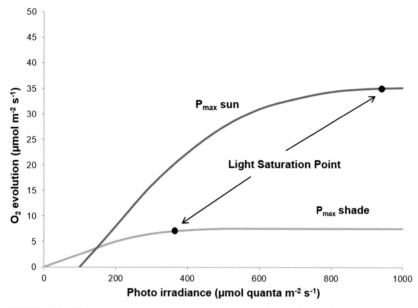

FIGURE 5.3 Light response curve of a sun-loving (purple) and shade-tolerant (green) plant, illustrating the interplay of photosynthetic rate (O_2 evolution) and light intensity (photon irradiance). *Reproduced with permission from Plants in Action [33].*

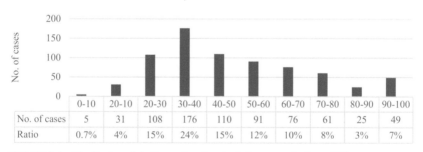

	0-10	20-10	20-30	30-40	40-50	50-60	60-70	70-80	80-90	90-100
No. of cases	5	31	108	176	110	91	76	61	25	49
Ratio	0.7%	4%	15%	24%	15%	12%	10%	8%	3%	7%

FIGURE 5.4 Distribution of shading ratios of agrivoltaic systems in Japan [37].

growth, shading is not necessarily a hindrance, and may even be beneficial by reducing water demand.

Simulating the impact of agrivoltaic systems on the availability of direct solar irradiation on the crop level can be done by sun ray projections of the respective position of the sun over time. To also assess the impact on diffuse light, different models of the sky hemisphere have been employed in the past. If reflections to the rear side of the PV modules or parts of the plants should be considered, two approaches are the view-factor method and the ray-tracing method [36]. Akira Nagashima suggests a simplified method to approximate the shading rate as the ratio of module surface area and total area [2]. Fig. 5.4 shows an analysis by Makoto et al. [38], visualizing the shading ratio distribution among agrivoltaic systems in Japan based on statistics by the Ministry of Agriculture, Forestry and Fishery (MAFF) in 2018.

5.2.2 Light management

According to the sun's changing position, the shade of PV modules of an agrivoltaic system varies over a day and a year. Homogeneous light distribution is desirable for uniform plant growth and ripening especially in arable farming applications. To ensure this, there are different possibilities for light management:

- Deviating the module row orientation from the electrical optimum of pure south or north (depending on the hemisphere) towards east or west achieves a more homogenous shading pattern on the field. This avoids "shade trenches" where some crops receive significantly less light than others. In a case study in Southern Germany, such an orientation deviation resulted in a reduction of electrical yield of less than 5% [39].
- Smaller modules can be used so that shade is minimized and light distributes more equally. This method is frequently applied in Japan [2].
- In glass-glass modules, the PV cells can be spaced out. Such semitransparent modules allow for partial shade below the PV modules and facilitate smaller row distances while still maintaining the same level of light availably. This is particularly interesting in horticulture applications when plants benefit from the shelter of larger PV modules.
- Likewise, modules can be arranged in a checkerboard pattern to increase the homogeneity of shade.
- The maximal movement of shade over a day is accomplished with an east-west alignment of the module rows. This can be combined with single or dual-axis tracking to deliberately adjust the light conditions according to crop-specific needs. If opaque PV modules follow an electrically

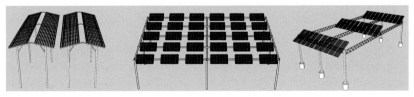

FIGURE 5.5 Illustration of different types of agrivoltaic systems with east—west, south- and south—east orientations [18].

optimized tracking pattern or are installed in a fixed east-west system, the width of the module rows should be considerably less than the height of the system to avoid that too less light reaches the ground below the module rows.
- The larger vertical clearance of the PV modules also increases the availability of light on the field [39].

Fig. 5.5 illustrates three types of module arrangements for agrivoltaics systems.

5.2.3 Microclimatic impact

Agrivoltaics can also alter crop growth by modifying the microclimatic conditions. The design of the agrivoltaic system, particularly the ground coverage ratio, determines how the microclimate parameters, such as air and soil temperature, humidity, soil moisture, wind speed, and evapotranspiration, is changed.

Table 5.2 presents selected findings of changes to microclimatic conditions due to agrivoltaics [45]. Soil and air moisture levels were found to be higher in agrivoltaic systems, whereas the soil and the air temperature were found to be more balanced out throughout the day. As the distance between rows is larger, there is an increase in radiation at the ground level. Consequently, there is higher irradiation in the summer months compared to winter [39].

5.2.4 Suitable crops

Studies have identified crops that, under given circumstances, are considered suitable for growing in agrivoltaic systems. These include leafy greens, fodder varieties such as clover grass, several fruits and berries, herbs, and spices. Lettuce, like other crops, responds to shade with an increased leaf area to overcome shade-related drawbacks. As the leaves are the product of the plant, lettuce cultivars are particularly attractive for agrivoltaics. Fruits and berries are suitable as the shelter of PV modules can protect crops that are relatively sensitive to extreme weather events. Thus, from an agricultural perspective, the benefits of stable yields achieved through crop protection measures outweigh potential shade-related yield reductions.

Table 5.3 represents a summary of different studies, showing the yield changes of various crops induced by partial shading and altered microclimatic conditions. In a case study from Germany, potatoes, winter wheat, and celery showed a decreasing yield of almost 20% in the first year of field trial and an increase in yield of more than 10% (in the case of potato and celery) in the second year [10]. As this season was exceptionally dry and hot, the results harmonize with the general expectations: If for light availabilities below the light saturation point, no other factors limit crop growth, shading will lead to a yield decline while in periods of drought stress with

TABLE 5.2 Affected microclimatic parameters in agrivoltaics.

Category	Panel elevation (m)	Location	Observed impact relative to the reference area	References
Soil and air temperature	3.3	Tucson, Arizona, USA	• Air temperature was 1.2°C + 0.3°C cooler during the day and 0.5°C + 0.4°C warmer during the night.	[40]
	4	Montpellier, France	• Crop temperature at midday decreased significantly while increasing during the night. • Average soil temperature decreased significantly. • Average air temperature remained steady.	[32]
	5.5	Heggelbach, Germany	• No major change in average air temperature. • Lower soil temperature in summer.	[10]
	3	Junagadh, India	• Average hourly air temperature decreased by 0.5°C–1.5°C between September and February.	[41]
	4	Santiago de Chile, Chile	• Monthly average air temperature decreased by 1°C during warm months.	[42]
Soil and air moisture	3.3	Tucson, Arizona, USA	• Vapor pressure deficit (VPD) was lower, by average −0.52 + 0.15 kPa, hence higher air moisture (to be found). • Soil moisture remained about 15% higher. • Soil moisture remained 5% greater (1.0% volumetric units) when irrigated daily.	[43]
	5.5	Heggelbach, Germany	• Slightly higher soil moisture in summer.	[10]
	4	Santiago de Chile, Chile	• Relative humidity was increased by 5%– 0%.	[42]
	3	Junagadh, India	• Higher relative air humidity in dry winter months.	[41]
Evapotranspiration	4	Montpellier, France	• Reduction of about 14%–29% of evapotranspiration. • Evapotranspiration daily average decreased by 19%–26%	[32,44]

TABLE 5.3 Crop yield changes with agrivoltaics.

Crop	Location	Shading rate	Yield change	References
Lettuce	Santiago, Chile	30%	(−) 8%	[42]
Broccoli	Santiago, Chile	30%	(−) 29%	[42]
Winter Wheat	Heggelbach, Germany	35%	(−) 19% (2017)	[10]
			(+) 3% (heat summer 2018)	
Potato	Heggelbach, Germany	35%	(−) 18%	[10]
			(+) 11%	
Celery	Heggelbach, Germany	35%	(−) 19%	[10]
			(+) 12%	
Clover Grass	Heggelbach, Germany	35%	(−) 5%	[10]
			(−) 8%	
Lettuce (varieties Kiribati and Madelona)	Montpellier, France	Half density, solar tracking, controlled tracking	(−) 5% to (−) 30% with fewer losses on controlled, that is, crop friendly tracking	[46]
Chiltepin pepper	Tucson, Arizona, USA	70%−80%	~ (+) 150%[a]	[40]
Jalapeno	Tucson, Arizona, USA	70%−80%	~ (−) 15%	[40]
Cherry Tomato	Tucson, Arizona, USA	70%−80%	~ (+) 90%	[40]
Lettuce	Montpellier, France	Half density	(−) 19% to (−) 1%[b]	[32]
		Full density	(−) 42% to (−) 21%	
Vine grapes	Piolenc, France	36%	~ (+) 25%	[47]
		66%	~ (−) 25%	
Apple	Mallemort, France	~ 50%	Similar growth rates whiles less water demand. However lower yields due to reduced fruit drop	[48,49]
Rice	Chiba, Japan	20%	(−) 20%[c]	[50]

(Continued)

TABLE 5.3 (Continued)

Crop	Location	Shading rate	Yield change	References
Corn	Kyoto, Japan	Low density	(+) 4.9%[d]	[51]
		High density	(−) 3.1%	
Lettuce	Japan	50%	(−) 10% to (−) 40% depending on season.	[52]
Swiss Chard	South Deerfield, Massachusetts, USA	38%	~ (+) 70% (2016, hot dry summer)	[53,54]
			~ (−) 25% (2017 cold summer)	
			~ (−) 60% (2018)	
Broccoli	Massachusetts, USA	38%	~ (+) 40%	[53,54]
			~ (−) 40%	
			~ (−) 45%	
Kale	South Deerfield, Massachusetts, USA	38%	~ (+) 25%	[53,54]
			~ (−) 50%	
			~ (−) 45%	
Bell Pepper	Massachusetts, USA	38%	~ (+) 40%	[53,54]
			~ (−) 40%	
			~ (−) 70%	
Common Bean	South Deerfield, Massachusetts, USA	38%	~ (+) 350%	[53,54]
			~ (−) 65%	
Cabbage[e]	Massachusetts, USA	38%	~ (−) 30%	[53]

[a]Yields given in the paper were described as "Fruit production (cumulative, per individual)." All crops achived their growth with significantly less water losses through transpiration.
[b]For summer and spring harvest.
[c]Further simulations of the studied showed a proportional relationship. Shading rate hence equaled yield reduction percentage.
[d]Quite unusual findings for a sun-loving crop like corn. A potential explanation can be hypothesized concerning the small size of the experimental plot. The size of the agrivoltaic installation was only 100 m² which could explain only marginal changes in microclimate and high availability of diffused sunlight despite high direct shading density.
[e]Fresh weight and harvest of 2017.

temperature and water availability as limiting factors, shading might be beneficial. The decreased water stress in the heat period probably resulted from reduced evapotranspiration by the shading of the plants.

In Arizona, chiltepin pepper and tomato showed a yield increase of 150% and 90%, respectively, with high shading rates of 70%−80%. The results of a study conducted with corn are somewhat surprising as it would be expected that shading of corn usually limits biomass production. Being a so-called *C4 plant*, the light saturation point of corn lies higher than *non-C4 plants* [55−57].

Table 5.3 depicts a selection of empirical studies of crop growth and yield impacts of agrivoltaics. It must be noted, however, that the stated results can serve only as orientation since the research was conducted in a wide range of contexts and locations. Most results are also lacking robust long-term observation.

As climatic conditions differ from region to region, it is hard to define universally "suitable" crops. But especially in hot and dry climates, shading is expected to have a positive impact on crop yield. Looking at Germany, the precipitation average of the past 30 years shows a clear downward trend [58]. The climatic conditions in spring, which are crucial for the development of many plants, have been too dry over the past years. Increasing warmth promotes a moisture deficit in the soil already in April, which cannot be compensated for in summer [59]. Simultaneously increasing global radiation [58] confirms that agrivoltaics is a suitable measure to increase the resilience of agricultural systems against climate change and at the same time to benefit from its effects both in Germany and Europe. Moreover, agrivoltaic systems may potentially replace

TABLE 5.4 Documents for identification of suitable crops.

References	Description
[60]	Literature review of the most important crops in Germany, categorized into suitable, neutral, and less suitable for agrivoltaics.
[61]	List of important crops in Japan, their light saturation point, and advice on cultivation, published by the Japanese "solar sharing" Association.
[62]	A list of the most common crops cultivated under agrivoltaic systems in Japan. The survey on the status of "solar sharing" in Japan was conducted by Chiba University and provides also data on suitable shading rates for selected crops in Japan
[10]	A literature review conducted by the University of Hohenheim, analyzing empirical findings regarding shading effects for important crops on an international scale. The review also focuses on changes in the chemistry and quality of the produced crops.
[63]	Assessment of selected greenhouse crops for a variety of PV greenhouses with different shading regimes within the Mediterranean−European context.

protective measures, shielding the susceptible crops from excessive radiation and hail. Particularly the combination of permanent crops and agrivoltaics has caught the attention of many solar developers in Europe. The perennial nature of these crops is a further advantage for agrivoltaics as focusing on the preferred light and microclimatic conditions on one specific crop allows for a well-tailored system while crop rotations of arable farming make it difficult to target clear shading levels and coverage ratios. Table 5.4 provides a selection of documents that provide lists of potentially suitable crops.

5.3   Typical systems and applications

To ensure suitable growing conditions for the crops, agrivoltaic systems require design adjustments to the PV system beyond what is seen in GM-PV. This section provides an overview of frequently applied approaches to PV system designs in agrivoltaics. According to the classification presented in Fig. 5.1, we differentiate between interspace PV and overhead PV systems. Generally, the land use efficiency of interspace PV systems is lower than that of overhead PV designs. While the cost-effectiveness speaks in favor of interspace PV systems, the social acceptance might be lower compared to overhead PV systems as the difference to conventional GM-PV is more distinct. Also, the synergetic land use might seem more obvious in overhead PV systems as the added value from the agricultural activities—typically arable farming or horticulture—tends to be higher compared to those of interspace PV which are usually applied in grassland and arable farming.

In the installation of open agrivoltaic systems, the impact of soil compaction should be considered. To reduce the negative impact of soil compaction on soil fertility, installation work should take place in periods of low soil moisture. Lighter installation machinery and the use of mobile construction roads can further help to reduce soil compaction.

5.3.1   Interspace photovoltaic

Interspace PV refers to a system with agricultural activity between PV module rows of tracked or fixed PV modules. In most cases, the mounting structure is similar to GM-PV. Even though these systems focus mostly on electricity generation, agricultural yields can also be of significant relevance. Due to the large variety of agricultural applications, it is difficult to draw a clear line between the technical designs of GM-PV and interspace PV agrivoltaic systems. The German DIN SPEC presented in Section 5.6 addresses this issue setting criteria for a prioritized agricultural use of the land for agrivoltaic systems.

Interspace cropping systems typically differ from overhead PV agrivoltaic approaches by having zero or little vertical clearance. Hence, agricultural machinery usage is limited to the space between module rows [18].

Vertical PV is a special type of interspace PV, using bifacial PV modules, and east−west orientation. Most of the electricity is generated in the morning and evening hours [64]. Given a row distance of 6−15 m (depending on row height), the space between rows can be utilized for agricultural purposes. Some research papers have simulated potential impacts of such module arrangements, but only Germany has seen large-scale implementation [65]. So far, only fodder and forage are harvested at these German interspace PV sites.

The agricultural areas in a vertical agrivoltaic system have uneven light distribution with full shade in the mornings and evenings and not around midday. Hence, for some agricultural applications, heterogeneous crop growth might occur. Benefits for plant growth are expected mainly in windy areas, where the modules serve as windbreakers and thus might help to reduce wind-induced soil erosion and evapotranspiration [66].

The following two subsections shed light on grassland farming and arable farming, two main applications of interspace PV. The subsections do not include horticultural activities, which is more typical in overhead PV systems.

5.3.1.1 Grassland farming

Dual use of GM-PV systems with sheep husbandry is commonly practiced in many countries. With this approach, the systems are typically optimized only for the electrical yield. The expected synergy effects tend to be low compared to other agrivoltaic systems, in line with the agricultural value creation per unit area. However, concrete research results in this area are still pending. The installation of vertical agrivoltaic systems is a new approach that may have positive impacts on fodder production land (see Fig. 5.6). Reference systems have already been built in Germany, for instance in Donaueschingen (Baden-Württemberg) and Eppelborn (Saarland) [68].

FIGURE 5.6 Tractor while mowing through module rows ©Next2Sun GmbH [67].

5.3.1.2 Arable farming

In arable farming, the employment of large land machinery is a central component of the cultivation methods, particularly in more industrialized countries. Therefore, the distances between PV module rows of interspace PV systems should be large enough to facilitate machine operations. Also, the light distribution of interspace PV systems might be challenging as in most arable farming applications a simultaneous ripening of the crops is essential for facilitating a uniform harvest. The heterogeneity of light availability is typically highest in the south or north-facing interspace PV systems with similar constructional geometry to electrically optimized GM-PV. This contrasts with tracked or non-tracked east-west facing interspace PV systems, where the distribution of sunlight over the day on the plant canopy is more homogenous. Generally, the magnitude of these effects depends on the latitude of the system's location.

Difficulties of heterogeneous light distribution could be avoided by further increasing the distances between the PV module rows. This way, though, the already low land-use efficiency of interspace PV further decreases.

Considering the challenges of worldwide biodiversity losses, interspace PV systems could contribute to less intensive cultivation of land, creating biodiversity stripes for more sustainable farming methods. There are few interspace PV systems installed in which the arable farming activities were monitored. Some systems exist in India, for example in the Jodhpur province, where bean species were cultivated. In this interspace PV system, only 49% of the land was used for agricultural applications [43]. Vertical PV applications could also be suitable for arable farming, experimentation with food crops is currently being conducted [67].

5.3.2 Overhead photovoltaic

Overhead PV refers to the cultivation of crops below PV module rows of tracked or fixed PV modules. The vertical clearance of the agrivoltaic mounting structure depends on the height of the plants being harvested underneath and the size of agricultural machinery used. Typical vertical clearances range between 2 and 6 m. The land-use efficiency of overhead PV systems is usually much higher than those of interspace PV systems. While the higher cost of overhead PV systems due to the mounting structure remains a challenge in countries with insufficient agrivoltaic support schemes, there is a range of advantages of overhead PV compared to interspace PV systems. The higher vertical clearance of the systems enables the installation of the inverters and other electrical components at a height so that no fence or security measures are needed. This way, fauna permeability is fully maintained, and social acceptance has the potential to rise. Further benefits result from the sheltering effects of PV modules. Additionally, rainwater harvesting and irrigation management approaches can be integrated more efficiently than in interspace PV systems.

Some of the most prominent examples are the "solar sharing" approach in Japan and the Fraunhofer ISE pilot plant in south Germany. Shading rates range between 20% and 50% depending on row spacing. In Japan, the shading rate has even reached 70% with suitable shade-tolerant crop cultivation such as mushrooms [38].

The following two subsections focus on arable farming and horticulture as typical applications of overhead PV.

5.3.2.1 Arable farming

Technically, the largest obstacle for overhead PV systems in arable farming is the employment of large land machinery. As a benchmark for the dimensions of the mounting structure, the typical height of a large combine harvester is up to 5 m, and the width while spraying is 24 m. One approach to realize these dimensions are tensile structures typically used in thin-shell structures in which combinations of cables and masts can facilitate high clearances without large material usage. Due to the footprint of tension zones in which horizontal forces are transferred to the ground at the border areas, tensile structures seem less suited for small-scale installations as the ratio between the cultivable area and the tension zone would be relatively low. Examples of an agrivoltaic tensile structure system in arable farming are the "Agrovoltaico" pilots of the Italian company REM Tec [63].

Another approach is steel frame construction. While, with reasonable efforts of material and cost, the span widths of steel frame structures are lower compared to tensile structures, this approach can be used for a smaller system without losing land for tension zones.

Compared to interspace PV, heavier wind load must be considered due to the high elevation of overhead PV systems. This also includes suitable foundations which, in the case of tensile structures, must also absorb torsional forces. In contrast to the present state-of-the-art GM-PV and interspace PV systems in which foundations are typically rammed or drilled into the ground, the high static requirements of overhead systems might encourage the project developer to use concrete foundations. However, discussions with farmers indicate a preference for fully reversible foundations without concrete [39]. Similarly, social acceptance might be higher for systems that are easier to dismantle.

The average size of an overhead PV system is largely influenced by policy goals of agrivoltaic support schemes, as exemplified in Japan, where small systems are favored, and China, where massive systems are favored. Even though there is still a lack of experience from enough countries to see clear trends, there are some good reasons why systems in arable farming can be expected to be much larger compared to systems in horticulture. One argument is that arable fields are usually much larger than horticulture fields and that uniform cultivation in arable farming prohibits small agrivoltaic systems on only a portion of the field. From a practical point of view that considers uniform tramlines, light availability, and harvesting periods across the field, it is recommended to install an agrivoltaic system on an entire field instead of just on a part of it.

Another argument is grid connection. While, due to high labor intensity and lower spatial requirements, horticulture areas tend to be closer to residential areas, arable farming areas are typically further away from the grid, making grid connection much more expensive. Economically, the high costs of grid connection can be justified only by a sufficiently high PV capacity of the installation which, again, speaks in favor of larger systems.

Fig. 5.7 shows the pilot overhead PV system of Fraunhofer ISE at Lake of Constance in Germany with a vertical clearance of 5 m.

5.3.2.2 Horticulture

Agrivoltaics has significant synergies with specialty crops, specifically wine cultivation and orchards. The main reason is the high-value creation of these crops and the need for protective measures to cultivate them [69]. An appropriately designed agrivoltaic system can ensure direct protection against environmental factors such as rain, hail, and wind, additionally, nets and other protective gear can be integrated into the PV mounting structure [18]. In viticulture, an increased amount of solar radiation and heat could have adverse effects on the crop and might also lead to sunburn and the fruits drying out on the grapevine. Solar radiation also increases the sugar content in the grapes resulting in a declining quality of the wine [70]. Shading, therefore, can have a positive effect on the grape quality, helping to prevent premture aging. The clearance height for the cultivation of grapes is approximately 2−3 m, which significantly reduces the costs associated with mounting structures. The protected cultivation of berry bushes is another area of application for specialty crops.

FIGURE 5.7   Fraunhofer ISE pilot facility in Heggelbach, southern Germany [23].

In horticulture applications, the use of tracked PV modules is particularly relevant, enabling the adjustment of light availability according to crop needs. Unlike arable farming, the higher added value of horticulture activities makes it economically more feasible to deviate from the electrical optimum tracking and to sacrifice electrical yield for the benefit of the agricultural layer. Fig. 5.8 shows a single-axis tracking system combined with a grapevine plantation in France. Dual-axis systems need more complex steering technology but can further increase the electrical yield and, hence, the maximal shade level [72].

5.4 Photovoltaic modules

There are various module technologies currently deployed in agrivoltaic systems. The major market share of modules consists of crystalline silicon modules. Experiments and demonstrations with promising results have been conducted with emerging module technologies, including semitransparent materials.

5.4.1 Crystalline silicon modules

Crystalline silicon (c-Si) modules dominate the PV market with a 95% share [73]. The cells are available in multicrystalline (multi-Si) and mono-crystalline (mono-Si) variants, with mono-Si as the majority with a 70% share of the total c-Si modules manufactured in 2019. Changes in the overall production share of different crystalline technologies between 1980 and 2019 can be seen in Fig. 5.9. Due to technological advances on the cell and module level, the efficiency of average commercial c-Si modules climbed during the last 10 years, from around 12% to 17%. Highly efficient modules can even reach 21% efficiency [73]. Moreover, technological progress, learning curve, and economies of scale led to a substantial

FIGURE 5.8 4.5 ha vineyard by Sun'Agri in France [71].

price decrease during the last years, with typical module's cost ranging today between 0.24 and 0.41 USD/Wp [74].

The dominance of c-Si modules is also noticeable in agrivoltaic systems, with three different types usually applied: monofacial, bifacial, and semitransparent (see Fig. 5.10). Monofacial modules are the standard type, with solar radiation uti-

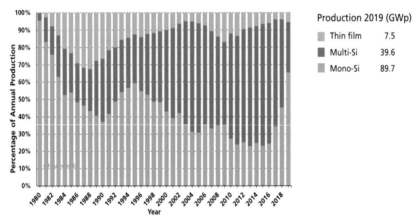

FIGURE 5.9   Share of photovoltaic production by technology [73].

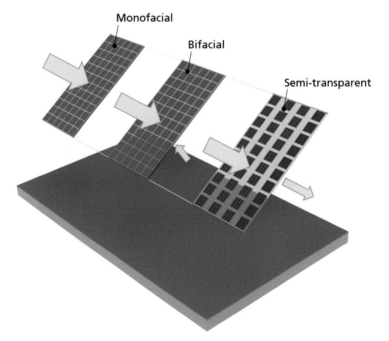

FIGURE 5.10   Illustration of the three main c-Si modules used in agrivoltaic systems.

lized only from the front side. On the other hand, bifacial modules can also utilize the incoming irradiance on the rear side. Bifacial modules evolved into a mature technology in the past years due to the progress of new solar cell concepts with diffused and passivated pn-junctions and passivated rear sides, like passivated emitter and rear contact (PERC) and passivated emitter rear totally diffused (PERT). In 2020, bifacial modules had a market share of around 20%, which is expected to reach 70% by 2030 [75]. Their price is already competitive with their monofacial equivalents [64] at US$0.37−0.43/Wp [74].

The additional yield of bifacial modules from the rear side (bifacial gain) depends on several parameters, among them the bifaciality factor (BF), the site configuration, and the ground albedo. The BF is the ratio between the rear side efficiency to the front side efficiency under standard testing conditions. Currently, typical BFs range between 70% and 80% for p-PERC modules, around 90% for n-PERT, and more than 95% for hetero-junction modules [76]. Regarding the site configuration, a larger row to row distance and a higher elevation of the PV modules leads to higher bifacial gains. Moreover, the bifacial gain can be increased by the incorporation of tracking systems [64]. Finally, the third important parameter is the albedo factor or the percentage of solar irradiation reflected by the ground. In agrivoltaics, an average value of 0.2 is assumed although the albedo factor varies spatially and temporally. Literature values are given for instance for bare soil (0.17), grass (0.25), and crops (0.17−0.3) [77]. Studies that measured albedo values during the growth process of crops identified a steep increase of albedo with increased leaf area index (LAI) of the crops (up from 0.15−0.2 to 0.24−0.28) [78]. In the Fraunhofer ISE pilot system located near Heggelbach, Germany, where bifacial modules were used, the bifacial gain was found to be around 8.7% [39].

Semitransparent c-Si modules have their cells spatially segmented, allowing part of the irradiation to pass through. The level of transparency, cell layout, color, and type of glass substrate can be tuned, allowing great flexibility in the design. On the other hand, there is a direct trade-off between transparency and efficiency, and their cost in terms of USD/Wp is higher than those of opaque modules. So far, semitransparent c-Si modules are mainly used in agrivoltaic systems over orchards where they increase light homogeneity on plant level [26] while utilizing the glass between the cells for the crop's protection.

5.4.2 Thin-film modules

Currently, commercial thin-film modules are typically based on cells like amorphous silicon (a-Si), copper-indium-gallium-diselenide (CIGS), and cadmium telluride (CdTe). They have prices comparable to c-Si modules [79], and offer great design flexibility regarding the module's size, shape, mechanical flexibility, and appearance [80]. Moreover, thin-film modules can achieve semitransparency by reducing the thickness of the absorbing layer while applying transparent electrodes. Still, their market penetration is low with a total share of 5% [73], also caused by their lower efficiency compared to wafer-based modules. Following the overall PV market, future developments of agrivoltaic systems with thin-film modules seem rather restricted.

5.4.3 Emerging module technologies

Apart from the options mentioned above, which are commercialized to a high level, there are several emerging PV module/cell technologies that could reach widespread application in agrivoltaics in the upcoming years. Here are some examples:

The Swiss start-up Insolight developed a concentrator module with integrated planar tracking. In this module, direct sunlight is concentrated through lenses on highly efficient multijunction solar cells while most of the diffuse sunlight is available for the plants [81]. Minor movements of the backplane of the module can shift the cells to the focal point of the concentrated sunlight. With this approach, the module can be installed in a fixed orientation. Fig. 5.11 shows the working principle of the Insolight modules.

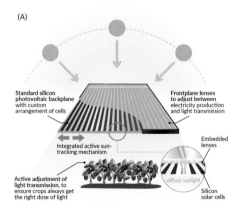

FIGURE 5.11 (A) Illustration of Insolight micro-concentrator module, (B) direct light is focused on the cells: Electricity mode or "E-mode," and (C) direct light is focused in between the cells: Maximum Light Transmission mode or "MLT-mode" [81].

Another promising option is offered by organic photovoltaics (OPV) which offers tailored absorption properties [82], alongside other unique features. Due to this, wavelength-selective transparency can be achieved, allowing also for high transparency in the solar spectrum between 400 and 700 nm where photosynthesis takes place. The Franco-German company ASCA manufactures and commercializes OPV modules, amongst others, for greenhouse applications [83].An example of their customizable and flexible OPV modules is depicted in Fig. Fig. 5.12.

In the USA, the company Soliculture developed a semitransparent luminescent solar collector (LSC), which consists of low-density crystalline cells on a special luminescent material (see Fig. 5.12) [84]. The working principle is that a portion of the incident light is absorbed by the luminescent dye to then be reemitted and waveguided to the solar cell while the rest of the light is transmitted to the plants [85].

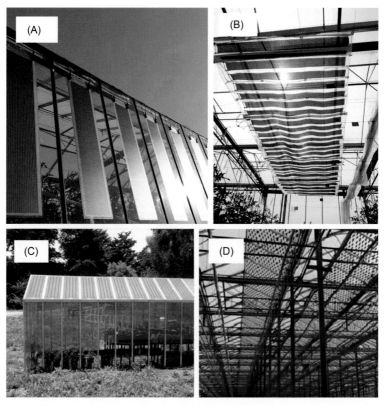

FIGURE 5.12   (A), (B) OPV modules developed and implemented in a greenhouse in France ©ARMOR ASCA [83], and (C), (D) Soliculture LSC modules and their integration in a greenhouse [84].

Another approach consists of wavelength-selective concentrating photo-voltaics (CPVs), where the concentrator is coupled with a dichroic mirror. These mirrors reflect the near-infrared spectrum to the focal point while allowing the visible spectrum to pass through. A group of scientists in China built and tested an agrivoltaic CPV system which showed promising results [86]. Fig. 5.13 illustrates the principle of wavelength separation in CPV.

The German company TubeSolar developed tubular modules based on the fluorescent tube production of OSRAM/LEDVANCE. The main advantage is the homogenous shading it provides along with lower wind loads and lower costs in terms of the mounting structure [87]. Fig. 5.14 shows an illustration of the TubeSolar module concept.

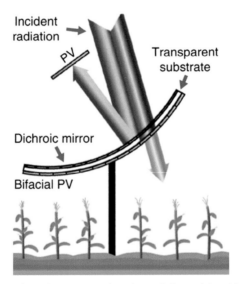

FIGURE 5.13 Illustration of a concentrating photovoltaic module with a dichroic mirror, allowing the PAR spectrum to reach the plants while NIR reflected on a PV cell [86].

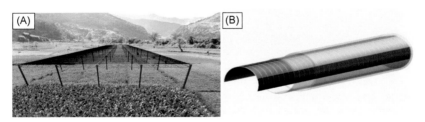

FIGURE 5.14 (A) Visualization of a tubular module facility and (B) a closer view of components [87].

5.5 Economics

The main factors of agrivoltaic system cost include installed capacity, agricultural management, facility location, and module technology. Due to the additional mounting structure necessary to increase vertical clearance, the capital expenditure (CAPEX) of the PV component of overhead agrivoltaics is higher than GM-PV. More material is used to elevate the PV modules, to increase the light availability on the agricultural field, and to facilitate the use of large agricultural machinery. Furthermore, increased complexity of site preparation and system installation must be considered.

To improve financial viability, therefore, the use of smaller agricultural machinery and a higher share of labor involved in agriculture seems more favorable. However, in contrast to GM-PV or interspace PV, overhead agrivoltaics usually does not require fencing which eliminates this cost factor. Also, additional income could be generated from the leased land [18].

In terms of operational expenditures (OPEX), in our calculations we assumed land cost per year to be lower for agrivoltaics than GM-PV due to the dual use of the land. The cost of maintenance and weed control is usually lower since it is already performed under the PV modules within the scope of agricultural activity. On the other hand, the cost of agricultural activity may increase due to more careful maneuvering of machinery around the mounting structure pillars. Repair services also result in a higher cost for overhead agrivoltaics [3]. In the case of permanent crops, consistent crop management is required to favorably adapt agrivoltaics to meet agricultural needs. With regards to crop rotation, the agrivoltaic design must be altered to meet the special crop requirements. Due to the site-specific adaption of agrivoltaic systems, economies of scale seem to be more difficult to achieve compared to more standardized GM-PV systems.

The following sections describe changes in CAPEX and OPEX as well as business models. The findings are mainly drawn from European and German experiences. Agrivoltaics in countries with significantly lower labor costs will lead to different cost structures.

5.5.1 Capital expenditure

The differences in CAPEX can be traced to three different major factors. Firstly, the module cost can increase since the type of modules is determined by crop requirements. With the use of bifacial glass-glass modules, the average price increases from US$260 to 326/kWp compared to standard monofacial modules. These expenses are, however, usually compensated by higher power generation per installed capacity.

Second, in arable farming, the average cost for mounting structure is expected to increase to US$491/kWp compared to US$74−98/kWp in the case of GM-PV. These costs range between US$393 and 737/kWp depending on the vertical and horizontal clearance and local static requirements like wind or snow loads. For specialty and permanent crops, the mounting structure costs are estimated lower at US$160−270/kWp.

The costs for site preparation and installation are considerably higher as well, being estimated at US$307−430/kWp in arable farming (GM-PV: US$86−123/ kWp). These costs arise from soil protection measures such as the potential use of temporary construction roads that spread the weight of heavy machines to avoid soil compression. For permanent crops or grassland applications, a lower cost increase is expected of about US$147 and 221/kWp [18]. Fig. 5.15 shows the cost differences of GM-PV and two agrivoltaic systems based on the German context. It can be seen that significant savings in mounting structure and site preparation and installation can be achieved by agrivoltaic systems for grasslands and permanent crops compared to arable farming agrivoltaics [3].

In the USA, benchmark studies on a variety of agrivoltaic systems have been conducted as well. The NREL concludes for a 500 kWp reference system a CAPEX range of US$1.53/Wp for GM-PV and up to US$2.33/Wp for arable farming agrivoltaics, with a maximum increase of around 52% [16].

5.5.2 Operational expenditure

In contrast to the CAPEX, the OPEX of agrivoltaics has the potential to be lower compared to GM-PV. The cost differences considered in the German context calculation are discussed in this section.

The cost for providing the land is assumed to drop from around US $3.7−1/kWp in arable farming and US$1.2/kWp for permanent and special crops. This estimate assumes that land costs for agrivoltaic systems are aligned with agricultural lease rates and are evenly shared between the farmer and the operator of the agrivoltaic system. However, this value can differ depending on the business model and the ownership structure [3]. On the one hand, agricultural use eliminates the PV-side costs for land maintenance, for example, weeding under the modules. Furthermore, in arable

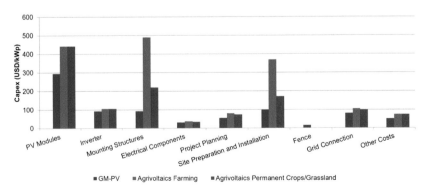

FIGURE 5.15 CAPEX comparison of ground-mounted photovoltaic and two different overhead agrivoltaics applications [18].

farming, loss of agricultural land due to mounting structure foundations may decrease crop yield per hectare. Agricultural subsidies for instance as part of the EU Common Agricultural Policy (CAP) may also not be paid and, hence, were not considered in the calculations.

Higher costs can be expected with regards to the cleaning and maintenance of the modules because the modules are elevated. As in the calculations, we refer to a location in southern Germany with sufficient precipitation, it was not accounted for costs of cleaning the PV modules. In places where regular cleaning is required, this cost might be up to and more than one-third of the operation and maintenance expenses, but this can also be mitigated by emerging cleaning techniques [18].

5.5.3 Levelized cost of electricity

On the bottom line, the generation of electricity in arable farming over a 20-year term with an averaged levelized cost of electricity (LCOE) of up to US$0.12/kWh is comparable on average with small rooftop systems in the German context. In the case of permanent crops with a lower clearance height, the LCOE is at an average of US$0.087/kWh which is about one-third higher compared to GM-PV [18].

The cost estimate does not consider that economies of scale in agrivoltaic systems with arable farming (and the tendency to larger field sizes) may result in a cost advantage, compared to systems in horticulture applications (Fig. 5.16).

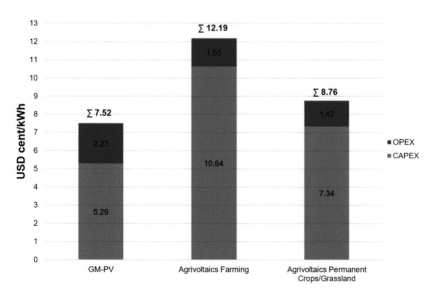

FIGURE 5.16 Levelized cost of electricity comparison of ground-mounted photovoltaic and two different agrivoltaics applications [18].

5.5.4 Business models

Despite a higher CAPEX, agrivoltaics can provide viable business models. In case the agrivoltaic system does not incorporate especially high-income crops and/or very significant crop yield increases or general production synergies, the ratio of crop revenues of the total agrivoltaics income is rather low. Hence, the main drivers of profitability are CAPEX, annual power generation, and Feed-in-Tariff (FiT) [88]. Concerning revenue streams and cash flows, the following conditions may lead to reasonable profitability:

- Provision of a FiT similar to rooftop PV. Unlike FiTs for GM-PV rooftop FiTs are usually sufficient to cover agrivoltaics CAPEX. The same applies to ceiling tariffs of tenders, as agrivoltaics cost structures cannot financially compete with GM-PV.
- Farmer self-consumption of electricity. High self-consumption rates can, for example, be reached by cold storage operations to store agricultural produce.
- A power purchase agreement with a nearby business is negotiated that supplies electricity to lower rates as commercial tariff similar to savings realized by self-consumption.

Options for financing may include bank loans, stock market financing, crowdfunding, financial leasing, and other finance models [88]. There is, however, no research examining preferred finance options of agrivoltaics facilities. As the sector evolves, such studies will be necessary to determine suitable models.

Financing structures are likely to be dependent on different stakeholder arrangements, that usually exceed the complexity of GM-PV due to agrivoltaics' inter-sectoral approach. There are various actors to be considered in the implementation of agrivoltaics. They can be distinguished based on four functions: (1) providing the land (ownership); (2) agricultural management of the land; (3) providing agrivoltaic systems (ownership/investment); and (4) operating the PV system.

In a simple base case, all these four functions can be managed by one party, namely the farmer. This model is expected for small, on-farm agrivoltaic systems where the farmers can take up the investment themselves and own the land. Aside from the low costs for project planning, ease of contracting, as well as high degree of decentralization, the main benefit of this business model is that the possible advantages and disadvantages of an agrivoltaic system can be considered more readily and dynamically when the interactions between the agricultural and PV levels impact the same economic unit.

In many cases, however, the land is not owned by the farmer. Ownership of the PV system is probably less common for larger agrivoltaic systems as well, increasing the likelihood of external investments. Partial ownership could help to maintain the incentive structure for the synergetic dual use of land in

TABLE 5.5 Possible stakeholder structures for agrivoltaics [18].

Business model arrangement	Function			
	Land provision	Agricultural management	PV system provision	PV system operation
Base case	Farmer			
External land ownership	Landowner	Farmer		
External PV investment	Farmer		PV investor	Farmer
Cultivation and operation only	Landowner	Farmer	PV investor	Farmer
Cultivation only	Landowner	Farmer	PV investor	PV operator

this case. However, the higher the proportion of outside capital, the more difficult it will be to keep in mind the benefits of both production levels during operation. Scaling opportunities and possible optimization through a greater division of labor speak in favor of this business model [18] (Table 5.5).

5.6 The German preliminary standard DIN SPEC 91434

In Germany, the preliminary standard DIN SPEC 91434 has been developed by a dedicated consortium to provide a basic framework guideline for prioritized agricultural land use in open agrivoltaic systems. The preliminary standard aims to assure that agriculture activities are not significantly hindered or restricted by the PV components and mounting structures, to avoid agricultural negligence, and to foster synergies between agricultural and PV land use. The preliminary standard will serve as a template for voluntary certification by third-party auditors and might serve as the basis for a full standard in a few years. The consortium included mostly PV-related organizations while agricultural stakeholders were represented by the University of Hohenheim, who had been leading the consortium together with the Fraunhofer ISE. The most important contents of DIN SPEC 91434 and the categorization of agrivoltaic systems are presented in Table 5.6.

For all defined categories, it is mandatory to keep the agricultural process ongoing below and in between the PV modules. This is to ensure that agrivoltaics can serve as a method to mitigate existing and/or emerging land-use conflicts in affected regions. A more precise description of agricultural

TABLE 5.6 Categorization of agrivoltaic systems according to DIN SPEC 91434 [89].

Agrivoltaic systems	Use	Examples
Category I: Elevation with vertical clearance >2.1 m Farming under PV modules	1A: Permanent and perennial crops	Fruit growing, soft fruit growing, viticulture, hops
	1B: Annual and perennial crops	Arable crops, vegetable crops, rotational grassland, arable forage
	1C: Permanent grassland for mowing	Intensive farm grassland, extensively used grassland
	1D: Permanent grassland with pasture use	Permanent pasture, portion pasture (e.g., cattle, poultry, sheep, pigs, and goats)
Category II: Elevation with vertical clearance <2.1 m Farming between PV Modules	2A: Permanent and perennial crops	Fruit growing, soft fruit growing, viticulture, hops
	2B: Annual and perennial crops	Arable crops, vegetable crops, rotational grassland, arable forage
	2C: Permanent grassland for mowing	Intensive farm grassland, extensively used grassland
	2D: Permanent grassland with pasture use	Permanent pasture, portion pasture (e.g., cattle, poultry, sheep, pigs, and goats)

activity is included in an agricultural cultivation proposal which addresses the core requirements and criteria for agrivoltaic systems, amongst others:

- It must be guaranteed that the simultaneous prioritized agricultural production of the land remains possible during the lifetime of the agrivoltaic system.
- The loss of land due to an agrivoltaic system must not exceed 10% of the total project area for category I and 15% for category II.
- Concepts of light availability, homogeneity, and water availability must be assessed, and, if required, suitable mitigation adaptations should be made.
- Soil erosion and damage must be prevented by choosing appropriate anchoring structures and ensuring control of water flow from the modules.

- The agricultural yield after the installation of the agrivoltaic system must amount to at least 66% of the reference yield. The reference yield is a three-year average value of the same agricultural area before the installation or comparable data from relevant publications and yields in similar regions [89].

Fig. 5.17 illustrates the main parameters of the DIN SPEC 91434.

5.7   Case study: agrivoltaics and society

Over the last decades, renewable energy developments could be widely observed across the globe. Along with their technological success and a multitude of political frameworks driving the development, numerous social and societal aspects accompanied the technology uptake. Bioenergy production, for instance, has provoked land-use conflicts between food and fuel productions and, consequently, exerted pressure on agriculture and farmers. At the same time, high rates of general acceptance for renewable energies do not necessarily translate into local acceptance—"Not In My Backyard" (NIMBY) has been observed regularly when implementing renewable energies [90–92]. Esthetic changes in the landscape caused by renewable energy installations greatly affect local residents and their emotional bonds to their homes and places for leisure activities [93]. Such individual values, moral and normative ideals, and preferences are manifold and often drive citizens' argumentation in critical attitude or even rejection. Therefore, as a huge transformation project, the global energy transition—in this study the German Energiewende (Energy Transition)—is a complex challenge that can only be met in a cooperative mode with all relevant stakeholders (e.g., farmers) and the society (inter alia the rural population). Consequently, the energy transition process can be regarded as a field of social action in which Responsible Research and Innovation (RRI) is particularly needed. RRI is associated with the claim of being able to realize better innovations by shaping the innovation process as "a transparent, interactive process by which societal actors and innovators become mutually responsive to each other with a view on the (ethical) acceptability, sustainability, and societal desirability of the innovation process and its marketable products (to allow a proper embedding of scientific and technological progress in our society)" [94]. Thus, RRI strives for a possibly necessary adaptation or even realignment of research questions and innovation paths, and here, the identification of potential (normative) ideals and moral controversies [95,96].

This section of the chapter sheds light on the RRI approach in a specific context of innovation, the publicly funded German project APV-RESOLA,[1] to achieve the stated goal of "science for society, with society" [97].

---

1. Funded by the Federal Ministry of Education and Research, funding code 033L098G.

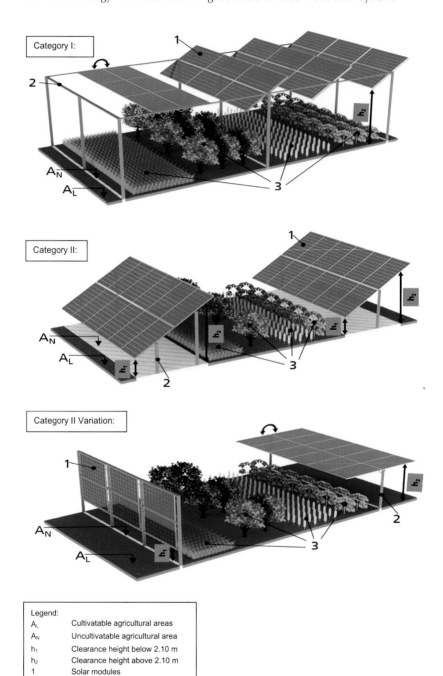

FIGURE 5.17   Main agrivoltaic systems according to DIN SPEC 91434 (own illustration). *Copyright: Illustration of crops, shutterstock, Ulvur, BlueRingMedia, Pisut trading, Ice AisberG.*

Accordingly, in the case study described below, the views of society were solicited in an attempt to make the scientific and technological developments more socially responsible [98]. Even more concretely, the field of anticipation was addressed in the context of RRI as stakeholders, and specifically, the rural population had to imagine future technologies to assess their potential desirability based on this [99]. Besides this, underlying expectations, fears, as well as individual ideas and values, among others, should be explored to identify conflict lines and possible conditions for success and to be able to mitigate these challenges that might be faced when agrivoltaics is introduced. Based on the findings of the transdisciplinary research, practical and practicable solution proposals can be developed.

5.7.1   Anticipation framework of APV-RESOLA

For a holistic understanding of the acceptance effects of solar power production in agrivoltaic systems, it is essential to reflect that technologies are always embedded in a socio-technical human-technology-environment system, that is, interact with both the groups of actors involved and the regional setting. In this context, the acceptance effects can be considered on different levels: On the socio-political level, it is about the overall societal discourse on solar power generation with GM-PV or agrivoltaic systems, which is strongly related to higher-level discourses such as energy transition and nuclear phase-out as well as the increase of organic food production. Relevant topics here are, for example, the role of renewable energies in general and agrivoltaic systems in particular (for the energy system), possible effects on food production as well as on the price of electricity. At the local level, the focus is on the direct impacts associated with agrivoltaic systems, which result from changes in the landscape, effects on local flora and fauna, and perceived visual effects at the level of residents, among other things.

  To better understand the local effects of agrivoltaic systems, within the context of the APV-RESOLA project, a multistage citizens and stakeholder participation design was developed and implemented in the farming community of Heggelbach in the Lake Constance region of southwestern Germany. Such a project approach taking place in a kind of living lab outside of researchers' premises offers a unique opportunity to assess a pilot facility in a real-life environment and to contrast preconceptions of the technology developers with the genuine expectations and tacit knowledge of citizens and knowledge outside the realm of science. RRI allows for a deliberative dimension for open provision of visions, goals, questions, and dilemmas that gives space for engagement and debate. Clear and open communication is required to successfully conduct such research projects and avoid misunderstandings about the process of citizen involvement and interest groups, especially on their role in science and research [100]. If the project does not meet these criteria, disappointment and conflicts may arise from unfulfilled expectations. Bringing together different stakeholders allows overcoming their own logic and interests set by science with its methodological targets and practice aiming at tailor-made, easy-to-implement solutions for marketable solutions.

To do so, the Institute for Technology Assessment and Systems Analysis (ITAS) at the Karlsruhe Institute of Technology (KIT) engaged societal actors early in the process and analyzed the local residents' expectations and fears [101]. For this research target, a framework for participation based on RRI (cf. Fig. 5.18) was developed based on a shared understanding of the technological vision. In this setting of an ex-ante (2015, before building the pilot facility) and an ex-post (2017, one year after the pilot facility started operation) citizens' workshop, different social views and perspectives were recorded and matched. Therefore, the study involved citizens aged 18−80 years (approximately 2100) affected by the technology to participate in the citizen research approach [101]. These workshops were accompanied by a survey at the pilot's opening ceremony in 2016 and a final stakeholder workshop summarizing the overall findings in 2018. In the workshops, a codesign approach of methods and instruments for empirical social sciences ranged from citizen workshops to stakeholder workshops.

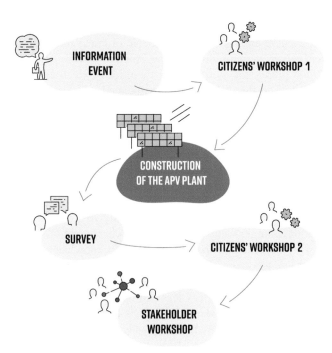

FIGURE 5.18   Multistage transdisciplinary agrivoltaics research approach. © *ITAS*.

With a group of 26 and 17 citizens in the first (in 2015) and second workshop (2017), respectively, social and societal questions of acceptance criteria for future technologies in a decentralized energy supply were discussed and success criteria synthesized. An open brainstorming of ideas, opinions, and expectations on agrivoltaics was analyzed in the first citizens' workshop to set up a world café (see Fig. 5.19) for continuity of ideas [102] with four main topics: (1) agrivoltaics and energy transition; (2) agrivoltaics and the landscape; (3) agrivoltaics and its profiteers; (4) agrivoltaics and politics. In qualitative content analysis, the statements were categorized and analyzed and the main findings were extracted [103]. The second citizens' workshop assessed possible changes in the appraisal of the technology and the local population's assessment patterns. As additional input, the results of the opening ceremony survey were also presented here to derive site criteria. To stress their rules of decision, participants were asked to apply the elaborated site criteria for agrivoltaics in a simulation game and argue for the identified sites within the region. In the final workshop, stakeholders from technology development, tourism marketing, administration at municipal, regional, and state levels, agricultural chamber, nature conservation, energy cooperatives, and citizen representatives jointly reflected and discussed the findings. To summarize, this approach organized around this on-site testing of a pilot-scale technology allowed (1) testing at a small scale in a real-life environment and (2) identifying acceptance criteria for the application of agrivoltaics and the surrounding legal framework early in the technology

FIGURE 5.19   Scene during the World Café at the first citizens' workshop. © *ITAS*.

designing process. This generates knowledge that helps to shape and accelerate transformation processes in the long term in a way that is compatible with society.

5.7.2 Optimization problems

The findings of the workshops and the studies suggest some optimization problems in the technology designing process that must be looked at in detail. Based on the application-related identification and analysis of opinions and normative value patterns, a system dynamics approach was applied to identify possible obstacles and drivers [104]. The efficiency of agrivoltaic systems highly depends on local climate conditions and may vary greatly from one year to another [105]. Acceptance levels closely relate to landscape impacts, similar to other renewables [106]. It was stressed to make agricultural cultivation below PV modules mandatory for agrivoltaic systems (cf. key factor 4). This finding proves to be of high importance when looking at optimization studies towards more power production within agrivoltaic systems. Putting up more PV modules significantly reduces radiation for arable land [107]. Heterogeneity of radiation reaching the ground is reduced with the height of the PV modules [108] while installation costs increase with more elevated and wind-exposed sites. Furthermore, participants in the workshops claimed more benefits for agrivoltaics. Combatting desertification or bringing desert areas back into cultivation would be of great value, as well as protecting plants against hail or other weather-related hazards. From a planning perspective, large sizes and accumulation of systems in the same region are limiting societal acceptance. The better agrivoltaic systems are integrated into local landscapes and the more they are attached to existing infrastructure (e.g., powering farms, industry, or communities), the more will citizens argue in favor of such technologies in a decentralized set-up. Such high demands on integrity and customizability require modular and adaptable set-ups, which contradicts to some extent a standardized mass-market product. For the planning process, the involvement of all relevant stakeholders supports a holistic analysis of planning criteria and needs but limits the transferability and a uniform planning process.

5.7.3 Factors of local acceptance

While the energy transition is a topic for society as a whole and, in principle, systems in the neighborhood are predominantly approved, a variety of factors of local acceptance are relevant for the evaluation of projects in the expansion of solar power production with agrivoltaic systems. On one hand, these are effects on people and nature. On the other hand, the factors include social norms and participation-related issues. Economic and climate protection reasons are joined by individual views on technology development and environmental protection measures in general. The respective importance of the individual aspects can be locally very different depending on various socio-economic, ecological factors as well as specific emotional evaluation. To adequately address these issues for the further development and commercial implementation of agrivoltaic systems, ten key factors were identified [18].

1. Utilizing the existing PV potential on rooftops, industrial buildings, and parking lots should take priority over identifying sites for agrivoltaic systems.
2. The agrivoltaic systems should be integrated into the decentralized energy supply to use solar power on-site or for processes with higher value creation, such as irrigation, cooling, or processing agricultural products.
3. The agrivoltaic systems should be combined with an energy storage system to increase resource efficiency, so the available electricity can be used to meet local demand.
4. Farming within agrivoltaic systems must be mandatory to prevent the one-sided optimization of power generation with "pseudo-agriculture" under the PV modules.
5. Agrivoltaic systems should be built on sites where synergies can be realized through the dual use of the land, such as shading to reduce heat stress for cultivated plants, generating electricity to irrigate the crops, or digital land cultivation with electrified and future autonomous systems.
6. The size and concentration of the agrivoltaic systems should be limited and, similar to wind power plants, minimum distances to residential areas should be established under consideration of local site characteristics and social preferences.
7. Agrivoltaic systems should not negatively alter the quality of local recreation, tourism, and the attractiveness of the landscape. Sites that are naturally screened from view (e.g., at the edge of a forest) or flat sites should be preferred to ensure the best possible integration of the systems into the landscape, making them less invisible
8. Approvals for agrivoltaic systems should be issued according to strict legal regulations and with citizen participation to avoid the uncontrolled growth of agrivoltaics, as seen with biogas plants due to privileged building laws for the agriculture sector. Municipalities and citizens should have codetermination rights in the identification of sites for agrivoltaic systems.
9. Agrivoltaic systems should preferably be constructed, owned, and operated by local farmers, energy cooperatives, or regional investors.
10. The strips along the agrivoltaic systems that cannot be farmed should be used for erosion protection and serve as corridor biotopes to maintain biodiversity in the agricultural landscape.

These findings show that acceptance is at least partly an expression of the possible negative effects of agrivoltaic systems on-site and the different spatial conflicts that can be identified in the following areas: Conflict between energy and food production, protection of the residential environment, tourism, nature and landscape conservation, sovereign decision-making regarding the selection of appropriate sites, governance and regulations, and operation models for agrivoltaic systems.

In addition, it must be examined to what extent a large-scale distribution of agrivoltaic system expansion that is perceived as fair can also have a positive effect on the general acceptance, for agrivoltaics specifically and for the energy transition in general. To date, the benefits, and burdens of large-scale solar power production on arable land have often been distributed unevenly across regions, with rural areas often bearing burdens for the more urban centers of consumption. However, spatial trade-offs between cost-optimal and equitable spatial distribution must also be considered in this context [109].

The findings of the RRI approach prove that reflective, anticipatory, and participatory technology development processes allow considering societal values, needs, expectations, and concerns that, when grasped at an early stage, can increase the responsible design of technological innovation and thus public acceptance [101]. Attaining high levels of acceptance might not be reached by political decisions based on democratic processes, while participation processes can provide a holistic picture [110]. Serious worries that may trigger opposition after the market introduction was observed by Hübner and Pohl whose analysis went well beyond the NIMBY hypothesis [111].

5.7.4 Responsible research and innovation of agrivoltaics

For agrivoltaic technology, stakeholder engagement and citizen involvement in renewable energy planning have proven to be a suitable approach for more societal integration of renewable energy projects. Rural development has been closely connected to renewable energy strategies and plans. Consequently, early involvement and trustful cooperation with various stakeholders and "affected" citizens early in the planning process are of crucial importance to identify societal implications of research and innovation outcomes in addition to technical, economic, and ecological aspects. For technology development processes, the involvement of stakeholders and future users offers important insights on local criteria and intangible knowledge assets. Practitioners provide other perspectives than researchers and technology developers, while citizens and other stakeholders can relate to individual normative ideals and moral perspectives, as well as to local and infrastructural conditions and thus to acceptability criteria for planning frameworks. Learning from previous renewable energy projects revealed the conflict between general acceptance levels for the technology as such, while local acceptance levels for neighboring facilities were significantly lower. It can be concluded that the procedural participation opportunities for the public and local added value chances are relevant acceptance factors. It is also important to strengthen the municipal implementation level and to provide municipal actors as implementers of the energy transition in the specific area with tools to deal constructively with the conflicts that arise. Administrations and political decision-makers are often not trained for the moderation of conflict and participation processes, and there are also limited time and personnel resources. In addition to structural changes such as targeted capacity building, specific tools, and other support services, a facilitated dialog at the local level is an important approach. Involving citizens both in the planning process and operation of renewable energies provides benefits for the entire energy system. Engaging stakeholders from rural communities early in the technology development process may thus offer insights and learning processes for more responsible socially shaped technology innovations and frameworks based on local characteristics.

5.8   Agrivoltaic policies

Recent governmental support has been seen with regards to supporting agrivoltaic systems worldwide. This section provides brief country profiles of the existing policies around the world.

5.8.1   Japan

Japan was the first country to introduce a support scheme for agrivoltaics. The concept of "solar sharing" was first developed here and in March 2019 there were almost 2000 "solar sharing" farms in the country accounting for about 0.6%−0.8% of the overall PV capacity. The "solar sharing" policy focuses on small-scale installations with 89% having the size of up to 0.3 ha and only 3% larger than 1 ha [38]. This reflects the conceptualization of "solar sharing" as a vitalization program for farmers and to respond to stagnating farming incomes and the rapid aging of the countryside population.

A large part of the land used (31%) is either abandoned, fallow, or devastated farmlands. Most of the new installations have a shading rate of 20%−50%. About one-third of these installations have a shading rate of 50%−80% and the major crops cultivated are mushrooms, different tea shrubs, or Myoga (Japanese ginger) [62].

The Ministry for Agriculture, Forests, and Fishery has set technical guidelines to approve the concept of "solar sharing" and its eligibility for an increased FiT. This regulatory framework was first introduced in 2013 and adjusted several times, including an update in 2020. It provides a clear concept of integrating agrivoltaics into the farmlands [112−115].

- There must be a clear farming plan and the continuation of farming activities must be ensured.
- The PV modules must be chosen in a way to ensure ample light for crop production and the mounting structure must have an elevation of about 2 m to make sure it does not hinder the use of agricultural machinery.
- There must be no effect on the neighboring farmlands.
- The crop yield must reach 80% of local standard yield and an annual report must be provided as evidence to prove the same.
- Lastly, if these conditions are not met, the agrivoltaic systems must be removed and the land must be restored.

5.8.2 South Korea

South Korea faces a similar problem to Japan, which includes lack of arable land, rapidly aging population, stagnating farmers' income, and the decreasing population in the countryside. About 63% of rural agrivoltaic systems were installed in farming areas [116]. The government has planned to promote small-scale agrivoltaics of 100 kWp to tackle the problem of farmland abandonment. Despite the ambitious plans, only 19 demonstrators have been installed by 2020, with a total capacity of 2 MWp representing a national research program to elaborate best practice [117].

Various organizations such as Green Energy Institute, the Korean Agrivoltaic Association (KAVA), and the Korean South-East Power Co. (KOEN) are trying to work on training to support the development of agrivoltaics in Korea [116].

5.8.3 China

In terms of agrivoltaic capacity, China is by far the most advanced country. Researchers predicted that by 2018, the annual installed capacity of agrivoltaics in China would have reached 3.26 GWp and the cumulative installed capacity more than 12 GWp [118]. At this stage, agrivoltaics in China is supported by dual policy support from the PV field and the agricultural field. The development prospect of agrivoltaics is very broad in China, it not only promotes the development of the PV industry but also the transformation of agricultural development [119].

The main companies involved in the installations of the large-scale agrivoltaic systems were Huawei, Jinko Solar, Longi Solar, Tongwei Group, and the Baofeng Group. The colocation of agriculture and PV could serve as a useful tool to fight against poverty in the rural areas in the Chinese context. Examples of corresponding policies are the "Administrative Measures for Poverty Alleviation with PV Power Plants" (2017) and the "Photovoltaic Power Station Project Land Use Area Control Indicators" [118,120]. The location of agrivoltaic systems is determined by the high share of the population living below the poverty line and the local authorities calculate the capacity requirement and tender projects to solar developers. This helps to diversify the income of families by either land leasing or employment in the systems or rents from the electricity generated [88].

5.8.4   France

The Agency for Ecological Transmission (ADEME) leads the definition and drafting of the application guidelines, while the Energy Regulatory Commission (CRE) collects the bids and selects the winners. Three tender rounds were conducted in 2017 (15 MWp), 2019, and 2020. A bidding process was undertaken, and the bidders had to submit two reports describing the innovative nature of the projects and justifying the synergies. The first round included a bidding field of 57 MWp whereas only 15 MWp was approved for innovative PV applications on crop and PV colocation. Much of the capacity was given to PV greenhouses and only 11% was allocated towards open field agrivoltaics. The third round saw a higher overall capacity allocated towards agrivoltaics [121]. The latest round of bidding allocated about 146.2 MWp towards innovations wherein agrivoltaics accounted for 80 MWp. The French company Sun'Agri alone received about 58 MWp and won 22 out of 31 projects. A fourth tender round for 140 MWp (projects up to 3 MWp) will receive bids by the end of 2021 [122−124].

5.8.5   Germany

Since January 2021, agrivoltaics entered the German legislation within the recent revision of the German Renewable Energy Sources Act (Erneuerbare-Energien-Gesetz, EEG). In April 2022, the EEG foresees a joint tender together with floating photovoltaics (FPVs) and PV roofs over parking.[2] The tendered capacity was increased in June 2021 from 50 to 150 MWp. In the same decision, the eligible areas—originally only encompassing arable farming land—were extended to permanent agricultural areas. Permanent grassland areas still remain excluded from the tender. The tender is specified in the German innovation tender regulation (InnAusV) [125]. The German Federal Network Agency was assigned to set the criteria for eligible agrivoltaic systems for the tender due to October 2021. A compliance to the requirements of the DIN SPEC 91434 is expected to serve as a precondition for joining the tender.

A detailed analysis of relevant policies in Germany as well as reform proposals can be found in the guidelines on agrivoltaics published by Fraunhofer ISE [18].

---

2. More on floating photovoltaics is discussed in Chapter 6.

5.8.6 United States of America

The past few years have seen a growing interest in agrivoltaics in the USA. The land utility for agrivoltaics is estimated to be over 800,000 ha by the NREL until 2030 [126]. Within the Innovative Site Preparation and Impact Reductions on the Environment (InSPIRE) project, data on the biodiversity impact of GM-PV are collected to assess and promote mitigation strategies for low-impact solar development opportunities [126]. The Rural Energy for America Program has been set up on the federal level providing incentives for PV development on farms. However, nothing explicit has been mentioned about agrivoltaics concerning dual-land use. The Department of Energy (DOE) also awarded 7 million USD as research funds for solar-agricultural colocation projects. The first exclusive policy for agrivoltaics was introduced in Massachusetts. The Solar Massachusetts Renewable Target (SMART) program provides for solar development with incentive payments [127]. In addition to current SMART categories, the Massachusetts Department of Energy Resources recently proposed a US$0.06/kWh rate adder for Agriculture Solar Tariff Generation Units [128]. Colorado has also experienced growing agrivoltaics support. In June 2021, as part of Colorado's renewable energy and energy efficiency program (ACRE^3), $150,000 of stimulus funding was set aside for "research, guidance, technical assistance, feasibility studies, and projects related to agrivoltaics" [129].

5.8.7 Israel

In unison with the Israeli government's goal to increase electricity generation from renewable energies to 30% until 2030, the Ministry of Energy and the Ministry of Agriculture and Rural Development have recently released a unique tender for agrivoltaics. The purpose of this application is to examine the feasibility of dual-use of agricultural land for electricity generation from renewable energies while maintaining optimal agricultural productivity in the selected sites. The first round of submissions for preliminary plans has been concluded on July 21, 2021. All submissions for this tender need to include a developer who presents the exact location of the facility, an agronomist that accompanies the project for 5 years to ensure the best possible agricultural productivity under the modules, and a research team that prepares and submits a detailed agricultural research plan. The total area of all approved facilities should not exceed 100 ha, with each facility not exceeding 1.5 ha and the total installed capacity should not exceed 111 MWp [130,131].

5.9 Conclusion and prospects

In the upcoming decades, PV is projected to become one of the main technologies meeting the global energy demand. Climate change and increasing water scarcity result in the need for mitigation and adaptation measures in agriculture as the sector is a large contributor of greenhouse gas emissions and suffering from increasingly extreme weather events. Concerning land-use competition between PV systems and agriculture, agrivoltaics enables an expansion of PV capacity while conserving farmland as a resource for food production. A dual-use of farmland considerably increases land-use efficiency. Additionally, PV modules can protect soil and crops that are exposed to increasing and more frequent severe weather events such as heat, heavy rain, or drought. Agrivoltaics can also provide climate-friendly energy to cover the demand of agricultural operations and diversify the income of farmers by providing additional revenues through the power generated which, in turn, can strengthen and vitalize rural economies.

In recent years, agrivoltaics has experienced a dynamic development in almost all regions of the world. The installed capacity increased from approximately 5 MWp in 2012 to more than 14 GWp in 2021, with government subsidy programs in Japan (since 2013), China (approx. 2014), France (since 2017), the USA and Korea (both since 2018) as well as Israel and Germany (both since 2021). Looking at the trends of land scarcity, renewable energy expansion, costs of PV systems, and the need to increase resilience in agriculture against weather extremes and water scarcity, there is little doubt that the total agrivoltaic capacity will continue to expand. High solar radiation and harsh temperatures can negatively affect the quality of the harvest in wine-growing by sunburn or drying out of the grapes. Partial shading has been seen to be beneficial to provide protection to the harvest and prevent it from premature ripening. Agrivoltaics have been increasingly funded and implemented in southern regions of France for this very reason.

Agrivoltaics can be classified as open and closed systems. This chapter only addressed open systems while closed systems are addressed in Chapter 3. Published in 2021, the German DIN SPEC 91434 defines criteria to assure a sufficient level of agriculture activities and to maintain the agriculture-focused character of open agrivoltaics considering both overhead and inter-space systems. Complying with the criteria of the DIN SPEC is expected to serve as a prerequisite to participate in the first German agrivoltaics tender in April 2022. If it will prove to be suitable and successful, in the future, the DIN SPEC 91434 might serve as a basis for a definition of agrivoltaics also at an EU or worldwide level.

Agricultural applications can be distinguished into three agricultural activities: grassland farming, arable farming, and horticulture. Depending on the agricultural activity, some trends can be observed regarding the technical approach of integrating PV into agricultural areas. Typical systems applied for permanent grassland are interspace systems where the added value for the agricultural layer is usually low. In contrast, typical horticulture applications benefit from the shelter of overhead systems with semitransparent PV modules.

Considering the typical low area potential on the one hand side and the overall potential benefits of climate adaptation on the other hand side, horticulture appears as a well-suited application for a first market launch of agrivoltaics in many countries. Reasons include the frequent close physical proximity of the growing area to the farmyard, the high synergy potential of the cultivated plants, the lower cost of mounting-structures, and the comparatively simple integration into the management methods for permanent crops. Also, horticulture areas might face fewer challenges regarding building permits as the impact on the landscape tends to be lower due to the existence of foil tunnels or hail protection nets. Regarding area potential, arable farming applications provide much larger surfaces for the installation of agrivoltaics compared to horticulture applications.

Regarding PV module arrangement, there is a wide range of approaches that aim to optimize the distribution and amount of sunlight as well as the use of the available solar spectrum in agrivoltaic applications. Despite their opaqueness crystalline silicon modules are by far the dominating PV cell technology in agrivoltaic systems mainly due to economic reasons.

Concerning general economic performance, today, agrivoltaics is competitive with rooftop PV and other renewable energy sources in some countries. However, an economically viable implementation of agrivoltaics depends on a corresponding regulatory framework which is still lacking in most countries. Adapting the regulatory framework to the technical developments of agrivoltaics could encompass easier permitting requirements, FiTs, and alignment with existing agriculture subsidies, amongst others.

The early involvement of local citizens is a key criterion for success in the implementation of agrivoltaics. Landscape impact by agrivoltaic systems is seen as a barrier that can be reduced by site-specific planning criteria and the involvement of stakeholders. Highlighting local added value opportunities and providing platforms for affected stakeholders such as neighbors, political decision-makers, farmers, and investors are crucial for societal acceptance towards agrivoltaics projects.

A general increase in agricultural added value could be another benefit of applications in horticulture. This is because horticulture is highly productive. Looking at numbers from Germany, with only about 1.3% of the farmland, horticulture contributes more than 10% of the value added in agriculture. Creating incentives for agricultural operations to become more active in this sector by subsidizing agrivoltaics in horticulture could therefore serve as leverage for the total agricultural production, even with a very small proportion of land used for agrivoltaics.

Further research could address the combination with energy storage, organic PV foil, employment of electrical agricultural machines, rainwater harvest, agroforestry, and solar water treatment and distribution. Another vision is "swarm farming" with smaller, automated, solar-powered agricultural machines working under agrivoltaic systems. Machines for autonomous weeding or the elimination of pests using lasers already exist today—without using chemicals or contaminating soils or groundwater. The mounting structure and availability of electricity of agrivoltaic systems facilitate the integration of such smart farming elements.

To tackle the challenges of climate change, it is essential to take effective steps towards our energy and climate policy goals and to foster the resilience of food production. To do so, a wide range of tools must be employed. Considering the opportunities of agrivoltaics for climate mitigation and adaptation, there is little doubt that the technology is among the most promising of these tools.

References

[1]   Goetzberger A, Zastrow A. On the coexistence of solar-energy conversion and plant cultivation. Int J Sol Energy 1982;1:55−69. Available from: https://doi.org/10.1080/01425918208909875.

[2]   Nagashima A. Solar sharing—changing the world and life. Tokyo: Access International Ltd.; 2015.

[3]   Schindele S, Trommsdorff M, Schlaak A, Obergfell T, Bopp G, Reise C, et al. Implementation of agrophotovoltaics: techno-economic analysis of the price-performance ratio and its policy implications. Appl Energy 2020;265:114737. Available from: https://doi.org/10.1016/j.apenergy.2020.114737.

[4]   Sun'Agri. Sun' Agri3: Le Programme de Recherche Sun'Agri 3. https://sunagri.fr/; 2020 [accessed 20.07.20].

[5]   Goetzberger A, Zastrow A. Kartoffeln unter dem Kollektor. Sonnenenergie 1981;3.

[6]   Lasta C, Konrad G, editors. Agro-photovoltaik: 15. Symposium energieinnovation; 2018.

[7]   Willockx B. A standardized classification and performance indicators of agrivoltaics systems; 2019.

[8]   Gorjian S, Minaei S, Maleh M, Ladan, Trommsdorff M, Shamshiri RR. Applications of solar PV systems in agricultural automation and robotics. Photovoltaic solar energy conversion. Elsevier; 2020. p. 191−235. Available from: http://doi.org/10.1016/B978-0-12-819610-6.00007-7.

[9]   Gorjian S, Bousi E, Özdemir ÖE, Trommsdorff M, Kumar NM, Chopra S, et al. Progress and challenges of crop production and electricity generation in agrivoltaic systems using semitransparent photovoltaic technology. Renew Sustain Energy Rev 2021; ARTICLE IN REVIEW (07/2021).

[10] Weselek A, Ehmann A, Zikeli S, Lewandowski I, Schindele S, Högy P. Agrophotovoltaic systems: applications, challenges, and opportunities. A review. Agron Sustain Dev 2019;39:545. Available from: https://doi.org/10.1007/s13593-019-0581-3.

[11] Aroca-Delgado R, Pérez-Alonso J, Callejón-Ferre Á, Velázquez-Martí B. Compatibility between crops and solar panels: an overview from shading systems. Sustainability 2018;10:743. Available from: https://doi.org/10.3390/su10030743.

[12] Jain P, Raina G, Sinha S, Malik P, Mathur S. Agrovoltaics: step towards sustainable energy-food combination. Biosource Technol Rep 2021;15.

[13] Hernandez RR, Armstrong A, Burney J, Ryan G, Moore-O'Leary K, Diédhiou I, et al. Techno−ecological synergies of solar energy for global sustainability. Nat Sustain 2019;2:560−8. Available from: https://doi.org/10.1038/s41893-019-0309-z.

[14] Toledo C, Scognamiglio A. Agrivoltaic systems design and assessment: a critical review, and a descriptive model towards a sustainable landscape vision (three-dimensional agrivoltaic patterns). Sustainability 2021;13:6871. Available from: https://doi.org/10.3390/su13126871.

[15] Macknick J, Beatty B, Hill G. Overview of opportunities for co-location of solar energy technologies and vegetation: NREL report no. TP-6A20-60240; 2013.

[16] Horowitz K, Ramasamy V, Macknick J, Margolis R. Capital costs for dual-use photovoltaic installations: 2020 benchmark for ground-mounted PV systems with pollinator-friendly vegetation, grazing, and crops; 2020.

[17] Macknick J, Caspari L, Davis R. Webinar co-location of solar and agriculture: benefits and tradeoffs of low-impact solar development; 2017.

[18] Fraunhofer ISE. Agrivoltaics: opportunities for agriculture and the energy transition; 2020.

[19] Fraunhofer ISE. Feasibility and economic viability of horticulture photovoltaics in paras, Maharashtra, India; 2019.

[20] Wang D, Zhang Y, Sun Y, editors. A criterion of crop selection based on the novel concept of an agrivoltaic unit and m-matrix for agrivoltaic systems: (a joint conference of 45th IEEE PVSC, 28th PVSEC & 34th EU PVSEC): 10−15 June 2018. Piscataway, NJ: IEEE; 2018.

[21] itk. Dynamic Agrivoltaism: Sun'Agri and ITK strengthen their partnership for resilient agriculture; 2020.

[22] Chiba Ecological Energy Inc. Website. https://www.chiba-eco.co.jp/; 2021 [accessed 08.04.21].

[23] Fraunhofer ISE. Website Agrivoltaics. https://www.ise.fraunhofer.de/de/leitthemen/integrierte-photovoltaik/agri-photovoltaik-agri-pv.html; 2022 [accessed 28.01.2022].

[24] Universität of Hohenheim. University of Hohenheim. https://www.uni-hohenheim.de/startseite; 2021 [accessed 30.07.21].

[25] ITAS. APV-RESOLA − Innovationsgruppe Agrophotovoltaik: Beitrag zur ressourceneffizienten Landnutzung. https://www.itas.kit.edu/projekte_roes15_apvres.php; 2021 [accessed 05.08.21].

[26] Willockx B, Herteleer B, Cappelle J, editors. Combining photovoltaic modules and food crops: first agrivoltaic prototype in Belgium; 2020.

[27] UMassAmherst. Dual-use solar & agriculture. https://ag.umass.edu/clean-energy/research-new-initiatives/dual-use-solar-agriculture [accessed 08.04.21].

[28] University of Arizona. What is agrivoltaics? https://research.arizona.edu/stories/what-is-agrivoltaics; 2021 [accessed 30.07.21].

[29] InSPIRE. Innovative site preparation and impact reductions on the environment. https://openei.org/wiki/InSPIRE/Resources; 2021 [accessed 30.07.21].

[30] Università Cattolica del Sacro Cuore. Stefano Amaducci - Docente. https://docenti.unicatt.it/ppd2/it/docenti/14925/stefano-amaducci/profilo; 2021 [accessed 30.07.21].

[31] CAZRI. Agri-voltaic system inaugurated at CAZRI, Jodhpur. https://icar.org.in/content/agri-voltaic-system-inaugurated-cazri-jodhpur; 2017 [accessed 30.07.21].

[32] Marrou H, Guilioni L, Dufour L, Dupraz C, Wery J. Microclimate under agrivoltaic systems: Is crop growth rate affected in the partial shade of solar panels? Agric For Meteorol 2013;177:117−32. Available from: https://doi.org/10.1016/j.agrformet.2013.04.012.

[33] Reproduced with permission from Plants in Action, http://plantsinaction.science.uq.edu.au, published by the Australian Society of Plant Scientists.

[34] Lambers H, Chapin FS, Pons TL. Plant physiological ecology. New York: Springer New York; 2008.

[35] Chapin FS, Matson PA, Vitousek PM. Principles of terrestrial ecosystem ecology. New York: Springer New York; 2011.

[36] Kang J, Jang J, Reise C, Lee K. Practical comparison between view factor method and ray-tracing method for bifacial PV system yield prediction; 2019. doi:10.4229/EUPVSEC20192019-5CV.3.30.

[37] Tajima M, Iida T. Evolution of agrivoltaic farms in Japan. In: Agrivoltaics2020 Conference: Launching Agrivoltaics World-wide, 14−16 October 2020. Perpignan, France, Online: AIP Publishing; 2021. p. 30002. doi:10.1063/5.0054674.

[38] Tajima M, Iida T. Evolution of agrivoltaic farms in Japan; 2020.

[39] Trommsdorff M, Kang J, Reise C, Schindele S, Bopp G, Ehmann A, et al. Combining food and energy production: design of an agrivoltaic system applied in arable and vegetable farming in Germany. Renew Sustain Energy Rev 2021;140.

[40] Barron-Gafford GA, Pavao-Zuckerman MA, Minor RL, Sutter LF, Barnett-Moreno I, Blackett DT, et al. Agrivoltaics provide mutual benefits across the food−energy−water nexus in drylands. Nat Sustain 2019;2:848−55. Available from: https://doi.org/10.1038/s41893-019-0364-5.

[41] Patel U, Chauhan PM. Studies of climatic parameters under agrivoltaic structure. Renew Sustain Energy: An Int J (RSEJ) 2018;1.

[42] Ayala P, Munos I, Acuna I. AgroPV: Mancomunión energía solar y agricultura; 2017.

[43] Barron-Gafford GA, Minor RL, Allen NA, Cronin AD, Brooks AE, Pavao-Zuckerman MA. The photovoltaic heat island effect: larger solar power plants increase local temperatures. Sci Rep 2016;6:35070. Available from: https://doi.org/10.1038/srep35070.

[44] Elamri Y, Cheviron B, Lopez J-M, Dejean C, Belaud G. Water budget and crop modelling for agrivoltaic systems: application to irrigated lettuces. Agric Water Manag 2018; 208:440−53. Available from: https://doi.org/10.1016/j.agwat.2018.07.001.

[45] Jung D. Agrivoltaic—a sustainable business case for small farmers in Chile [Masterthesis]. TU München; 2020.

[46] Valle B, Simonneau T, Sourd F, Pechier P, Hamard P, Frisson T, et al. Increasing the total productivity of a land by combining mobile photovoltaic panels and food crops. Appl Energy 2017;206:1495−507. Available from: https://doi.org/10.1016/j.apenergy.2017.09.113.

[47] Tiffon-Terrade B. Effect of transient shading on phenology and berry growth in grapevine; 2020.

[48] Juillion P, Lopez G, Fumey D, Génard M, Lesniak V, Vercambre G. Water status, irrigation requirements and fruit growth of apple trees grown under photovoltaic panels; 2020.

[49] Juillion P, Lopez G, Fumey D, Génard M, Lesniak V, Vercambre G. Impact of full sun tracking with photovoltaic panels on subsequent year bloom density and fruit drop in apple trees.

[50] Homma M, Doi T, Yoshida Y. A field experiment and the simulation on agrivoltaic-systems regarding to rice in a paddy field. J Jpn Soc Energy Resour 2016;37.

[51] Takashi S, Nagashima A. Solar sharing for both food and clean energy production: performance of agrivoltaic systems for corn, a typical shade-intolerant crop. Environments 2019;6:65. Available from: https://doi.org/10.3390/environments6060065.

[52] Tani A, Shiina S, Nakashima K, Hayashi M. Improvement in lettuce growth by light diffusion under solar panels. J Agric Meteorol 2014;70:139−49. Available from: https://doi.org/10.2480/agrmet.D-14-00005.

[53] Herbert SJ. Yield comparisons UMass farm NREL Co-Location Project 2016-17; 2018.

[54] Herbert SJ, Oleskewicz K. UMass dual-use solar agricultural report and final report summary; 2019.

[55]   Mayer DG, Butler DG. Statistical validation. Ecol Model 1993;68:21−32. Available from: https://doi.org/10.1016/0304-3800(93)90105-2.

[56]   Mbewe DMN, Hunter RB. The effect of shade stress on the performance of corn for silage vs. grain; 1985.

[57]   Reed AJ, Singletary GW, Schussler JR, Williamson DR, Christy AL. Shading effects on dry matter and nitrogen partitioning, kernel number, and yield of maize. Crop Sci 1988;28:819−25. Available from: https://doi.org/10.2135/cropsci1988.0011183X002800050020x.

[58]   DWD. Deutscher Wetterdienst: zeitreihen und trends. https://www.dwd.de/DE/leistungen/zeitreihen/zeitreihen.html?nn = 344886#buehneTop; 2021 [accessed 30.07.21].

[59]   Ionita M, Nagavciuc V, Kumar R, Rakovec O. On the curious case of the recent decade, mid-spring precipitation deficit in central Europe. npj Clim Atmos Sci 2020;. Available from: https://doi.org/10.1038/s41612-020-00153-8.

[60]   Obergfell T. Wie sieht eine Energiewende mit erhörter gesellschaftlicher Akzeptanz aus?; 2016.

[61]   APC Group. Solar sharing for fun! https://solar-sharing.org/sswp/wp-content/uploads/2018/08/1feb7cf46b50eece5b23cc808768e8c0.pdf; 2015 [accessed 03.10.19].

[62]   Chiba University. Solar sharing national survey results report; 2019.

[63]   Cossu M, Yano A, Solinas S, Deligios PA, Tiloca MT, Cossu A, et al. Agricultural sustainability estimation of the European photovoltaic greenhouses. Eur J Agron 2020;118:126074. Available from: https://doi.org/10.1016/j.eja.2020.126074.

[64]   Kopecek R, Libal J. Bifacial photovoltaics 2021: status, opportunities and challenges. Energies 2021;14:2076. Available from: https://doi.org/10.3390/en14082076.

[65]   Khan MR, Hanna A, Sun X, Alam MA. Vertical bifacial solar farms: physics, design, and global optimization. Appl Energy 2017;206:240−8. Available from: https://doi.org/10.1016/j.apenergy.2017.08.042.

[66]   Wu Z, Hou A, Chang C, Huang X, Shi D, Wang Z. Environmental impacts of large-scale CSP plants in northwestern China. Env Sci Process Impacts 2014;16:2432−41. Available from: https://doi.org/10.1039/c4em00235k.

[67]   Next2Sun. Reference projects. https://www.next2sun.de/en/references/; 2021 [accessed 11.08.21].

[68]   Krause-Tünker S. Next2Sun: experiences with vertical agro-photovoltaic; 2019.

[69]   Greer DH, Abeysinghe SK, Rogiers SY. The effect of light intensity and temperature on berry growth and sugar accumulation in Vitis vinifera 'Shiraz' under vineyard conditions; 2019. doi: 10.5073/VITIS.2019.58.7-16.

[70]   Mira de Orduña R. Climate change associated effects on grape and wine quality and production. Food Res Int 2010;43:1844−55. Available from: https://doi.org/10.1016/j.foodres.2010.05.001.

[71]   Sun'Agri. Agrivoltaics viticulture. https://sunagri.fr/en/project/nidoleres-estate/; 2021 [accessed 11.08.21].

[72]   Agostini A, Colauzzi M, Amaducci S. Innovative agrivoltaic systems to produce sustainable energy: an economic and environmental assessment. Appl Energy 2021;281:116102. Available from: https://doi.org/10.1016/j.apenergy.2020.116102.

[73]   Fraunhofer ISE. Photovoltaics report; 2020.

[74]   pvxchange. Solar price index. https://www.pvxchange.com/price-index; 2021 [accessed 30.07.21].

[75]   VDMA. Photovoltaic roadmap (ITRPV): Eleventh edition online.

[76]   PI Berlin. Bifacial PV technology: ready for mass deployment; 2019.

[77] Breuer L, Eckhardt K, Frede H-G. Plant parameter values for models in temperate climates. Ecol Model 2003;169:237−93. Available from: https://doi.org/10.1016/S0304-3800 (03)00274-6.

[78] Peng J, Fan W, Xu X, Wang L, Liu Q, Li J, et al. Estimating crop Albedo in the application of a physical model based on the law of energy conservation and spectral invariants. Remote Sens 2015;7:15536−60. Available from: https://doi.org/10.3390/rs71115536.

[79] PVinsights. 2021. http://pvinsights.com/ [accessed 30.07.21].

[80] Heinrich M, Kuhn T, Dimroth F, Würfel U, Goldschmied JC, Powalla M, et al., editors. A comparison of different solar cell technologies for integrated photovoltaics; 2020.

[81] Insolight. Technical features. https://insolight.ch/technology/; 2021 [accessed 30.07.21].

[82] Emmott CJM, Röhr JA, Campoy-Quiles M, Kirchartz T, Urbina A, Ekins-Daukes NJ, et al. Organic photovoltaic greenhouses: a unique application for semi-transparent PV? Energy Env Sci 2015;8:1317−28. Available from: https://doi.org/10.1039/C4EE03132F.

[83] ASCA. Die transparente und flexible organische Photovoltaik Folie https://de.asca.com/; 2021 [accessed 30.07.21].

[84] Soliculture. Soliculture - greenhouse integrated photovoltaics. http://www.soliculture.com/ ; 2021 [accessed 30.07.21].

[85] Loik ME, Carter SA, Alers G, Wade CE, Shugar D, Corrado C, et al. Wavelength-selective solar photovoltaic systems: powering greenhouses for plant growth at the food-energy-water nexus. Earth's Future 2017;5:1044−53. Available from: https://doi.org/10.1002/2016EF000531.

[86] Miskin CK, Li Y, Perna A, Ellis RG, Grubbs EK, Bermel P, et al. Sustainable co-production of food and solar power to relax land-use constraints. Nat Sustain 2019; 972−80.

[87] Tubesolar. Tubesolar - the future of photovoltaics. https://tubesolar.de/en/the-future-of-photovoltaics/; 2021 [accessed 30.07.21].

[88] APEC. APEC low carbon model town solar photovoltaic agricultural development mode study final report: APEC Project: EWG 11 2015A; 2016.

[89] DIN. Agri-photovoltaic systems − Requirements for primary agricultural use: English translation of DIN SPEC 91434:2021-05; 2021.

[90] Sütterlin B, Siegrist M. Public acceptance of renewable energy technologies from an abstract vs concrete perspective and the positive imagery of solar power. Energy Policy 2017;106:356−66. Available from: https://doi.org/10.1016/j.enpol.2017.03.061.

[91] Devine-Wright P. Rethinking NIMBYism: the role of place attachment and place identity in explaining place-protective action. J Commun. Appl Soc Psychol 2009;19:426−41. Available from: https://doi.org/10.1002/casp.1004.

[92] Devine-Wright P. Beyond NIMBYism: towards an integrated framework for understanding public perceptions of wind energy. Wind Energ 2005;8:125−39. Available from: https://doi.org/10.1002/we.124.

[93] Smyth E, Vanclay F. The Social framework for projects: a conceptual but practical model to assist in assessing, planning and managing the social impacts of projects. Impact Assess Proj Appraisal 2017;35:65−80. Available from: https://doi.org/10.1080/14615517.2016.1271539.

[94] Schomberg von R. Towards responsible research and innovation in the information and communication technologies and security technologies fields.

[95] Linder R, Goos K, Güth S, Som O, Schröder T. "Responsible research and innovation" als ansatz für die forschungs- und technologie- und innovationspolitik - hintergründe und entwicklungen; 2016.

[96]   Pacifico Silva H, Lehoux P, Miller FA, Denis J-L. Introducing responsible innovation in health: a policy-oriented framework. Health Res Policy Syst 2018;16:90. Available from: https://doi.org/10.1186/s12961-018-0362-5.

[97]   Owen R, Macnaghten P, Stilgoe J. Responsible research and innovation: from science in society to science for society, with society. Sci Public Policy 2012;39:751−60. Available from: https://doi.org/10.1093/scipol/scs093.

[98]   van Oudheusden M. Where are the politics in responsible innovation? European governance, technology assessments, and beyond. J Responsible Innov 2014;1:67−86. Available from: https://doi.org/10.1080/23299460.2014.882097.

[99]   Stilgoe J, Owen R, Macnaghten P. Developing a framework for responsible innovation. Res Policy 2013;42:1568−80. Available from: https://doi.org/10.1016/j.respol.2013.05.008.

[100] Rösch C, Gölz S, Bisevic A, Hildebrand J, Witte K, Venghaus S. Transdisziplinäre ansätze zur erforschung gesellschaftlicher akzeptanz; 2019.

[101] Ketzer D, Weinberger N, Rösch C, Seitz SB. Land use conflicts between biomass and power production − citizens' participation in the technology development of Agrophotovoltaics. J Responsible Innov 2019;15:1−24. Available from: https://doi.org/10.1080/23299460.2019.1647085.

[102] Jorgenson J, Steier F. Frames, framing, and designed conversational processes. J Appl Behav Sci 2013;49:388−405. Available from: https://doi.org/10.1177/0021886313484511.

[103] Mayring P. Qualitative content analysis. Forum Qualitative Social Research (FQS) 2000;1.

[104] Ketzer D, Schlyter P, Weinberger N, Rösch C. Driving and restraining forces for the implementation of the Agrophotovoltaics system technology − a system dynamics analysis. J Env Manage 2020;270:110864. Available from: https://doi.org/10.1016/j.jenvman.2020.110864.

[105] Fraunhofer ISE. Agrophotovoltaik: hohe Ernteerträge im Hitzesommer; 2019.

[106] Devine-Wright P, Batel S, Aas O, Sovacool B, Labelle MC, Ruud A. A conceptual framework for understanding the social acceptance of energy infrastructure: insights from energy storage. Energy Policy 2017;107:27−31. Available from: https://doi.org/10.1016/j.enpol.2017.04.020.

[107] Amaducci S, Yin X, Colauzzi M. Agrivoltaic systems to optimise land use for electric energy production. Appl Energy 2018;220:545−61. Available from: https://doi.org/10.1016/j.apenergy.2018.03.081.

[108] Dupraz C, Marrou H, Talbot G, Dufour L, Nogier A, Ferard Y. Combining solar photovoltaic panels and food crops for optimising land use: Towards new agrivoltaic schemes. Renew Energy 2011;36:2725−32. Available from: https://doi.org/10.1016/j.renene.2011.03.005.

[109] Drechsler M, Egerer J, Lange M, Masurowski F, Meyerhoff J, Oehlmann M. Efficient and equitable spatial allocation of renewable power plants at the country scale. Nat Energy 2017;. Available from: https://doi.org/10.1038/nenergy.2017.124.

[110] Schweizer P-J, Renn O, Köck W, Bovet J, Benighaus C, Scheel O, et al. Public participation for infrastructure planning in the context of the German "Energiewende.". Uti Policy 2016;43:206−9. Available from: https://doi.org/10.1016/j.jup.2014.07.005.

[111] Durstewitz M, Lange B. Meer - wind - strom: forschung am ersten deutschen offshore-windpark alpha ventus/Michael Durstewitz, Bernhard Lange (Hrsg.). Wiesbaden: Springer; 2016.

[112] MAFF. About farm power generation. http://www.maff.go.jp/j/shokusan/renewable/energy/attach/pdf/einou-5.pdf; 2018 [accessed 03.10.19].

[113] MAFF. Farming photovoltaics (FPV): guidelines; 2018.

[114] MAFF. Solar sharing case studies. https://www.maff.go.jp/j/shokusan/renewable/energy/attach/pdf/einou-31.pdf; 2019 [accessed 20.06.20].

[115] MAFF. About farm type solar power generation: Ministry of Agriculture, Forestry and Fisheries. https://www.maff.go.jp/j/shokusan/renewable/energy/einou.html; 2020 [accessed 20.06.20].

[116] Min-hee J. KHNP receives Korea's first patent on photovoltaic power generation on paddy fields.

[117] Lim C-H, Kim TY, Gim GH, Park JS, Kim WR, Kim DS. Recent R&D trends and status of agri- photovoltaic system in South Korea; 2020.

[118] Meng J, Jia W, Zhang H, Fan T, He J. Solar photovoltaic + multi-scene application to help combat climate change. China Environ 2021;44−7.

[119] Xue J. Photovoltaic agriculture - new opportunity for photovoltaic applications in China. Renew Sustain Energy Rev 2017;73:1−9. Available from: https://doi.org/10.1016/j.rser.2017.01.098.

[120] pv magazine, Huawei. Smart solar: convergence powers PV: *PV Magazine* special. https://16iwyl195vvfgoqu3136p2ly-wpengine.netdna-ssl.com/wp-content/uploads/2017/05/pvi_HuaweiEdition_170512.pdf; 2017 [accessed 17.06.20].

[121] Spaes J. Agrivoltaics prevail in France's tender for innovative PV technologies. https://www.pv-magazine.com/2021/01/05/agrivoltaics-prevail-in-frances-tender-for-innovative-pv-technologies/?utm_source = dlvr.it&utm_medium = linkedin; 2021 [accessed 08.04.21].

[122] Bénard S. Lastenheft der frz. Ausschreibung für die Errichtung und den Betrieb innovativer Stromerzeugungsanlagen mittels Sonnenenergie. https://energie-fr-de.eu/files/ofaenr/05-traductions/02-traductions-payantes/180123_Uebersetzung_Lastenheft_Ausschreibung_Innovation_DE.pdf; 2017 [accessed 20.07.20].

[123] Bénard S. Förderstrategie der APV in Frankreich; 2019.

[124] Enerdata. France opens several renewable tenders for wind, solar, and hydropower. https://www.enerdata.net/publications/daily-energy-news/france-opens-several-renewable-tenders-wind-solar-and-hydropower.html; 2021 [accessed 22.08.21].

[125] InnAusV. Verordnung zu den Innovationsausschreibungen (Innovationsausschreibungsverordnung - InnAusV). http://www.gesetze-im-internet.de/innausv/BJNR010610020.html; 2020 [accessed 30.07.21].

[126] NREL. Beneath solar panels, the seeds of opportunity sprout. https://www.nrel.gov/news/features/2019/beneath-solar-panels-the-seeds-of-opportunity-sprout.html; 2021 [accessed 30.07.21].

[127] MassGov. Guideline on establishing SMART compensation rates. https://www.mass.gov/doc/guideline-on-establishing-smart-compensation-rates/download; 2018 [accessed 18.07.20].

[128] MassGov. Guideline regarding in definition of agricultural solar tariff generation units. https://www.mass.gov/doc/agricultural-solar-tariff-generation-units-guideline-final; 2018 [accessed 17.06.20].

[129] General Assembly of the State of Colorado. Senate Bill 21-235 - Concerning additional funding for programs of the department of agriculture to support increased efficiency in agricultural operations, and, in connection therewith, making an appropriation. http://leg.colorado.gov/bills/sb21-235; 2021 [accessed 28.07.21].

[130] GovIL. A unique collaboration between the Ministry of Energy and the Ministry of Agriculture and Rural Development will increase the production of solar energy in Israel and will lead to the promotion of integrated energy production in agriculture. https://www.gov.il/he/departments/news/press_030521; 2021 [accessed 05.08.21].

[131] GovIL. Request for submission of preliminary plans for the dual use of renewable energies in agricultural areas. https://www.gov.il/he/Departments/publications/reports/agro_-voltai; 2021 [accessed 05.08.21].

Chapter 6

Aquavoltaics: dual use of natural and artificial water bodies for aquaculture and solar power generation

Charis Hermann[1], Flemming Dahlke[2], Ulfert Focken[2] and Max Trommsdorff[1,3]

[1]*Fraunhofer Institute for Solar Energy Systems ISE, Freiburg im Breisgau, Germany,*
[2]*Thünen-Institute of Fisheries Ecology, Bremerhaven, Germany,* [3]*Wilfried Guth Chair, Department of Economics, University of Freiburg, Wilhelmstr, Freiburg im Breisgau, Germany*

6.1 Introduction

Fish and seafood play an important role in the global food supply, especially in the provision of essential amino and fatty acids. The demand for fish and seafood is not only increased with population growth but also with increasing urbanization and income at an estimated global rate of 1.5% per annum [1]. The demand for fish and seafood is expected to increase from 154 Mton in 2011 to 186 Mton in 2030 [2]. Fish and seafood for human consumption originate mainly from capture fishery and aquaculture. Capture fishery is the exploitation of common property resources, which is sometimes poorly managed, resulting in overexploitation. Aquaculture is defined as the husbandry of aquatic organisms under private or corporate ownership, offering better options for management. Currently, capture fisheries and aquaculture respectively provide about 48% and 52% of fish and seafood for human consumption [3]. While the overall supply from capture fisheries is expected to be stagnant at best, and supply from aquaculture is estimated to be increased (Fig. 6.1). Aquatic plants, mostly algae from aquaculture, are predominantly used for industrial purposes. Between the years 2001 and 2018, global production of aquaculture products has been annually increased by 5.3%. This makes fish and fish products one of the most traded food products in the world, with Asia being the dominant continent. The number of employed people in this sector has also been increased so that in 2018, 20.5 million

Solar Energy Advancements in Agriculture and Food Production Systems.
DOI: https://doi.org/10.1016/B978-0-323-89866-9.00009-2

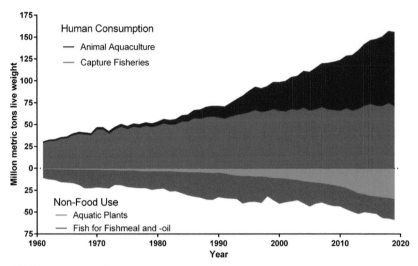

FIGURE 6.1   Yields of capture fisheries and aquaculture for both human consumption and non-food purposes [4,5].

people in the aquaculture sector and 29 million people in the fish capture sector were employed, specifically in the global south countries with the highest share of employment for small farmers and women [3]. The expansion of aquaculture production can be partly achieved by intensification in existing facilities, but it also requires the set-up of new equipment which increases the competition for resources, space, water, energy, and feed. To mitigate this issue, new smart approaches including aquavoltaics are placed at the focus of attention. Aquavoltaics describes the combination of photovoltaic (PV) technology and the production of aquatic animals and plants. Aquaculture and photovoltaics are two sectors with a decisive role in providing food and sustainable energy security.

Following the diverse possibilities of using water areas, the integration approaches of PV modules have also a wide range of potential applications. Floating photovoltaics (FPVs) is one technical approach employed to generate solar electricity on water surfaces. Apart from floating systems, water bodies can also be covered with elevated PV modules using a fixed substructure. PV modules can be installed on various types of water bodies such as pit lakes, gravel pits, natural lakes, reservoirs, irrigation ponds, and open seas. In 2007, a 20 kWp FPV plant was installed in Aichi, Japan, while in 2018, the global power generation capacity of FPV systems could reach 1.3 GWp (Fig. 6.2) which is estimated to reach 118 GWp by 2025 [6−9]. Since 2013, large-scale installations in China, Japan, and the Netherlands, in particular, have driven the global FPV market [7]. This massive expansion confirms the promising potential of FPVs resulted from the high availability of water bodies and cobenefit of profitable synergies for the aquatic ecosystem [10,11].

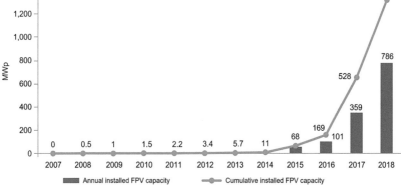

FIGURE 6.2 Floating photovoltaic development from 2007 to 2018 in MWp [6].

FIGURE 6.3 Aquavoltaic system installed in Les Cedres, La Réunion [12].

This chapter focuses on the integration of PV systems with aquaculture, representing the coproduction of electricity and aquatic animals in a dual-use approach. Till now, standard FPV systems and their integration with aquaculture systems have been less investigated. In terms of global deployment, Asia has developed several aquavoltaic systems, representing their feasibility and potential. Additionally, in recent years, India, Africa, and European countries have had an increasing number of implemented aquavoltaic projects. Fig. 6.3 shows an example of an installed aquavoltaic system in La Réunion island installed by Akuo Energy company. The system consists of a PV roof with a generation capacity of 1.5 MWp, covering 14 fish rearing ponds mainly stocked with tilapia.

6.2 Aquaculture

Aquaculture systems and practices are quite diverse and utilize different aquatic habitats, including natural and artificial ponds, offshore farms, and indoor facilities [13]. A basic classification of aquaculture is according to distinguishes between freshwater, brackish water, and marine systems [3]. Freshwater aquaculture refers to the production of fish and shrimp in facilities located in or connected to freshwater reservoirs, rivers, lakes, canals, or groundwater. Brackish aquaculture systems (fish, shrimp crabs, mussels) are usually installed in estuaries, bays, and lagoons. Marine systems (mariculture) may be land-based, located in fjords or open water. In total, around 600 different species of aquatic animals are cultivated in aquaculture systems [14]. Globally, the largest volume of aquatic animal production (85.5 Mton) comes from freshwater systems (61%), followed by marine water (27%), and brackish water (12%) aquacultures [3]. Freshwater systems contribute to 58% of the total annual production value of US$260 billion, followed by marine systems (25%) and brackish water aquaculture (17%) (see Table 6.1).

Aquaculture operations also vary in terms of production intensity and are categorized as extensive, semiintensive, or intensive systems [13]. Extensive systems, such as shellfish aquaculture and integrated rice-shrimp production, provide relatively low yields (per area) but require minimal or no supplemental feeding and aeration of the water body. The sunlight dependency of aquaculture ponds is decreased from extensive systems to intensive systems. The latter produce more yield per area but require additional feed and energy inputs for aeration devices. Open pond systems draw freshwater from surrounding rivers, lakes, or lagoons and discharge most of the wastewater directly into the surrounding environment. While closed pond systems rely on complex water treatment practices to minimize effluents, and increasingly employ roof structures to be protected against external contaminants, diseases, and predators. With increasing intensification, higher external inputs of feed and energy (for aeration and water quality management) are required to realize high stocking densities and animal growth rates [15] (see Fig. 6.4). For example, intensive shrimp farms can produce more than 50 tons/ha/year

TABLE 6.1 Global production of aquatic animals in different water habitats by volume and value in 2019 [3].

Environment	Production (Mton)	Value (bill. USD)	Percentage of total production	Percentage of the total value
Freshwater	52.3	150.9	61.2	58.1
Marine water	23.3	64.4	27.3	24.8
Brackish water	9.9	44.5	11.6	17.1
Total	85.5	259.8	100	100

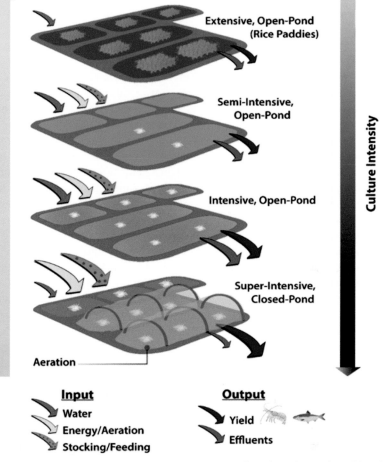

FIGURE 6.4 Relationship between culture intensity and sunlight dependency of pond-based aquaculture systems [16].

of shrimps with an average energy demand of 5500 kWh/ton for aeration [17], while extensive farms without aeration typically produce about 5 tons/ha/year.

6.2.1 Land use, energy demand, and greenhouse gas emissions

Aquaculture includes various production systems that are increasingly competing for land with other agricultural sectors such as crop and livestock production, forestry, and exclusive energy production [14,18]. In many regions, the rapid growth of aquaculture is causing major environmental problems, including the degradation of landscapes and water bodies with subsequent loss of ecosystem services and biodiversity [19]. A poorly managed production system especially with a high water input, escaping individuals, use of antibiotics and heavily polluted wastewater causes severe environmental impacts [20]. The largest volume of aquaculture production comes from pond systems, which are estimated to cover an area of more than 110,000 km^2 worldwide [21]. Most of the global pond areas are in Asia, especially along the coasts and river deltas of China and Southeast Asia where aquaculture has tremendously grown in recent years (Fig. 6.5). Hotspots of pond aquaculture in Asia and other parts of the world are characterized by the availability of highly productive farmland, abundant marine and freshwater resources, and therefore are often densely populated. In many of these regions, the growing international demand for aquatic products has led to increased production of profitable (export-oriented) products such as shrimp and pangasius catfish [23], putting pressure on ecosystems due to the requirement for new lands to build more pond systems [14]. As an example, the rapid growth of aquaculture production in coastal Bangladesh, Vietnam, and Indonesia has already altered or displaced traditional agriculture by converting coastal rice paddies and mangrove forests into shrimp ponds [24]. Land-use changes from agriculture to more profitable aquaculture systems can damage adjacent ecosystems if they are not properly managed. Poor construction planning and wastewater treatment have affected the health of lakes, rivers, reefs, lagoons, and salt marshes in many places, sometimes with serious consequences for local communities [19]. Innovative technologies and more efficient land-use strategies are therefore required to ensure the sustainable coexistence of natural habitats with both productive agricultural and aquaculture operations.

The energy consumption of an aquaculture system highly depends on the location, reared species, and system design. Approximately 0.5% of total

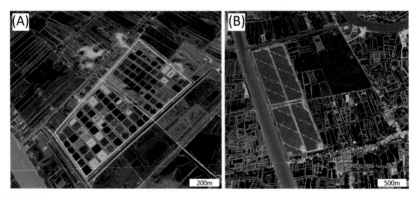

FIGURE 6.5 Satellite images showing land-use patterns in the Vietnamese Mekong Delta with: (A) intensive aquaculture ponds (*yellow outlined area*) surrounded by extensive mangrove-shrimp farms, and (B) a PV power plant (*red outlined area*) surrounded by aquaculture ponds, rice paddies, and mangrove forest [22].

global greenhouse gas (GHG) emissions is allocated to aquaculture [25]. Industrial-scale farming of shrimp and fish species involves high stocking (animal) densities, intensive feeding, and mechanical aeration (oxygen enrichment) of the culture ponds or tanks [18]. In addition to the energy required for feed production, water pumping, temperature control, and artificial light sources also require high energy input. Specifically, water aeration consumes a significant amount of energy, accounting for 50% of production costs. In shrimp aquaculture, about 70% of the global yield (5 Mton/year) comes from aerated ponds with an average energy consumption of 5500 kWh/ton, adding up to 20 TW for shrimp pond aeration worldwide [3,17]. In a superintensive shrimp production facility in Belize, the aerators consume 447 kWh/day on an area of 1 ha with 75% aerator's operation time. With an average production time of 132.5 days, it has been calculated that aerators alone consume 59,227.5 kWh during one production cycle. Hence, a shrimp harvest of 13,600 kg/ha/crop results in energy consumption of 435 kWh/kg of shrimp [26]. In another life cycle assessment (LCA) study for a shrimp production facility in the USA [recirculation aquaculture system (RAS)], an energy input of 99 MJ/kg shrimp was calculated for the entire life cycle, including feed production, construction material, and energy use. The energy consumption was also calculated as 4.2 kWh/kg for shrimp, representing the share of water pumps as 59%, followed by aerators as 24%, and foam fractionator pumps as 17% [27].

According to released statistics, annual GHG emissions from all aquaculture operations (260 Mton CO_2-eq/year) are rather small compared to that from cattle farming (3000 Mton) or pig farming (800 Mton) as depicted in Fig. 6.6 [25]. However, to meet the growing aquatic food demand for a world population of 10-billion in 2050 [28], energy consumption and GHG emissions from the aquaculture industry are likely to be significantly increased in the coming decades. In this regard, PV-integrated aquaculture systems with the coproduction of food and electricity would be an important contribution to sustainable land use and climate change mitigation.

6.2.2 Essential parameters of aquaculture

In this section, the most important aquaculture production parameters considering yield and animal welfare are presented and described. Parameters such as water and air temperature, light availability, water pH, dissolved oxygen (DO), feeding system, and predator pressure not only play major roles in a successful farming process but also must be given special consideration when PV modules are integrated to avoid production losses. The aquavoltaic approach aims to maintain these parameters and improve the system by exploiting synergies between the aquaculture and PV systems. First, the relevance and the management difficulty from an aquaculture perspective are explained, and then possible methods of improvement or complications

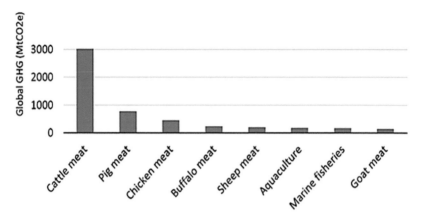

FIGURE 6.6 Global greenhouse gas emissions from the production process of terrestrial meat (2010), aquaculture (2010), and marine fisheries (2011) [25].

arising from combination with PV systems are discussed. It is worthy to be noted that cultured species have different requirements, confirming the need for variation of essential parameters as a function of species type and farming systems.

6.2.2.1 Water and air temperature

Fish are ectothermic animals, which means that their body temperature completely depends on the temperature of the surrounding. This implies that temperature fluctuations in the water are directly transmitted to the animals, affecting biochemical processes taking place in their bodies [29]. Each species has a different optimal temperature profile, and excessively high or low water temperatures and fluctuations impair vital functions such as growth and reproduction. The growth curve is increased by rising temperatures until an optimum level is reached. Temperature values exceeding the optimum value will result in a decreased growth rate due to a reduced feed intake and food conversion ratio [30]. Specifically, high-temperature values lead to stress reactions in fish, which in the worst case has lethal consequences. It should also be noted that optimum temperatures and tolerance ranges differ according to the development stage of the fish [31]. Fig. 6.7 depicts the relation between specific growth rate and temperature for Atlantic salmon. As shown in this figure, the optimum temperature varies with the initial weight of the fish. Accordingly, the management goal is to reduce water temperature fluctuations and maintain the optimum temperature range consistently.

Various experiments show that shading influences water temperature. In many regions, it is a common practice to install shading nets during hot periods to take advantage of their cooling effect (see Fig. 6.8). In a research study, milkfish were bred under different shading scenarios, in which a

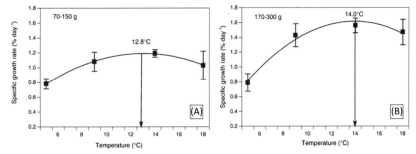

FIGURE 6.7   The effect of temperature and fish weight on the growth rate of Atlantic salmon: (A) fish with a bodyweight of 70−150 g, (B) fish with a bodyweight of 170−300 g [32].

FIGURE 6.8   Outdoor trout hatchery with shading nets installed in Western Australia [33]. © State of Western Australia [Department of Primary Industries and Regional Development, WA].

reduced temperature fluctuation between day and night was observed at a shading level of 40%, creating more stable breeding conditions [34]. In another similar study, shading experiments were conducted for *Lateolabrax japonicus* and the results were compared with a control case without shading. It was found that the water temperature of 35.8°C can be obtained for the control case, while for the pond with 60% shading, the water temperature was reported as 31.5°C. In addition, reductions in feed intake and growth rate were observed right after the water temperature was exceeded from 30°C. Although shading nets are often used in hot climates, low water temperatures in shaded ponds during cold seasons have also been reported as an impairing factor for fish growth [35]. With the integration of PV modules, the shading nets can be substituted, and therefore providing a moderate temperature in the summertime will eliminate the need for additional cooling facilities. In this case, the stress caused by high temperatures is mitigated and the day/night variation is minimized. The biggest trade-off for PV integration occurs in winter when a high solar energy reception is required to provide favorable temperature levels. In open systems with no heating equipment, an optimal degree of shading should be provided for the reception of adequate irradiation to heat the water. Therefore it can be concluded that the positive cooling impact of shading that compensates day/night temperature fluctuations in hot seasons is dominant compared to its slight negative impact in cold seasons.

6.2.2.2 Light availability

The vital role of light on life in water ecosystems is well-known. In both closed and open aquaculture systems, the light factor is decisive for many process interrelationships. The influence occurs mainly at three levels: (1) light quality, (2) light quantity, and (3) light periodicity [36]. Light quality refers to the wavelength that is absorbed by water, influencing life under water. The light quantity level reflects different light intensities, while light periodicity represents the day and night as well as summer and winter cycles. Each fish species has its light requirements, while some species are more sensitive to light [37]. In addition, fish differently react to light depending on their stage of development. For most species, an increasing light period promotes fish growth, confirming a direct correlation between food intake and light availability. Nevertheless, too much light negatively affects fish development and a certain dark period is important for the health of the aquatic animals [37]. In addition to light intensity and light periodicity, distinct wavelengths affect fish behavior differently [38,39]. This responsiveness arises as different wavelengths are absorbed in different water depths and aquatic animals have highly sensitive retinas. In most species, wavelengths in the range of 495−570 nm (green) and 450−495 nm (blue) affect fish behavior and consequentially the fish growth [38]. However, all species do not have light-dependent growth [40]. Not only fish are sensitive to different light conditions, but also various species of microorganisms and algae such as phytoplankton are directly or indirectly affected by the light factor [41]. The chosen degree of coverage is a function of location, type of the aquaculture system, and fish species. Since many organisms in aquaculture systems or natural water bodies require light, shading can lead to competitive behavior among the species. This shift within the ecosystem can lead to an alteration in the composition of species, and consequently disrupt the natural food chain [10]. Thus the compatibility of aquaculture and PV modules will likely depend on the extent to which the respective culture system relies on natural sunlight [10]. Shading of water bodies due to the installation of PV modules may be problematic in extensive and semiintensive systems where photosynthetic plants and algae represent essential components of animal feed and water quality management. Aquaculture ponds act as an ecosystem that includes photosynthetic microorganisms and algae, which are responsible for nutrient recycling (e.g., nitrogen and phosphorus), biomass production (natural feed), and oxygenation of the water body. There is a negative correlation between shading degree and microalgae biomass production since the photosynthetic efficiency of algae becomes limited in the shade. In addition to photosynthesis, solar radiation also affects water temperature, which in turn has an impact on the biology of aquatic organisms and their influence on water chemistry (Fig. 6.9). Extensive culture systems are often entirely based on natural feeds and photosynthetic oxygen production. With supplemental feeding and oxygen supply, intensive systems become less dependent on the activity of photosynthetic organisms. For example, biofloc technology (BFT) which is mainly applied in superintensive closed ponds, solves the dependency on microalgae. Biofloc is produced from feed residues and excrements, which then again acts as a protein-rich feed [42]. BFT is described as a blue revolution in the aquaculture sector, enabling sustainable water management. Mostly applied in shrimp farming, BFT is also applicable in the farming of other species such as tilapia or carp [43].

Once an aquaculture system becomes independent of microalgae, reduced light availability will have fewer negative impacts on the whole system, allowing for partial or complete shading of the water body by PV modules. However, applying good management can make PV integration possible even in systems that depend on algae growth. As an alternative, tracked PV modules can benefit from adapted light conditions, resulting in a more productive system. In comparison with fixed modules, tracked PV modules are costly and their maintenance is labor-intensive, especially when they are installed on water areas.

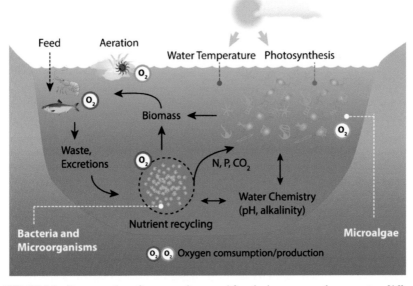

FIGURE 6.9 Representation of an aquaculture pond functioning as a complex ecosystem [16].

An important factor, particularly in the context of highly technologized systems, is the utilization of artificial light sources. The use of light-emitting diodes (LEDs) enables precise control of lighting conditions. As an example, the use of artificial lighting can prolong daylight cycles, or compensate for high cloudiness periods to optimize fish growth [10]. Likewise, artificial lights could have a compensatory effect in cases the PV modules block a large amount of light from entering the system. LEDs can be operated with relatively low energy input. Therefore allocating a part of the generated electricity by PV modules directly to supply light, no external electricity source would be required. In conclusion, it can be underlined that light availability, periodicity, and wavelength play crucial roles when PV modules are integrated with an aquaculture system because of two reasons: (1) fish growth and their feeding behavior are light-dependent and (2) many aquaculture systems depend on microalgae growth.

6.2.2.3 Dissolved oxygen content

The oxygen content of water is another crucial factor required for successful fish breeding. Insufficient oxygen causes aquatic animals to have difficulty in breathing, making them gasp for air at the water surface. This causes increased stress levels or even death of individuals. Stressed animals experience a decrease in appetite and feed conversion ratio and an increase in susceptibility to diseases [44]. DO levels are influenced by physical and biological processes. As the pressure in air is higher than in water, the water will accumulate oxygen from the atmosphere. This process eliminates when the pressure in both mediums is equal when the water is saturated with DO. Accordingly, this physical process strongly depends on the exchange between the water surface and the atmosphere. This exchange is largely affected by the turbulence derived from the wind stream flowing directly above the water surface. This water−atmosphere interaction mostly occurs in open systems. Therefore increasing the PV coverage leads to a decrease in air turbulence and consequently DO values of the water in the pond [37]. In addition to physical processes, biological processes also affect DO levels. Phytoplankton organisms release oxygen as a photosynthetic byproduct which confirms that light-dependent photosynthetic activity can affect DO content. During daytime and under sunny conditions, DO content can rise above saturation level when photosynthesis is higher than respiration processes. Decreasing and fluctuating DO levels cause problems not only in closed and highly industrialized aquatic systems but also in less industrialized aquaculture environments [45].

Hitherto, the literature has provided contradictory results regarding DO levels in aquaculture ponds integrated with PV modules in a way that some studies reported increased DO levels in shaded systems [34,46], while others revealed that DO content is decreased with an increase in the coverage ratio of PV modules [34,37]. These conflicting results can be explained by different PV integration approaches since space can be created between the modules and water surface when PV modules are installed above the water areas in which the air stream can flow, or described by the light availability and favorable temperature levels that affect photosynthesis ratios. Different effects on DO levels due to shading also depend on the aquatic production system. In closed indoor systems using BFT, photosynthetic processes and air fluxes above the water surface play a less important role. In these systems, oxygen is artificially added 24 h a day. Paddlewheels and oxygen pumps account for a significant share of energy consumption in aquaculture systems (see Fig. 6.10). In this context, the integration of PV modules would be beneficial due to providing the power demand of the oxygen management system. When there is no additional supply for oxygen, the installation of PV modules should be done in a way to ensure that the physical and biological DO production processes are not negatively affected.

6.2.2.4 Further water quality parameters

The assessment of water quality includes the analysis of a wide range of parameters. As an example, depending on the nitrogen content, the adaptive abilities of aquatic animals are affected. The pH value affects toxic processes of specific compounds in water and thus causes a strong effect on the life of aquatic organisms. Although other water quality parameters such as water salinity and turbidity, and CO_2 content are important when monitoring the water quality [47], they are not discussed in this chapter since they can be indirectly affected by the PV integration and changed by water temperature, DO content, and light availability.

6.2.2.5 Feeding system

The way that PV integration affects the above-mentioned parameters strongly depends on the aquaculture system management. Besides, the layout of the system (open vs closed), the stocking density (intensive vs extensive),

FIGURE 6.10   (A) Long-arm aerator (paddlewheel system), and (B) vertical pump aerator [17].

the degree of technologization (high vs low), and the feeding system are crucial when the consequences of PV integration are evaluated. According to Boyd and Tucker [47], a modified classification shows different feeding approaches as follows:

- *Autotrophic feeding system*: Feed is directly available to aquatic animals from plants growing in the pond.
- *Heterotrophic feeding system*: Feed consists of organic matter added to the pond. It is used to feed organisms, which in turn is consumed by target species.
- *The feed of organic origin*: It is consumed directly by target species.
- *Biofloc*.

When PV modules are integrated, especially in the case of the first two feeding systems, the requirements of target species as well as the intermediate organisms, either of plant or animal origin should be considered.

6.2.2.6 Predator pressure

Predominantly in open systems, other external influences are of importance in parallel with the controllable water quality parameters. The loss of cultivated animals through predatory birds is a major problem. Aquatic animals are food for many bird species and are particularly easily obtained prey if they are maintained in high densities. Thus, with limiting bird management, large losses can occur depending on the species, farming system, and location. For example, a loss of fish ranging from 33% to 95% has been recorded in catfish ponds in Mississippi [48]. Price et al. [49] described less drastic figures, however, their findings also show that fish predation by birds is an issue that should not be neglected. To address this problem, equipment, money, and workforce have already been put into the development of bird exclusion devices. Fig. 6.11 shows a bird-excluding net draped over a

FIGURE 6.11 Predator net to protect cultivated fish from bird attacks [50].

cultivated fish pond which is considered a common way of protection against bird attacks.

With the integration of PV modules over the water surface, aquatic animals are automatically given protection from aerial attacks. The modules can thus serve as a safety net substitute and reduce losses due to fish predation by birds [10]. In addition to the aspects that have already been listed, other factors do not directly affect the growth and performance of aquatic animals but are nevertheless essential for aquaculture operations.

6.2.2.7 Harvesting method

When PV modules are installed either directly on the water surface or elevated above the water surfaces where the substructure is anchored in the pond, the issue of harvesting methods becomes important. For many farmers, it is a common practice to pull a large net through the pond to gather the aquatic animals and harvest them. If either modules or substructure parts are installed in the water, this operation becomes complicated. Modules that can be moved to one side of the pond at harvest time or a substructure that is fixed to the shore could be feasible approaches to minimize the disturbing factor. Additionally, rope systems without anchors at the pond level could be a solution.

6.2.2.8 Occupational safety

The installation of a PV system is accompanied by the task of maintenance. In large companies with financial flexibility, the provision of technical know-how and workforce is a minor problem, while medium-sized to smaller companies and farmers can face difficulties. Smaller farmers may already have very limited capital to make sufficient investments for aquaculture operations [51]. Consequently, the farmer who manages the pond does not necessarily own the farm. This could lead to a conflict of interest as the owner aims to achieve a second income with the PV integration, whereas the farmer faces possible losses in the harvest as well as more difficult working conditions. In this regard, a fair agreement and financial distribution must be realized [37]. Although the possibility of increased working demands cannot be ignored, some system designs can have positive effects on workers. In the case of elevated modules, shading can lead to improved working conditions, especially on hot summer days.

6.3 Aquavoltaics

The combination of PV and aquaculture is possible to achieve through different approaches. FPV modules installed on water bodies and elevated modules mounted on a fixed substructure are common types of how to simultaneously cultivate fish and generate electricity in the same area. The technical specifications of these approaches will be explained here, highlighting their advantages and disadvantages from the PV perspective.

6.3.1 Floating photovoltaic modules installed on inland water bodies and seas

The basic components of a typical FPV system are a supporting substructure, PV modules, and floating elements [52] as shown in Fig. 6.12. Different external conditions require different FPV technologies. Pringle et al. [10] classified FPV systems into four main types of (1) thin-film, (2) submerged, (3) surface-mounted, and (4) microencapsulated. To prevent the installation from drifting away and moving out of control, mooring systems are essential. Depending on the properties of the water body which mainly depends on waves and tide, the mooring system must buffer strong forces from water and wind to ensure stability and intended module orientation. Mooring can be done by ropes fixed to the shore or by an anchor system on an open water surface.

Thin-film systems are characterized by their lightweight and high flexibility. For the FPV approach, thin-film modules are encapsulated by a transparent polymer. The encapsulation usually includes air to increase buoyancy (Fig. 6.13) [53]. Advantages of thin-film FPV systems are as follows:

- *High flexibility*: They are less vulnerable to high waves and in emergencies, they can even be rolled up [55]. This feature can be considered an advantage when PV modules are integrated with aquaculture systems. In this case, for certain aquaculture operations such as harvesting, PV modules can be rolled up to minimize obstacles.
- *Cooling and cleaning effect*: Since the modules are in direct contact with the water surface, a cooling effect is efficiently achieved, resulting in an

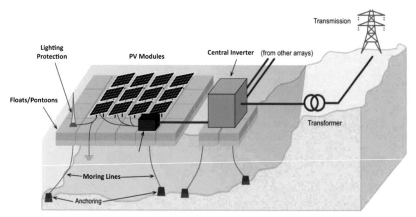

FIGURE 6.12 A standard floating photovoltaic system including photovoltaic modules, substructure, and floating elements [6].

FIGURE 6.13 (A) Prototype of a thin-film floating photovoltaic system (570 Wp) installed in Sudbury, Canada, and (B) A thin-film floating photovoltaic system installed in Kyrholmen, Norway (floating photovoltaic modules on a polymer membrane) [53,54].

FIGURE 6.14 (A) A submerged system with a sinking system and (B) a surface-mounted system on a pontoon [56,57].

increased power conversion efficiency [53]. But any contamination on the surface of the modules is flushed into the water, which can adversely affect water quality and fish performance.

- *Weight*: Thin-film FPV systems are lighter and thus require a less massive mounting structure. Additionally, cost savings due to material savings are achieved as no extra floating bodies are required.

A relatively new approach integrates thin-film PV modules on a floating polymer membrane as shown in Fig. 6.13B. The flexibility of the membrane layer ensures robustness in stormy weather. The main benefit of the system is the reduced material usage and a direct cooling possibility that results in increased module efficiency. The two remaining approaches use floating bodies, called pontoons, which are usually made of polyvinyl chloride (PVC) to keep the modules floating on the water. In the case of submerged systems, PV modules are positioned slightly below the water surface while being attached to the pontoons at their edges (Fig. 6.14A). In this configuration, the cooling effect is maximal and the loss of efficiency due to soiling is minimal [58]. Additionally, the characteristics of the surrounding water in this type must be taken into account. Depending on the weather and season, increasing turbulence in the water ecosystem can lead to increased sediment resuspension. Furthermore, water is a light absorber, and therefore, incoming radiation, especially wavelengths between 350 and 550 nm will be decreased with the water depth [56]. Surface-mounted systems consist of pontoons supporting either flat, tilted, or tracked modules as shown in Fig. 6.14B. Except for tilted modules in which the bottom side may be in contact with the water surface, these modules do not have direct contact with water.

FPV systems installed at sea are often large-scale installations. In comparison with offshore FPV systems, the plants installed on inland water bodies are of smaller capacities. The majority of these plants are installed on abandoned or active quarry lakes and pit lakes [59]. A combination of hydropower and photovoltaics is achieved by PV installation on reservoirs. Insufficient research has been done to determine the feasibility of fish farming under these FPV installations and a minority of studies have reported their negative impacts on fisheries [59]. However, some other studies indicate benefits to fish populations and biodiversity increase in lakes covered with FPVs [60,61]. Since each lake has its unique feature and each ecosystem reacts to interventions with specific sensitivity, making a precise conclusion about the environmental impacts of FPVs installed on these environments is difficult. However, to minimize the ecological impacts, covering the entire water surface with PV modules have not been suggested [6]. Additionally, the impact of FPVs on the aquatic fauna and flora is not well understood. Since the open sea displays high heterogeneity, it is difficult to make specific statements about the impact on fish and plant breeding.

Generally, installation of infrastructures such as drilling platforms for oil and gas, wind parks, or FPV systems are obstacles for large-scale fishing operations such as trawling. However, these installations typically attract a large variety of aquatic organisms, including high-value species of finfish and crustaceans, which can be exploited by fisheries with gill nets, trammel nets, or traps, offering new perspectives for income generation in connection with FPV systems.

6.3.2 Photovoltaic modules installed on closed aquaculture systems

Closed aquavoltaic systems are highly technologized aquatic animal and plant production facilities. Closed systems consist of water ponds installed onshore, which have a constant barrier between water within the system and the environment [62]. In these systems, there is no uncontrolled exchange between the agricultural unit and the natural environment, providing conditions of air and water to be easily controlled and managed. Aquaculture in greenhouses also offers the possibility of PV integration (see Fig. 6.15). By using the

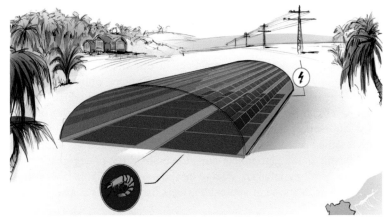

FIGURE 6.15 Photovoltaic integration into a shrimp greenhouse [63].

existing substructure of the fish-greenhouse, it is possible to replace the PVC or glass roof with PV modules without much additional material. An on-roof construction is also possible, however, in both solutions, the additional weight must be coordinated with the load capacity of the already existing substructure. Depending on the target organism, shading effects must also be considered. However, due to the high level of technologization, it can be assumed that LED lighting is already implemented or represents a suitable addition. The power of LEDs and in general the high energy input of these intensive breeding systems can be provided by the generated PV power. In these structures, the distance between the water surface and PV modules can be several meters, maintaining the temperature of the air inside the greenhouse at an optimal fish- or plant-specific level for growth. As for many species, the optimal temperature range is rather high, and it is questionable whether the cooling effect and hence an increased module efficiency occurs. The main advantages of PV integration in closed aquaculture systems are:

- Dual land use.
- Overheating of greenhouses especially during hot months of the year can be fatal for target organisms. Therefore shade nets and/or cooling systems must be utilized to control the climate in fish greenhouses. Integrated PV modules can be a substitution for cooling/shading devices.
- The high energy demand of intensive aquaculture operations can be supplied by electricity generated by PV modules. With an integrated storage system, power outages that are critical to the oxygen supply can be alleviated.
- Closed systems offer a high level of monitoring possibilities and alternative feeding methods which are not exclusively based on algae. Therefore they can highly benefit from PV integration due to being (partially) independent of light availability.

6.3.3 Photovoltaic modules installed on a substructure above the water surface

In addition to FPV integration and closed aquavoltaic approaches, the use of elevated modules is a common practice. In this method, PV modules are mounted on a substructure above the water surface, similar to a ground-mounted system. The modules have no direct water contact which can reduce the cooling effect. While generating electricity in the upper layer of the system, fish cultivation can be carried out below the panels. The EPC company Renhui installed a 133 MWp aquavoltaic system integrating with elevated PV modules in 2018. The project aims to combine PV power generation and fish farming at Mengjiafangzi Village in Tuanbo town, China. Scientific evaluations on how this system approach affects fish are not yet available.

6.3.4 General aspects of aquavoltaic systems

Regardless of the system design, when integrating PV modules with aquaculture systems, several aspects need to be considered. As already discussed, module efficiency decreases as temperature increases. In the aquavoltaic approach, a positive cooling effect can be achieved both by water and by increased wind speeds, especially at sea. Depending on the system design and the distance between modules and water, an additional cooling system may also be required. Trapani and Miller [53] demonstrated the efficiency improvement of a thin-film FPV array by 5% for 3 months due to the cooling effect. Generally, an efficiency increase of 10%−12% [56] or 10%−15% [60] has been reported for FPV systems. Another advantage related to the proximity to water arrives when considering soiling effects. First, particles are being washed from the module surface more regularly, and second, especially for offshore FPV systems, a spatial distance makes dust swirls possible. However, soiling of the surface of PV modules can also occur by other sources such as bird droppings (see Fig. 6.16) or biofouling [10]. Biofouling describes the settling of organisms such as algae or mussels on PV surfaces, by which not only the modules but also mounting

FIGURE 6.16 Bird droppings on floating photovoltaic modules [65].

systems and cables can be affected. Another problem occurring when PV modules are installed on saltwater bodies is the deposition of salt on the module surface that occurs over time. In a study by Gretkowska [66], a decrease of 13%−28% in power performance within 30 days was reported due to salt deposition on the surface of the PV module. Apart from salt residues on PV modules, salt contact and an increased moisture factor can lead to more severe corrosion. Although the modules are made of noncorrosive materials, after a certain time, oxidation may occur [6]. Furthermore, the mechanical stress would also be high because of tides, increased wind speeds, and waves, especially in stormy conditions. To compensate lateral forces, a stable anchoring is essential [57], while a flexible mounting of PV modules offers the advantage of floating with the waves, protecting the system against external forces [56]. With increased flexibility, microcracks can also be prevented [67]. On the other hand, flexible systems are more exposed to wind forces and thus rotations are more likely to occur. Depending on the rotation degree, the amount of received incoming radiation at the module surface may change [68].

Particularly for PV systems directly installed on or close to water surfaces, special working conditions during maintenance arise. Depending on the location, the maintenance of the system can be more difficult since the work has to be done out of boats or from the moving pontoons. However, since accessibility is difficult, it can be assumed that vandalism and theft are declined [6]. The compliance with the food−energy−water nexus represents the main target of every aquavoltaic approach. One of the most important synergistic effects resulting from coupling the PV systems with aquaculture is saving water. Especially in the light of climate change, when dry periods becoming more frequent, a reduction of evaporation is a major success [69]. Previous studies have found substantial decreases in evaporation rates due to module coverage. Santafé et al. [70] reported a 25% decrease in water loss in a reservoir in Spain. Sahu et al. [67] reported an average drop of 33% in water loss in natural lakes and ponds and a 50% reduction in artificial water storage reservoirs. Especially in aquaculture systems, where high water flow rates are observed, the prevention of water loss is a great benefit from both economic and ecological points of view. Pringle et al. [10] reported that a typical aquavoltaic system reduces water consumption by up to 85% due to first, preventing the water evaporation because of the coverage of water body with PV modules and second, due to providing the required power by PV modules. The latter is because, despite the generation of electricity by burning coal in fossil fuel-based power plants, PV systems consume significantly less amount of water to generate the same amount of electricity.

6.3.5   Economics and environmental impacts

The degree of economic incentive for aquavoltaic systems is multifactorial. One problem the aquaculture sector is facing is the constant increase in production costs. However, with more technical advanced farming systems, higher yields can be achieved, which is particularly beneficial when fish prices rise [3,71]. The implementation of aquavoltaic technology should be economically possible for large companies as well as small farmers. Considering only the figures of the FPV sector, the average capital expenditure (CAPEX) is still 25% higher than that of ground-mounted systems [72]. Especially the mooring system accounts for a significant share of total costs. The operating expenditure (OPEX) values of FPV systems are also increased due to more difficult maintenance conditions [60]. However, in recent years, there has been a strong trend towards reducing CAPEX and OPEX costs. The main reason is the massive expansion of large-scale FPV facilities mainly in Asia. The levelized cost of energy (LCOE) for large-scale FPV systems thus equalizes that of ground-mounted systems. A major reason for the massive expansion is the feed-in-tariff incentive and various governmental plans promoting the utilization of renewable energies. Due to decreasing investment costs for PV integration, it becomes continuously easier for small farmers to implement aquavoltaic approaches. Especially in periods of crop failures, the electricity yields can provide a second income source and ensure a secure and regular income for smaller farmers. Even in the scenario where the aquatic food production experiences a slight decline due to shading, the dual land use creates a higher output. Chateau et al. [37] performed a simple calculation to demonstrate the financial impact of an aquavoltaic system in Taiwan. In 2018, the feed-in tariff for FPV systems in the country was at US$0.155/kWh and the selling price of milkfish, cultivated in the aquavoltaic system was US$2.692/kg. At 60% coverage, which corresponded to a 452 kWp facility, the researchers were still able to observe more than 70% of fish yields. A simple calculation demonstrated a fivefold increase in economic value due to the dual land-use approach, even if slight losses in fish production have to be accepted. In a well-designed system, taking full advantage of synergistic effects, it can be expected to achieve an increase in fish production, at least during the summer months. Economic details strongly depend on space and time. Depending on climatic conditions, different LCOE values are obtained. Especially in tropical and arid climates, decreasing LCOE has been noted [6]. Due to the developing level of technology in aquaculture systems, the cost of electricity consumption is also increasing. In particular, the permanent operation of aerators affects the energy balance. For shrimp farms, up to 80% of the total energy demand can be due to the operation of aerators [44]. Tien et al. [44] compared the levelized costs of gasoline (US$1.3 /kWh) and PV electricity (US$0.6 /kWh) in Vietnam, indicating a positive economic effect by using PV modules for energy supply.

The extent of the negative impacts of aquavoltaic systems on the environment has not been sufficiently investigated. Concerning closed aquavoltaic systems, a direct negative environmental impact of PV installations is unlikely to occur. However, in open systems where modules are placed on a natural water

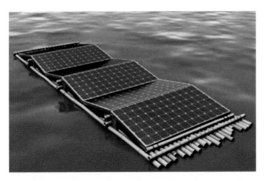

FIGURE 6.17 Floating substructure made of bamboo for floating photovoltaic modules [11].

surface, negative impacts cannot be excluded. Gorjian et al. [60] reported temporary disturbance factors which mainly occur during installation. Anchors are secured to the lake or seabed, and ropes must be fixed to the shore where sensitive coastal areas may be affected. Furthermore, the same study mentioned the inhibition of algae growth due to the shading effect. This is positive, especially in waters facing the risk of eutrophication. However, a strongly reduced light availability results in a competition for the light which may cause an undesirable shift in species. Sahu et al. [67] drew attention to the increased risk of contamination of water. If a module is severely damaged or improperly maintained, heavy metals and other environmentally harmful substances could contaminate the ecosystem and affect aquatic animals' production. Since there is a high material input with the installation of almost all PV systems, the literature suggests bamboo substructures as a feasible alternative for FPV systems. Bamboo is hollow and floats on the water surface and has high durability of over 10 years even in contact with water (see Fig. 6.17) [60]. Although the production of PV equipment emits CO_2, the substitution of fossil fuels can lead to an overall CO_2-saving effect. Especially considering the high energy input for aquaculture production, CO_2 saving becomes an important aspect [60]. A developed sustainable energy model for shrimp farms in the Mekong delta indicated that each ton of produced shrimp emits 10 tons of CO_2 only due to the use of energy from coal-fired power generation [44].

6.4 Conclusions and outlook

Floating PV approaches are gaining popularity in Europe, while Asia is leading the way in installing mega-systems. The technical feasibility of integrating PV modules with water areas has been demonstrated, but still, there is a lack of robust studies concerning fish farming. Although some best practice examples show that PV and aquaculture can be combined without significant drawbacks, there is little material on shading effects and fish behavior. Since the integration of PV into the aquaculture sector is still in its early stage, further studies are required, especially regarding environmental impacts and financial feasibility for small farmers.

With an increasing demand for food and feed, the consumption of aquatic animals and plants will continue to rise. To relieve pressure on the world's oceans, aquaculture offers a way to produce large quantities of food with a relatively small footprint on land and water surfaces. However, aquaculture facilities need to be operated sustainably, otherwise, the environmental impacts become immense. Due to the high energy consumption of intensive aquaculture facilities, it is essential to obtain energy from renewable sources. To avoid increasing land use, the approach of aquavoltaics offers a solution in the double use of land. Especially in countries with long periods of drought, the aquavoltaics concept offers many synergies. Besides the beneficial shading for animals in summer, the strong reduction of water loss through reduced evaporation rates is of particular interest. With an appropriate system approach, aquavoltaics can contribute to the sustainable use of water, fulfilling the concept of the food-water-energy nexus.

References

[1] Delgado CL, Wada N, Rosegrant MW, Meijer S, Ahmed M. Fish to 2020: supply and demand in changing global markets. Washington, DC; 2003.

[2] Kobayashi M, Msangi S, Batka M, Vannuccini S, Dey MM, Anderson JL. Aquaculture Econ & Manag 2015;19:282.

[3] FAO. The state of world fisheries and aquaculture 2020. Rome: FAO; 2020.

[4] FAO. FishStatJ-software for fishery and aquaculture statistical time series, <http://www.fao.org/fishery/statistics/software/fishstatj/en>; 2021.

[5] FAO. New food balances, <http://www.fao.org/faostat/en/#data/FBS/metadata>; 2020.

[6] World Bank Group, ESMAP, SERIS, Where Sun Meets Water. Floating solar market report. Washington, DC: World Bank; 2019.

[7] Ortmann B. An introduction to the rise in Floating-PV, the challenges and the opportunities it presents; 2020.

[8] Y. Ueda, T. Sakurai, S. Tatebe, A. Itoh, and K. Kurokawa. Performance analysis of PV systems on the water; eupvsec-proceedings; 2008, p. 2670−2673.

[9] Cazzaniga R, Rosa-Clot M. Sol Energy 2021;219:3.

[10] Pringle AM, Handler RM, Pearce JM. Renew Sustain Energy Rev 2017;80:572.

[11] Rosa-Clot M, Tina GM. In: Rosa-Clot M, Tina GM, editors. Submerged and floating photovoltaic systems. Elsevier; 2018, p. 185.

[12] Akuo Energy. Aquanergie application in La Réunion, Les Cedres, <https://www.akuoenergy.com/en/aquanergie>; 2021.

[13] Bostock J, McAndrew B, Richards R, Jauncey K, Telfer T, Lorenzen K, et al. Philosoph Trans Roy Soc Lond. Ser B, Biol Sci 2010;365:2897.

[14] Ottinger M, Clauss K, Kuenzer C. Ocean & Coast Manag 2016;119:244.

[15] Hargreaves JA. Aquacultural Eng 2006;34:344.

[16] Graphical objects were partly provided by the Integration and Application Network, University of Maryland, center for environmental science, integration and application network <https://ian.umces.edu/>; 2021.

[17] Boyd CE, McNevin AA. J World Aquac Soc 2021;52:6.
[18] Naylor RL, Hardy RW, Buschmann AH, Bush SR, Cao L, Klinger DH, et al. Nature 2021;591:551.
[19] Ahmed N, Thompson S, Glaser M. Environ Manag 2019;63:159.
[20] Páez-Osuna F. Environ Manag 2001;28:131.
[21] Verdegem MCJ, Bosma RH. Water Policy 2009;11:52.
[22] Google earth, satellite images of the Vietnamese Mekong Delta. imageID:2021; Google Maxar Technologies; 2021.
[23] Troell M, Naylor RL, Metian M, Beveridge M, Tyedmers PH, Folke C, et al. Proc Natl Acad Sci USA 2014;111:13257.
[24] Renaud FG, Le TTH, Lindener C, Guong VT, Sebesvari Z. Climatic Change 2015; 133:69.
[25] MacLeod MJ, Hasan MR, Robb DHF, Mamun-Ur-Rashid M. Sci Rep 2020;10:11679.
[26] Boyd CE, Clay JW; Evaluation of Belize Aquaculture, Ltd: A Superintensive Shrimp Aquaculture System. Report prepared under the World Bank, NACA, WWF and FAO Consortium Program on Shrimp Farming and the Environment; 2002.
[27] Sun W. Life cycle assessment of indoor recirculating shrimp aquaculture system; 2009.
[28] UN. World Population Prospects 2019, in Highlights. United Nations, Department of Economic and Social Affairs, Population Division; 2019.
[29] Pörtner HO, Peck MA. J fish Biol 2010;77:1745.
[30] Jutfelt F, Norin T, Åsheim ER, Rowsey LE, Andreassen AH, Morgan R, et al. Funct Ecol 2021;35:1397.
[31] Dahlke FT, Wohlrab S, Butzin M, Pörtner H-O. Sci (N York, NY) 2020;369:65.
[32] Handeland SO, Imsland AK, Stefansson SO. Aquaculture 2008;283:36.
[33] Department of Primary Industries and Regional Development, Government of Western Australia. Starting an aquaculture operation in Western Australia. Outdoor trout hatchery with shading nets <https://www.agric.wa.gov.au/small-landholders-western-australia/starting-aquaculture-operation-western-australia>; 2017.
[34] Chang P-H, Yeh S-L, Lee Y-C, Huang W-L, Li S-Y, Yeh C-Y, et al. J Fish Soc Taiwan 2020;95.
[35] Lin T-S, Kuo I-P, Der Yang S. J Fish Soc Taiwan 2020;105.
[36] Boeuf G, Le Bail P-Y. Aquaculture 1999;177:129.
[37] Château P-A, Wunderlich RF, Wang T-W, Lai H-T, Chen C-C, Chang F-J. Sci Total Environ 2019;687:654.
[38] Yeh N, Ding TJ, Yeh P. Renew Sustain Energy Rev 2015;51:55.
[39] Koike T, Matsuike K. Nippon Suisan Gakkaishi 1988;54:829.
[40] Ruchin AB. Fish Physiol Biochem 2004;30:175.
[41] Hama T, Miyazaki T, Ogawa Y, Iwakuma T, Takahashi M, Otsuki A, et al. Mar Biol 1983;73:31.
[42] Krummenauer D, Abreu PC, Poersch L, Reis PACP, Suita SM, dos Reis WG, et al. Aquaculture 2020;529:735635.
[43] Emerenciano MGC, Martínez-Córdova LR, Martínez-Porchas M, Miranda-Baeza A, Tutu H, editors. Water Quality. InTech; 2017.
[44] Tien NN, Matsuhashi R, Bich Chau VTT. Energy Procedia 2019;157:926.
[45] Boyd CE, Torrans EL, Tucker CS. J World Aquac Soc 2018;49:7.
[46] Li P, Gao X, Jiang J, Yang L, Li Y. Energies 2020;13:4822.
[47] Boyd CE, Tucker CS. Pond aquaculture water quality management. Boston: Kluwer Academic; 1998.

[48] Kumar G, Hegde S, Wise D, Mischke C, Dorr B. N. Am. J. Aquac 2020.
[49] Price IM, Nickum JG. Colonial Waterbirds 1995;33.
[50] Gareware-Wall Ropes Ltd. Sapphire ultracore predator nets <https://www.garwarefibres.com/product/sapphire-ultracore/>; 2021.
[51] Asche F., Khatun F. Aquaculture: issues and opportunities for sustainable production and trade, 5; 2006.
[52] Lee Y-G, Joo H-J, Yoon S-J. Sol Energy 2014;108:13.
[53] Trapani K, Millar DL. Renew Energy 2014;71:43.
[54] Ocean Sun. Floating PV demonstrator system for fish farm operator Leroy Seafood <https://oceansun.no/project/kyrholmen/>; 2021.
[55] Trapani K, Redón Santafé M. Prog Photovolt: Res Appl 2015;23:524.
[56] Cazzaniga R, Cicu M, Rosa-Clot M, Rosa-Clot P, Tina GM, Ventura C. Renew Sustain Energy Rev 2018;81:1730.
[57] Ferrer-Gisbert C, Ferrán-Gozálvez JJ, Redón-Santafé M, Ferrer-Gisbert P, Sánchez-Romero FJ, Torregrosa-Soler JB. Renew Energy 2013;60:63.
[58] Rosa-Clot M, Rosa-Clot P, Tina GM, Scandura PF. Renew Energy 2010;35:1862.
[59] Da Pimentel Silva GD, Branco DAC. Impact Assess Proj Appraisal 2018;36:390.
[60] Gorjian S, Sharon H, Ebadi H, Kant K, Scavo FB, Tina GM. J Clean Prod 2021;278:124285.
[61] Lima R, Boogaard F, Sazonov V. Assessing the Influence of Floating Constructions on Water Quality and Ecology. Paving the Waves Conference proceedings 2020; WCFS 2020.
[62] Pahri SDR, Mohamed AF, Samat A. Int J Life Cycle Assess 2015;20:1324.
[63] Fraunhofer ISE. SHRIMPS − solar-aquaculture habitats as resource-efficient and integrated multilayer production systems <https://www.ise.fraunhofer.de/en/research-projects/shrimps.html>; 2021.
[64] Rosa P. Clot in floating PV plants. Elsevier; 2020, p. 101.
[65] Gretkowska L. A study on the performance of solar photovoltaic modules exposed to salt water; 2018.
[66] Sahu A, Yadav N, Sudhakar K. Renew Sustain Energy Rev 2016;66:815.
[67] Choi Y-K. IJSEIA 2014;8:75.
[68] Campana PE, Wästhage L, Nookuea W, Tan Y, Yan J. Sol Energy 2019;177:782.
[69] Santafé MR, Ferrer Gisbert PS, Sánchez Romero FJ, Torregrosa Soler JB, Ferrán Gozálvez JJ, Ferrer Gisbert CM. J Clean Prod 2014;66:568.
[70] Ioakeimidis C, Polatidis H, Haralambopoulos D. Glob NEST J 2013;282.
[71] Oliveira-Pinto S, Stokkermans J. Energy Convers Manag 2020;211:112747.

Chapter 7

Solar heating and cooling applications in agriculture and food processing systems

Amir Vadiee

Division of Sustainable Environment and Construction, School of Business, Society and Engineering, Mälardalen University, Västerås, Sweden

7.1 Solar heating technologies

7.1.1 Solar water heating systems

Solar water heaters are among the most common solar thermal systems. Solar water heaters are mainly used in residential applications, although they can also be used in large-scale industrial applications [1]. In principle, solar water heaters are used anywhere hot water is required for purposes such as sanitary hot water; space heating or supplying preheated water for heating systems; and supplying heat required by absorption systems, air conditioning and cooling applications, and desalination plants [2]. Fig. 7.1A and B shows the principle of solar water heating systems integrated for direct heating and preheating applications that can be used in any industrial heat process sector, including the agricultural and food processing industries. In most modern solar water heaters, the operating fluid, which is the water−glycol mixture, flows in a closed cycle between the tank and the collector through a piping circuit [3,4]. The collector absorbs the solar energy and transfers it to the operating fluid, which is used in a heat process and, when cold, returns to the collector. This type of solar water heater is known as an indirect-circulation system. There is another type of system in which the heated water is used directly, called the direct-circulation or open-cycle solar water heater. In the open-cycle solar water heater, a water−glycol mixture cannot be used as the hot fluid needs to be directly used. Both types of solar water heaters can operate in passive mode (i.e., thermosiphon) or active mode using a pump to circulate the working fluid through the system [3,4].

Solar Energy Advancements in Agriculture and Food Production Systems.
DOI: https://doi.org/10.1016/B978-0-323-89866-9.00001-8

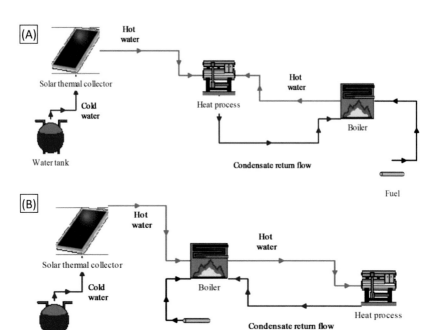

FIGURE 7.1   Simplified principles of solar thermal systems: (A) for direct heating processes, and (B) for preheating in indirect heating processes.

In direct-circulation systems, the pump power equals the total friction pressure losses inside the pipes plus the height difference between the collector and storage tank, while in indirect-circulation systems, the pump only needs to compensate for the pressure loss due to the internal friction of the pipes. The main challenge of solar heating systems is that they periodically produce more heat than is needed, especially in the summer [5]. The control units of the solar circuit are generally set to switch off the pump in active indirect solar heating systems when the accumulator tank is filled with heat, which is often defined by the temperature at the bottom of the tank reaching 95°C. When the solar collector pump is switched off, the solar collector will continue to heat up until it reaches its stagnation temperature, which is the temperature at which the heat losses from the solar collector equal the solar radiation absorbed by the solar collector [6].

However, since recently developed solar collectors are designed to have small heat losses, the stagnation temperature is high, often up to 200°C. At the same time, as the temperature increases, the pressure in the solar collector circuit also increases. Stagnation can occur even if the tank is not charged, for example, in the event of a power failure, when the circulating pump connected to the solar collector stops. The solar collector, pipes, and insulation must therefore be designed to withstand the stagnant temperature of the solar collector. Although active solar heating systems are more robust, they are also more complicated than passive solar heating systems due to the integrated control system [2,7,8].

Solar heating systems integrated with heat pumps have been attracting attention in recent years. In both residential and industrial applications, solar water heaters are mainly used in one of two ways. First, solar water heaters can supply hot water in summer when the heat pumps can be switched off. In addition to saving energy, this also increases the heat pump lifetime due to shorter operating time and fewer starts and stops [9]. The second alternative, which has become more relevant in recent years, is for the regeneration (i.e., recharging) of boreholes for geothermal heat pumps [9]. Using the large supply of solar heat in the summer, the borehole temperature can be increased, resulting in higher heat pump efficiency throughout the rest of the year. In summary, hot water is among the most essential elements of a wide range of heat processes in various industries including food processing, livestock, and agriculture [10−20]. In the food processing and livestock industries, hot water is used mainly for washing, while in the horticultural industry, hot water is used to supply part of the heating demand. The solar water heaters used in the food processing, livestock, and horticultural industries are usually energy-efficient and cost-effective with low operation and maintenance costs in comparison with other thermal energy systems [21].

7.1.2 Solar air heating systems

Solar air heating is one of the most robust and reliable methods to supply hot air for space heating applications in residential and commercial buildings as well as in industrial systems such as the agricultural, dairy, food processing, and livestock industries. The solar air dryer, which is a specific type of solar air heater with a theoretical principle similar to that of space heating, is widely used for drying agricultural products, timber, fruits, and herbs [22−26].[1] The various design configurations and modifications to improve energy performance have been reviewed in detail by Arunkumar et al. [27,28]. In both these studies, solar air heating systems are categorized as of three main types, that is, active, passive, and hybrid, while there are several configurations of each type based on the energy storage integration, design layout, and tracking system. In its simplest form, a solar air heater consists of a channel for airflow and a transparent covering layer. Despite their ease of use and low cost, solar air heaters have lower thermal efficiency than other solar thermal solar energy systems. An effective method to improve the performance of a solar air heater is to use a finned plate in the air

---

1. Solar dryers and their applications in agriculture are presented in Chapter 9, Solar Applications for Drying of Agricultural and Marine Products.

channel to enhance the heat transfer coefficient. The higher surface roughness provided by a finned plate increases the turbulence of the airflow, which increases the mixing of cooler and warmer air near the channel wall. This allows for better heat transfer, although the use of a finned plate also increases the pressure drop [27]. Chan et al. [29] compared the thermal performance of glazed perforated-plate and glazed flat-plate air collectors, two of the most common types of commercial solar air heaters. They concluded that the highest thermal efficiency of a perforated plate collector was over 20% higher than that of a flat-plate collector (FPC) [29].

Solar air heaters can usually supply 20%−30% of average annual space heating requirements under a wide range of climatic conditions [27]. Transpired solar collectors, a type of unglazed air collector in the form of a perforated wall-cladding system, are a commercialized technology to supply part of the space heating requirements of commercial buildings where sufficient wall area is available for their installation. In such systems, fans are used to draw air through thousands of tiny holes in the air collector where it is exposed to solar irradiation, and the solar heat gain is transferred directly to the air that is circulated through the building [27].

7.1.3 Solar cooling systems

The temperature has a major impact on almost all agricultural and food processing industries. In commercial greenhouses, for example, the indoor temperature should be precisely controlled to maintain optimum growing conditions, and overheating is one of the main problems in these facilities [30,31]. Cold storage is also widely used to keep agricultural products fresh before and during transportation.

Solar-driven cooling systems can be categorized as of three types: thermal, electric, and hybrid energy source systems. In a review, Allouhi et al. [32] studied all types of solar-driven cooling systems with both thermal and electric paths. They categorized solar-thermal-driven cooling systems as either sorption or thermomechanical refrigeration systems and categorized solar-electric-driven cooling systems as vapor-compression systems, Stirling systems, and thermoelectric systems. Their study introduced and discussed the third group of solar-driven cooling systems called hybrid systems combining different refrigeration technologies, such as adsorption−desiccant, adsorption−ejector, absorption−ejector refrigeration, and hybrid desiccant−conventional cooling systems. They also comparatively studied different types of solar-driven cooling systems considering their environmental impacts, thermo-economic indicators, and technological barriers [32].

In solar-thermal-driven cooling systems, the refrigerant is heated directly by the solar radiation absorbed in the solar collectors, while in solar-electrical-driven cooling systems, the generated electricity runs the cooling system. All solar-thermal-driven cooling systems start with a solar collector that provides the energy required by the integrated cooling systems. Compared with concentrating solar collectors, FPCs can use both direct and diffuse solar radiation, generating heat even on cloudy days. FPCs are efficient at collector temperatures of 50°C above the ambient temperature, while evacuated tube collectors (ETCs) are more efficient at higher temperatures. However, the maximum operating temperatures in both FPCs and ETCs are similar at about 200°C [32]. The performance of solar-thermal-driven cooling systems is greatly influenced by the available solar radiation and the characteristics of the collectors. However, the typical output range of solar-thermal-driven cooling systems is about $130-170$ W/m^2 at solar radiation of 1000 W/m^2 [32-38]. The variation in the efficiency of solar collectors implemented in refrigeration systems depends on their specific properties and operating conditions (i.e., optical efficiency, heat loss, and angle dependence effects). In addition, the performance varies with the working temperature inside and outside the solar collector. The return temperature to the solar collector varies and depends on the solar radiation, the number and properties of solar panels, and the accumulator's capacity and heat load. High heat consumption and a larger accumulator tank on the heating side result in a lower return temperature to the solar panels, which increases their efficiency [33].

Solar thermal collectors installed in cooling systems are considered the heat-generating units used to meet the heating requirements of integrated facilities and systems during the months when the cooling demand is negligible. This is an important parameter when dimensioning a plant so that the solar heating system is operated as energy efficient as possible, using heat for other purposes, such as domestic hot water and space heating during spring and autumn [32].

Sorption cooling systems comprise absorption, adsorption, and desiccant cooling systems—the most common types of solar-thermal-driven cooling systems [32]. In general, the performance of absorption systems is similar to that of conventional compression refrigeration systems, although electricity as the main energy source in compression systems is replaced with a heat source. The working fluid in an absorption system is usually water mixed with ammonia or lithium bromide. This cooling cycle is properly understood by considering its four main components: absorber, generator, condenser, and evaporator. Fig. 7.2 shows a schematic of a simplified sorption solar cooling system. The working principle of a solar absorption cooling system is that the collectors absorb the solar energy and heat the vacuum tubes to a maximum temperature of 170°C (this temperature can vary depending on the type of collector) [39]. This heat is transferred to the working fluid and then to the absorption chiller generator. In the generator, the heat leads to the separation of water from the working fluid. The working fluid then enters the condenser where it is liquefied before entering the evaporator. The evaporator is a heat exchanger in which the working fluid is cooled, delivering

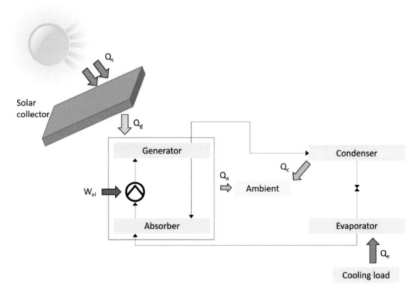

FIGURE 7.2 Working principle of a typical solar sorption cooling system [32].

cooling for any needed purposes (e.g., fan coils in commercial greenhouses to meet cooling needs). The working fluid is then absorbed in the absorber and the distilled working fluid is pumped to the generator to be reheated and separated. A gas or electric heater is also used as an auxiliary energy source for the sorption system.

Sorption cooling systems can be divided into two general groups based on the working fluid type, that is, ammonia and lithium bromide chillers, the first type being used usually for cooling below 0°C and the second for cooling above 5°C [32]. Absorption cooling systems can also be divided into single-effect and double-effect sorption systems, differing in the number of generators. Single-effect solar absorption systems operate with a limited coefficient of performance (COP) of 0.73−0.79; the COP rises to 1.22−1.42 in double-effect and to 1.62−1.90 in triple-effect absorption solar cooling systems [40]. One aspect to consider when the absorbing refrigeration system is powered by hot water from a solar collector is that a high return temperature from the refrigerator, which often works with a temperature difference of only 10°C, adversely affects the solar panel performance.

A solar-driven desiccant cooling system consists of both dehumidification and evaporative cooling processes. A desiccant cooling system has three main components, that is, desiccator, evaporative pad, and regenerator, and is an effective cooling system specifically for a commercial greenhouse requiring cool (18°C−20°C) and humid (RH 70%−85%) air [41]. In a simplified desiccant cooling system, fresh outside air is drawn into the desiccator where it is dehumidified by the desiccant, which absorbs the water content of the fresh air making it less humid. The lower moisture content of the air enhances the cooling effect of the evaporative cooling cycle, which is often used to provide cool humid air for greenhouses. To reuse the desiccant in the cycle, a solar collector provides the heat required in the regenerator to condense the water from the desiccant [42]. Fig. 7.3 presents the schematic of a solar-driven desiccant cooling system applied in a greenhouse. Lychnos and Davies [42] assessed a conventional evaporative cooling system and a solar-driven desiccant cooling system in a case study of a greenhouse in a hot climate, demonstrating that better thermal performance is achieved year-round using a solar-driven desiccant cooling system.

A relatively small amount of electricity is required in a solar sorption cooling system to operate electrically driven systems such as pumps. The energy efficiency ratio (kW_{th}/kW_{el}), defined as the ratio between the total thermal power capacity and the electric power demand in the system, can range from 20 to 40 kW_{th}/kW_{el} depending on the climatic conditions and integrated auxiliary energy systems [34]. Employing solar sorption-driven cooling systems reduces the electricity required for space cooling by 80% compared with conventional air conditioning systems [34]. The second way to use a solar thermal collector in the cooling system is in thermo-mechanical refrigeration cycles (i.e., the Rankine, Stirling, and ejector cycles). Through this concept, solar energy is used to generate part or all of the heating required by the heat engine to produce the electrical energy needed to run the conventional cooling cycle. Fig. 7.4 shows the principle of the thermo-mechanical refrigeration system.

Solar-electric-driven cooling systems are similar to thermo-mechanical cooling systems as both types use solar energy to directly or indirectly

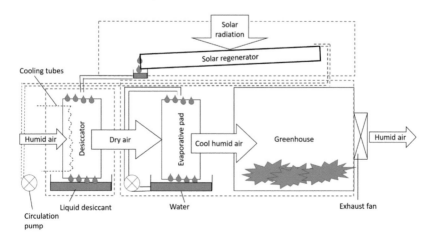

FIGURE 7.3 Schematic of a solar-driven desiccant cooling system for a greenhouse [42].

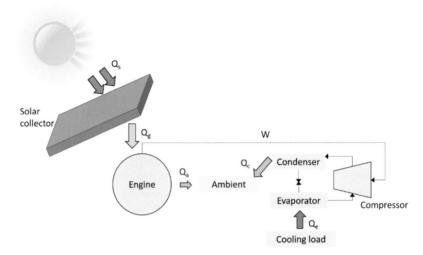

FIGURE 7.4 Schematic of a solar thermo-mechanical cooling system [32].

generate electrical energy to drive a conventional compressive cooling system. Solar electric refrigeration consists of series of photovoltaic (PV) modules that supply the required electricity for the refrigeration system. One specific type of solar electric cooling system is called the solar blind system, in which solar panels serve as a curtain blocking solar radiation from entering the greenhouse, reducing the cooling demand while supplying the electricity required by a conventional cooling system. A solar blind system operates in the shading position mainly when the light intensity and temperature inside the greenhouse are sufficiently high and simultaneously there is high availability of solar radiation, leading to high electricity and heat generation through hybrid photovoltaic−thermal (PVT) modules. The electricity generated by a solar blind system can be utilized directly in the integrated cooling system or be used as the energy source to run a heat pump for either heating or cooling purposes. Like other solar energy systems, the solar blind system must be integrated with energy storage systems to manage the mismatch between the generated electricity and heat and the total energy demand [31,43−45].

In addition to using active solar cooling systems, on sunny days, passive solar cooling systems can also be used. Passive cooling systems are mostly used in space cooling and ventilation applications incorporated into specific architectural features of buildings. Currently, the largest solar cooling system in operation has been installed in Arizona, USA since 2014, where 4865 m^2 of roof area is covered by solar thermal collectors to obtain a total of 3.4 MW$_{th}$ of heating capacity that supplies the heating required by a single-effect lithium bromide absorption cooling system with a total cooling capacity of 1.75 MW [46]. However, in other projects in Austria, Spain, Italy, Jordan, and Singapore, various types of solar collectors, such as ETCs or Fresnel reflectors, are used to provide the heat for absorption cooling systems commissioned in the food processing sector. A gigantic solar thermal cooling system is close to being in operation in Singapore where it may become the world's largest solar-powered cooling system, covering 6500 m^2 with solar collectors to supply 1.8 MW of cooling power [46].

7.2 Applications of solar thermal technologies in agriculture and food processing systems

The total operational solar thermal capacity worldwide at the end of 2019 was about 480 GW$_{th}$, corresponding to 390 TWh of annual energy yield and mitigating 135 Mtons of carbon dioxide (CO_2) emissions. The growth rate of the solar thermal market has not been as fast as that of PV systems, ranging from 2% (India) to 170% (Denmark) in 2019. The greatest potential for solar thermal market growth is in district heating systems and industrial thermal processes. By the end of 2019, at least 1 million m^2 of solar thermal collectors corresponding to 700 MW$_{th}$ were in operation in over 800 solar commercial thermal processing systems [46].

Their widely ranging operating temperatures and heating requirements make industrial thermal processes a promising market for solar thermal applications. Depending on the required operating temperatures of thermal processes, different types of solar thermal collectors can be used. At operating temperatures below 100°C, FPCs and ETCs can be considered, while at higher operating temperatures up to 400°C, concentrating collectors such as Fresnel reflectors and parabolic trough collectors (PTCs) are usually employed [47].

Solar energy technology has been used in agriculture since time immemorial when solar dryers were used to dry agricultural products for long-term storage [48]. The main barrier to solar energy system implementation in the agricultural sector is upfront costs. Therefore, a detailed and precise life-cycle cost analysis is crucial when proposing a cost-optimized system. Solar energy systems are more economically feasible in large farms, resulting in costs lower than those of conventional fossil fuel-based energy systems.

One of the biggest advantages of solar energy systems integrated with energy storage is that they can be used in rural areas where no electricity or thermal grid networks are available, providing independent energy systems with a clean energy source. Furthermore, the use of solar energy systems is an effective and strategic solution in areas suffering from water scarcity. Concerning environmental aspects, the agricultural sector is greatly affected by global warming, while solar-based technologies are the most efficient solution to mitigate climate change. Integrating solar energy with the agricultural industry, therefore, has a positive environmental impact and represents an effective way to prevent global warming.

Given current energy and environmental concerns and sustainable development strategies, the agricultural industry, like other industries, is trying to adopt solar energy technologies. By introducing high-efficiency, low-cost solar energy systems with a wide range of operating temperatures, more applications will become possible in the agricultural and food processing sectors. Currently, there are three main applications in which active and passive solar thermal energy systems can be implemented in the agricultural industry:

- drying agricultural products,
- heating and cooling greenhouses, livestock shelters, and other buildings, and
- food processing applications including dehydration and water heating.

By presenting some case studies, the following section of this chapter discusses the main applications of solar heating and cooling systems in the agricultural and food processing industries.

7.2.1 Solar air heaters for drying and dehydration processes

Drying or dehydration is a traditional method of preserving food for storage by removing excess moisture from agricultural products. Decreased humidity reduces or eliminates microbial activity and chemical interactions in fresh products. Furthermore, the reduced volume and weight of dried food means easier and cheaper transportation and storage in the food industry. Throughout human history, solar energy has been the main energy source for the dehydration and drying of agricultural products for long-term storage.

Solar dryers used in the food processing industry can be classified based on the method of air movement (i.e., natural or forced convection) and solar radiation utilization (i.e., direct or indirect). In a direct passive solar dryer, the products are exposed to direct solar radiation while the air moves through an insulated box based on the buoyancy force or pressure difference. The direct passive solar dryer is the simplest and cheapest type of solar dryer [49].

In an indirect solar dryer, the air is first heated in a solar air collector and then passes over the drying product, with the heat and mass being transferred either by natural convection through a ventilator or chimney, or being forced by convection using a fan.

There is a third type, the hybrid solar dryer, in which both direct and indirect heating systems are employed. In such dryers, the air is heated by passing through an absorber plate and then flows under the product trays. Additionally, the product tray is placed in a tilted glazed box that is exposed

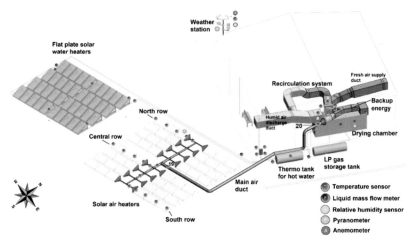

FIGURE 7.5 An example of a hybrid solar drying system [50].

to direct solar radiation. In hybrid solar dryers, an auxiliary heating system such as an electrical heater, biomass boiler, or fossil-fuel-based furnace integrated with sensible or latent thermal energy storage maintains the desired thermal performance of the system regardless of climatic conditions [50]. Hybrid solar dryers are usually used on an industrial scale where continuous operation and a secure energy supply are crucial.

Fig. 7.5 shows an example of a hybrid solar dryer system in which both solar air and water heating systems are indirectly used to supply the heating required for the dehydration and drying processes, while a backup energy system is incorporated to increase system resilience [50].[2]

7.2.2 Solar water heater in agriculture and food processing systems

Through each level of the food value chain, including the production of horticultural and livestock infrastructure, crop and food production and processing, transportation, marketing, and consumption, both electrical and thermal energy are utilized. However, crop and food processing contribute the most to global energy consumption along agricultural value chains. In particular, countries with rapidly growing agro-industry sectors have higher energy consumption in the food processing sector. Of food industry operations, sugar processing followed by dairy and meat processing are considered the most intensive water and energy users due to the highly controlled hygienic standards [51]. In this sector, most thermal energy is used for washing,

---

2. Solar drying systems and their applications in the agriculture sector are presented in Chapter 9, Solar Applications for Drying of Agricultural and Marine Products.

TABLE 7.1 Operating temperatures and specific thermal energy consumption levels of different heat processes on dairy farms [56–58].

Process	Temperature range (°C)		Specific thermal energy consumption (kW$_{th}$ ton^{-1} of final product)
Yogurt culturing	40–45	Low-temperature processes <80°C	17–24
Bottle washing	60		28–118
Pasteurization	70		12
Cleaning-in-place (CIP)	70–80		56–168
Bottle sterilization	100–140	High-temperature processes >100°C	92–140
UHT treatment (milk sterilization)			118
Multi-stage evaporation			7–168
Spray drying			123–179

pasteurizing, boiling, extraction, pureeing, brewing, baking, and cooking procedures [52–55]. Table 7.1 shows the usual temperature ranges of the hot water and steam required for various processes in dairy farms. As the operating temperatures required for these processes do not usually exceed 140°C, both FPCs and ETCs can be utilized for these applications.

The most common types of solar water heaters used in cleaning and washing applications are active direct (i.e., open loop) circulation systems, as the processed water contains pollutants and the temperature drops mainly within the cleaning/washing process [59].

The Centre for Renewable Energy Sources and Saving, a Greek organization, summarized several examples of solar water heating systems implemented in the dairy industry for different processes [58]. The first cited example was solar water heating used for the CIP washing process and the preheating system of the steam boilers by Mevgal S.A. in Thessaloniki, Greece. The hot water consumption of this dairy system is 120–150 m^3 day^{-1}; the required temperature is 20°C–80°C for the washing process but does not exceed 130°C in any of the other processes. Regarding the washing process, 403.2 m^2 of selective FPC together with a 5000-L storage tank meets the hot water requirements. A steam boiler as an auxiliary heating system was also considered in this system. Furthermore, 216 m^2 of FPC together with 108 m^2 of compound parabolic concentrator (CPC) collectors integrated with a 5000-L storage tank were proposed to be used for preheating the steam boiler inlet water [58].

In Alpino S.A., another dairy industry located in Thessaloniki, 576 m^2 of FPC was installed on the building's roof to be used as a preheating system for the steam boiler [58]. Another successful example of solar energy system implementation in the Greek dairy industry is located in Korinthos, where a 170-m^2 FPC is installed to support yogurt culturing. As this process requires a low temperature of 40°C−45°C, the excess collected heat is used for preheating purposes for the steam boiler [58]. There are other reported examples of solar water heating system applications in dairy factories in South Africa. For example, the Fairview Cheese Farm in Paarl meets its heating requirements with a boiler having a storage capacity of 4000 L and a 90-m^2, etc. [60]. Solar water heating systems are also available for use in small-scale applications. For example, Tally Ho Farm in North Carolina, USA installed a 30-m^2 FPC and a 1135-L water storage tank in 2007 with an electric heater as an auxiliary heating system; the total investment was paid back within three years. According to some commercial reports from the solar companies providing the technology, the payback period for such small-scale solar water heaters in agricultural and food systems is three to ten years, whereas larger systems have shorter payback times [61]. Liu et al. [62] studied the proposed application of a solar water heating system with an integrated heat pump in a pig farm in Yunlin County, Taiwan for evisceration, sanitation during processing, and daily cleanup. The proposed case consists of a 239.3-m^2 glazed FPC integrated with a total of 24,000 L of storage tanks. The results indicated that the proposed solar water heating system could supply up to 70% of the total required energy. The expected payback period for this system was calculated to be under 6.5 years [62].

Hess and Dinter [63] assessed a proposed solar process heating system for a sugar mill in South Africa, considering the application of different solar thermal collectors for various systems including live and exhaust steam, feedwater, and clear juice systems. Table 7.2 shows the estimated system performance of these heat processes.

As shown in Table 7.2, the PTC was considered suitable to produce live steam for use in the turbine and the preheating system in connection with exhaust steam and the feed water tank, while the etc was used to preheat the juice. The schematics of these applications in the proposed system are presented in Figs. 7.6−7.9 [63]. Furthermore, Hess and Dinter concluded in the same study that none of the proposed systems was economically viable given current market conditions. The levelized cost of heat (LCOH) and internal rate of return (IRR) are used as indicators in economic assessments. Regarding the LCOH, different scenarios were investigated and the results

TABLE 7.2 Solar energy system performance in a case study of a sugar mill in South Africa [63].

Heat process	Heating demand (MWth)	Max. field area (m²)	Required max. temperature (°C)	Collector type	Annual solar fraction (%)
Live steam	74.4	53,168	360	PTC	17
Exhaust steam	62.5	89,238	130	PTC/etc	18–21
Feed water	9.0	12,880	200	PTC	18–34
Clear juice	4.7	6739	114	etc	24

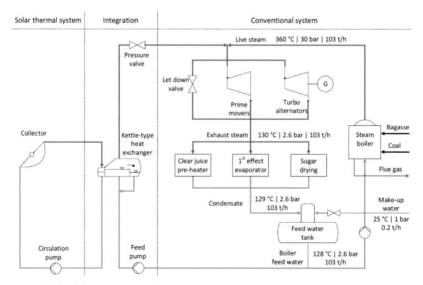

FIGURE 7.6 Schematic layout of a proposed integrated solar thermal system to produce live steam in a sugar mill [63].

were compared with heat generation using coal. The levelized cost of energy (LCOE) is another economic measure that gauges the feasibility of integrated renewable energy projects; however, LCOH was used in this study to capture the energy measured as heat and not electricity. The IRR is another indicator of the economic viability of projects in which conventional heating is replaced with integrated solar systems.

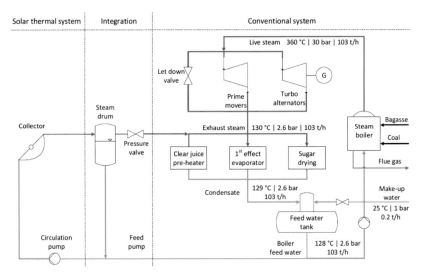

FIGURE 7.7 Schematic layout of a proposed solar thermal system integrated with exhaust steam in a sugar mill [63].

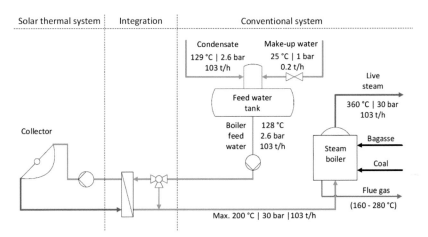

FIGURE 7.8 Schematic layout of a proposed solar thermal system integrated with a preheating system in a sugar mill [63].

Table 7.3 presents the LCOH and IRR of the above-mentioned systems in optimal operating conditions [63]. The LCOH values of conventional heat processes based on coal are 3.36 and 4.06 for exhaust and live steam, respectively. The cost of capital in the sugar milling industry is 10%−15% and to ensure economic viability, the corresponding IRR should exceed this cost of capital.

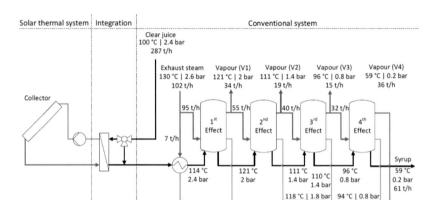

FIGURE 7.9 Schematic layout of a proposed integrated solar thermal system in a sugar mill to meet the heating requirements for the clear juice process [63].

TABLE 7.3 Economic assessment of solar thermal energy applications in sugar mill systems [63].

Heat process	Collector type	System costs (Euro/m^2)	Capital investment (EUR millions)	LCOH (EUR/ kWh)	IRR (%)
Live steam	PTC	378	20.10	4.57	4.6
Exhaust steam	etc	188	16.78	4.55	3.3
Feed water	PTC	378	8.46	4.57	2.6
Clear juice	etc	188	1.06	3.97	4.6

7.2.3 Solar thermal systems for cooking processes

Solar thermal energy systems mainly employing parabolic collectors can be used for cooking and pasteurizing foods and liquids. A solar cooker is a device that allows food to be cooked using solar energy to save time and fuel. Such systems can also be used to pasteurize water or sterilize various devices. Pasteurization and sterilization together with ultra-high-temperature treatments are considered the main food preservation methods involving heating. The temperatures required for these applications are 70°C–150°C depending on the food to be preserved, temperatures that can be supplied easily by different solar thermal energy systems [64].

In principle, all solar cooking systems concentrate solar radiation in a specific space using either reflectors or glazed enclosed space, and the solar radiation is turned into heat for utilization. The operating temperatures of solar cooking, pasteurization, and sterilization systems are affected by different practical parameters, although under standard conditions, the operating temperature can typically reach 150°C. Solar cooking, pasteurization, and sterilization systems are categorized based on the storage system integration, direct or indirect heating process, collector type, working fluid, etc. [64−67]. Fig. 7.10 shows a typical detailed classification of solar cooking systems and Table 7.4 summarizes a literature survey of different types of solar cooking systems.

More detailed reviews of recent advances in solar cooking systems, including parametric assessments of their thermal performance, have been presented by Aramesh et al. [64,86]. Aramesh et al. reviewed almost 40 case studies and compared the reported overall efficiencies of solar cookers, with the highest reported efficiency being 77% for a parabolic solar cooker and the lowest 10.2% for a box cooker. An economic assessment was also performed in the same study to determine the estimated equivalent costs for 2018, which were EUR 42-385. Furthermore, considering the fuel prices in countries such as India, Pakistan, Mexico, Tanzania, and Indonesia, the simple payback period reportedly ranged from six months to 18 years. However,

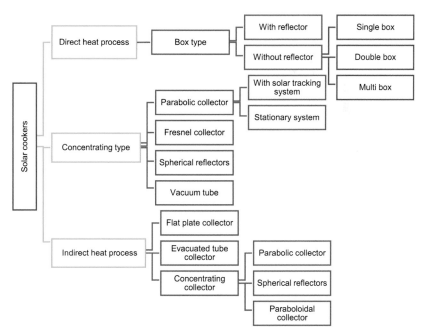

FIGURE 7.10 Broad classification of solar cookers [64−66].

TABLE 7.4 Selected studies of solar cooking systems.

Type of solar cooker	Type of study	Main features	Technical assessment	Economic assessment	Refs.
Solar box	Experimental	• Single-family solar cooker • Small size • Socially acceptable design • Lightweight • High affordability	• Max. temp. 144°C • Cooking power 103.5 W • Heat loss 1.474 W/°C	-	Mahavar et al. [68]
Solar box	• Theoretical • Experimental • Comparative study with and without internal reflector in both dry and boiling modes	• Higher performance with internal reflector • Higher standard stagnation temperature and cooking power with internal reflector	• Max. temp. 129.3°C and 106.2°C with and without internal reflector, respectively • Max. efficiency below 20%	-	Kahsay et al. [69]
Truncated pyramid solar box cooker	Experimental	• Utilized in farming household	• Max. temp. 140°C • Max. efficiency 54% • Heat-loss coefficient 3.75–5.75 W/cm²	-	Kumar et al. [70]
Solar box integrated with a parabolic mirror	Experimental Theoretical	• Improvement in both thermal and radiative performance of the solar cooker	• Overall efficiency of the system increased by 35.5%	-	Zamani et al. [71]
Solar box equipped with a fixed asymmetric compound parabolic concentrator (CPC)	Experimental Theoretical	• Improvement in maximum stagnation temperature • Greater economic viability on large scale for commercial applications	• Max. temp. 166°C • Cooking power 78.9 W	Overall fabrication cost estimated at about EUR 63	Harmim et al. [72,73]
Solar box with a trapezoidal air duct	Experimental Theoretical	• Enabled the cooker to operate even in poor irradiance	• Max. temp. 165.5°C • Cooking power 60.2 W • Overall efficiency 45.11%	-	Saxena and Agarwal [74]

Type	Method	Features	Results	Cost/Payback	Reference
Solar box	Experimental Theoretical	• Animal feed cooker • Short payback period • Save a lot of firewood and agricultural waste, producing corresponding environmental benefits	• Max. temp. 146°C • Overall efficiency 28.4%	• Overall cost EUR 670 • Payback period 6 months to 3.5 years, depending on initial fuel costs	Nahar et al. [75–77]
Parabolic solar cooker	Experimental	• Low cost • Simple design	• Max. efficiency 15.7% • Max. energy output 78.1 W	-	Öztürk [78]
Parabolic solar cooker	Experimental	• Two automatic tracking axes • Only for liquids	• Max. temp. 96°C	-	Al-Soud et al. [79]
Truncated pyramid parabolic solar cooker	Experimental	• Improved radiation reflection inside the cooker • Utilizable as a solar dryer as well	• Max. temp. 140°C	-	Kumar et al. [70]
Seven types of solar cookers	One-year comparative field test	• Three study areas in South Africa • Fuel savings of 38%	-	Pay-back periods range from 8 months to 5 years (average 2 years)	Biermann et al. [80]
Paraboloid solar cooker	Experimental	• Minimal heat loss due to wind shield provided	• Max. temp. 326°C • Max. efficiency 26%	-	Kalbande et al. [81]
Trapezoidal prism solar panel cookers	Experimental	• Cooker direction could be manually changed	• Max. temp. 267°C	-	Edmonds [82]

(Continued)

TABLE 7.4 (Continued)

Type of solar cooker	Type of study	Main features	Technical assessment	Economic assessment	Refs.
Indirect parabolic dish cookers	Comparative study	• Multifunctional application in both residential and commercial scales • Study of a domestic-size cooker (DSC) and a community-size cooker (CSC) • Vessel capacity 5 and 50 L for the DSC and CSC, respectively	• Max. power output 288.17 W (DSC) and 1026.66 W (CSC) • Max. efficiency 43.56% (DSC) and 21.97% (CSC)		Kaushik and Gupta [83]
Vacuum tube and linear Fresnel collectors	Experimental Theoretical	Study of effects of ambient temperature and water load on different system parameters	• Max. temp. 250°C • Max. efficiency 30% • Power 208 Wm^{-2}	-	Farooqui [84,85]

the most common types of solar cookers have a simple payback period of one to three years [86].

The environmental impact of replacing combustible fuels with solar energy in cooking systems has been assessed by Panwar et al. [87], who found that solar cookers could have a considerable climate mitigation impact, reducing emissions by $101-425$ kg CO_2 per year depending on the initial fuel type.

The world's largest solar cooking system is in operation in India, funded by the Indian government with a total investment of about €251,000. This system generates 3500 kg of steam per day and is designed to operate independently of the domestic electricity grid. This system consists of 73 solar dishes distributed over 16 m^2 that save 100,000 kg of LPG, costing about €37,700, and reducing CO_2 emissions by 366 kg $year^{-1}$ [88].

7.2.4 Solar heating and cooling in commercial greenhouses

Greenhouse horticulture is among the most important agricultural sectors with the highest energy consumption [89]. Operating costs, including electricity and fuel costs, have a direct impact on the final product price. Using renewable energy technologies to minimize energy cost is one of the most important issues in sustainable greenhouse operation [41]. A typical commercial greenhouse usually needs energy equivalent to $100-300$ kWh/m^2 per year for lighting, cooling, heating, and ventilation applications [31]. Solar thermal energy systems can be widely used in modern commercial greenhouses, and since the operating temperatures are mainly below 60°C, all types of solar thermal collectors can be employed for this application [47]. Different types of active solar thermal energy systems have been implemented and assessed in commercial greenhouses on both small and large scales [90]. Solar water heating integrated with root heating or underfloor heat distribution systems, solar air heating coupled with the air handling system, solar-powered air conditioning, and ventilation systems, solar-driven electric heating, and solar-driven heat pumps is the most common integrated solar systems in commercial greenhouses [91].

Besides active solar thermal systems, passive solutions are also an effective way to utilize solar energy in the agriculture industry. Closed and semiclosed greenhouses are examples of passive solutions [89]. The closed greenhouse is an innovative passive system in which the greenhouse acts as a big solar collector. In this concept, the excess heat collected inside the greenhouse is stored in either a short-term storage system (i.e., a water storage or Phase change material storage tank) or a seasonal storage system (e.g., borehole thermal energy storage). Therefore, the excess stored heat will ultimately be used to meet the heating requirements based on the actual demand. In conventional greenhouses, the excess heat (i.e., cooling demand) is managed through natural (i.e., windows) and forced ventilation systems, while external fuel (e.g., natural gas, oil, and biofuel) is used in conventional heating systems to meet the heating requirements [89]. Closed greenhouse systems are conceptually independent of external fuel for heating purposes, which leads to significant cost savings of over 20% of the product cost. The closed greenhouse concept is also considered a promising solution in cold climatic conditions where the windows need to be open for ventilation during the winter to control excess humidity while the heating system needs to operate simultaneously to compensate for the corresponding heat losses. Analytical assessment shows that a closed greenhouse can supply all the thermal energy needed to operate the greenhouse independently [30,89,91,92]. In an analytical study performed by Vadiee et al., the total annual heating requirements of a closed greenhouse were calculated to be 57 kWh/m^2, compared with 320 kWh/m^2 in a conventional greenhouse, while the total cooling requirements (to be stored for later use as a heat source) of the closed greenhouse were about 165 kWh/m^2 [41]. Therefore, the ideal closed greenhouse has the potential to reduce the energy requirements and corresponding CO_2 emissions of a commercial greenhouse by up to 80% and 75%, respectively. Vadiee and Martin [30] reviewed experimental and theoretical studies of closed greenhouses and corresponding assessments; their findings regarding system specifications and key outcomes are summarized in Table 7.5.

In a conventional greenhouse, there are several ways to meet cooling needs, for example, through ventilation, shading, and conventional thermodynamic cooling methods, such as evaporative, absorption, and compression cooling systems. Solar energy systems can be integrated with each of the aforementioned cooling methods. A solar absorption cooling application in a 300-m^2 experimental greenhouse at the University of Bari in Italy was studied by Vox et al. A chiller with lithium bromide—water as the refrigerant with a total cooling capacity of 35 kW and electrical consumption of 4 kW was used to supply 7°C cooling water; an 80-m^2 ETC was used to supply the heating required for the absorption cooling system [101].

Solar PV systems can be directly implemented in conventional ventilation systems to meet the electrical energy requirements. However, passive solar systems such as solar blinds, solar chimneys, and solar walls can be also used together with conventional ventilation systems to enhance air circulation and reduce cooling requirements [102,103]. The solar blind system, which consists of a series of smart-controlled PVT modules installed on the glazing of the greenhouse, is used as a shading device that also produces electricity, partially supplying the energy needed for cooling [31,43−45]. The concept of the solar blind system, proposed by Vadiee et al., has been assessed in several studies [31,44,45,89]. The overall assessment shows that a solar blind system can meet 66%−99% of cooling requirements, depending on the set point temperature, while the corresponding heating requirements can be reduced by 80% in a blind-equipped commercial greenhouse compared with a conventional greenhouse [31,44,45,89].

TABLE 7.5 Summary of system specifications and key outcomes for selected solar-heated greenhouses [30].

Name of greenhouse/project	System specifications	Key outcomes	Ref.
Aircokas	Experimental study of 700-ha semiclosed greenhouse with heat pump and active cooling system	• 30%–45% CO_2 emission reduction • Investment cost covered by 20% higher production	[93]
Zero-fossil-energy greenhouse	Experimental study of 1-ha semiclosed greenhouse with CHP, boiler, heat pump, and active cooling system	• 2% higher production • 10% higher cost than the conventional greenhouse	[94]
Themato	Experimental study of 1.4-ha closed greenhouse with CHP, heat pump, Air handling unit (AHU), and active cooling system	• 17% higher production • 30% lower fossil-fuel use • Cost competitive due to higher total yield	[95]
ECOFYS	Experimental study of 1400-m^2 partly closed greenhouse with boiler, CHP, heat pump, absorption chiller, and free cooling system	• 50% greater energy efficiency • 20% higher production • 50% less water consumption • Total benefit depends on energy demand, energy price, energy saving, production, and closed part fraction	[96]
Energy-producing greenhouse	Theoretical study of 1-ha partly closed greenhouse with natural gas, geothermal heat pump, active cooling, and free cooling	• 12% higher production	[97]
Watergy	Theoretical study of 1-ha closed greenhouse with natural convection established by a tower and a shading system for cooling purposes	• 75% lower water consumption • Minimum electrical cost	[98]
GESKAS PSKW & PCH	Theoretical study of 160- and 240-m^2 closed greenhouses with cogeneration, boiler, heating coil, AHU, and cooling coil	• 6% higher production and 34% lower primary energy consumption	[99]
ENEA	Theoretical study of 9000-ha closed greenhouse with a boiler, heater, AHU, and active cooling	• Reduced costs mainly for electricity for forced ventilation and active cooling	[100]

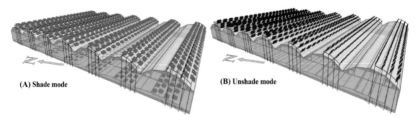

FIGURE 7.11 Solar blind system integrated with a commercial greenhouse in Shiraz: (A) Shade mode and (B) unshade mode [31].

Furthermore, the electricity generated by solar blinds in this concept is estimated to be 80 kwh/m^2 annually. The results indicate that a solar blind system in a closed greenhouse could reduce CO_2 emissions from 27.8 to 18.5 tonnes per 1000 m^2 [41,45].

The solar blind concept has also been studied for a commercial rose greenhouse located in Shiraz (Fig. 7.11) with a total cultivation area of 4081 m^2. The results indicate that by covering 19.2% of the greenhouse roof with a solar blind system, 42.7 kWh/m^2 of electricity would be generated annually while natural gas consumption, electricity demand, and CO_2 emission would be reduced by 3.57%, 45.5%, and 30.56 kg/m^2, respectively [31].

Table 7.6 presents a list of selected examples of PV-integrated cooling systems for experimental and commercial greenhouses.

A detailed description of solar-powered greenhouses is provided in Chapter 3, Advances in solar greenhouse systems for cultivation of agricultural products.

7.2.5 Solar cooling systems in agriculture and food processing systems

Cooling demand is increasing continuously in agriculture and food processing operations, where cooling is mainly used for cold storage as well as space air conditioning purposes [109]. The main aim of cooling systems in the food industry is to maximize the lifespan of both fresh and processed products by minimizing the biochemical and microbiological activity in the food products. Various processes in the food industry are defined based on the cooling technique, for example, in the dairy industry, meat storage, and processing, as well as fermentation and storage in the bread and beverage industries. The setpoint temperature in such cooling applications is quite variable ranging from $-20°C$ for the deep freezing to $18°C$ in space cooling applications [109].

One example of a cooling application in the food sector is that of extracting heat from production rooms to keep food products at a certain temperature. This also applies to storage rooms for agricultural products where the

TABLE 7.6 Summary of system specifications of selected photovoltaic-integrated cooling systems for experimental and commercial greenhouses.

Location	Type of solar thermal energy system	System specifications and performance	Ref.
Malaysia	PV-integrated with fan system	• System includes 48 PV modules of 18.75 W each • Cooling requirements supplied by two fans with a total power of 400 W • Total average daily load is 2 Wh day^{-1}	[104]
Italy	PV-integrated system	• 3842-m^2 two-span greenhouse • 246.16 kWp monocrystalline, textured, antireflective-layer PV array • Energy savings on average 30% for summer cooling and 11% for winter heating	[105]
India	PV-PEM fuel cell integrated with a fan-pad ventilated system	• 90-m^2 floriculture greenhouse • 51 PV modules of 75 W each • 3.3 kW electrolyzer with two PEM fuel cell stacks of 480 W each	[106]
Saudi Arabia	PV-integrated system for a hot arid area	• 351-m^2 greenhouse • 14.72 kW PV arrays • 15 kWp power conditioning system	[107]
Japan	Side-ventilation controller driven by PV energy	• 32-m^2 greenhouse covered with plastic film • 0·078 m^2 with a rated maximum power of 3.2 W • Indoor temperatures controlled between 20°C and 25°C	[108]

crops must be kept at a low temperature for a certain period as part of their growing procedure. Sadi et al. [38] conducted a case study of a proposed storage building in Anand, India where a combined solar-powered cooling system would be used to supply annual total cooling requirements as shown in Fig. 7.12.

The proposed system consisted of a single-effect lithium bromide--water absorption chiller, a heater, and three different types of ETCs (i.e., flat-plate, U-type, and heat-pipe ETCs) to supply the cooling required to

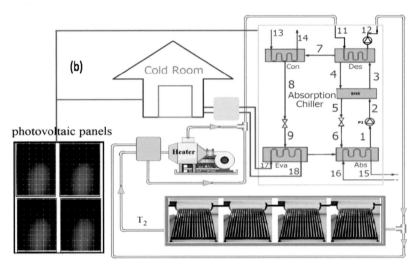

FIGURE 7.12 Schematic of a proposed combined solar-powered cooling system for use in an agricultural product storage facility in India [38].

store 180 and 85 tonnes of potatoes and apples, respectively, per year in a 3-m³ room. A 300 W PV was also used to supply the electricity required by electrical components in the system and the power consumption of the storage room. Results indicated that HP-ETCs would have the highest efficiency with the fewest collectors (i.e., 24 collectors) being needed to meet the heating requirements of the absorption system. However, in the case of using FP-ETCs, 33 collectors would be required. The shortest payback period in this study was calculated to be 11 years corresponding to the HP-etc collectors [38]. In another study by Eltawil and Samuel [110], a vapor compression cooling system integrated with a PV system for a 2.5-m³ potato storage facility was assessed. The average PV production and cooling load were 5.65 and 4.115 kWh, respectively, to keep the indoor temperature of the storage room below 10°C with 86% relative humidity. The average daily COP was also calculated to be 3.25 for this system [110].

Best et al. [111] studied the feasibility of implementing Fresnel concentrating collectors in a single-effect air-cooled lithium bromide−water absorption chiller system for the agricultural and food processing industries in Mexico (Fig. 7.13). The system located in Hermosillo would incorporate various processes, including livestock, feed production, and the processing of pork derivatives. In this study, two refrigeration subsystems were considered consisting of one low-temperature unit (i.e., −35°C to −20°C) and one medium-temperature unit (−2°C to 2°C). The results indicated that 330 m² of Fresnel collectors integrated with a 39 kW thermal chiller would provide 93,500 kWh of cooling for the proposed system. However, on a larger scale

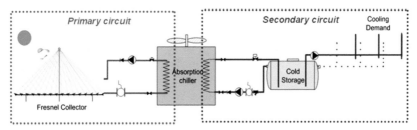

FIGURE 7.13 Schematic of a Fresnel solar collector integrated with an absorption chiller in an agro-food industry in Mexico [111].

using 2640 m^2 of collectors, the system could generate up to 758,400 kWh of cooling and the solar fraction would increase from 1.5% (for the first case study) to 14% for the scaled-up case. The authors also studied the energy saving and emission reduction potential of implementing solar cooling systems in the agricultural and food processing industries in Mexico, demonstrating that at least 95 GWh and 18.3 kt CO_2 could be saved annually.

Edwin and Sekhar [112] performed a techno-economic assessment of a hybrid air conditioning system integrated with ETCs. Their study showed that the total cooling requirements can be met by a combined solar and biomass-based energy system for which the payback period is 4−6 years in different studied configurations [112].

According to the aforementioned studies, a solar vapor compression cooling system has a lower LCOE, needs a smaller collector area, reduces CO_2 emissions more, and has higher thermal efficiency (COP) than does a solar absorption system integrated with an, etc. for the same cooling capacity. In general, solar thermal absorption refrigeration systems including both evacuated tube and Fresnel collector systems are not economically competitive in the case studies. This is mainly due to the high investment cost and low thermal efficiency of solar cooling systems. Furthermore, when electricity from the grid is highly subsidized, solar-electric-driven cooling systems are unrealistically uncompetitive. However, the ever-increasing cooling demand due to global warming as well as population growth and corresponding food demand mean that solar cooling will be the most promising solution in the future as initial costs decline and system performance improves.

7.3 Conclusions and prospects

Electricity and fuel price increases and environmental concerns have led us to use renewable energy technologies such as solar energy systems in various sectors, including the agricultural and food processing sectors. In recent years, with the declining total initial costs of solar collectors, solar thermal energy systems have become more competitive with shorter payback periods for different applications. The costs of solar thermal energy systems depend strongly on the required process temperature, thermal power capacity, and available solar radiation. For conventional flat-plate and ETCs, the investment costs are €250−1000 kW^{-1}, while the energy costs for concentrated solar thermal energy systems are €0.04−0.07 kWh^{-1}. Currently, solar thermal energy systems are mainly used for solar air drying, solar water heating, and solar space heating and cooling systems in the agricultural and food processing industries. Besides the economic benefits (especially considering available subsidies), solar energy systems can have a great impact on the carbon footprint of agricultural products. Solar cookers are another type of solar thermal system application that is still under development and has not yet been commercialized at a large scale in the industry. Currently, the main applications of solar cookers are in developing countries and rural areas where electricity and thermal grid networks are unavailable. Given recent advances, solar cooking systems are expected to be utilized widely at the industrial level, with considerable climate mitigation impacts.

Due to recent advances in the performance and economic viability of thermal energy storage systems, solar thermal energy systems now have greater potential than ever for all types of heat processes over a wide range of operating temperatures. Thermal energy storage is specifically used in both heating and cooling systems in both active and passive integrated solar greenhouses. The closed greenhouse is a promising and innovative concept that has not been commercialized yet, although recent studies have identified the great potential of such greenhouses to reduce energy costs and emissions, which will make them feasible for a new generation of agricultural production.

Several factors require further improvement to ensure the greater acceptance of solar thermal energy systems in agricultural and food processing industries, as follows:

- the techno-economic feasibility assessment of solar thermal energy systems applied in large-scale agricultural systems;
- specific in-depth analyses of the business models of innovative applications in agricultural and food processing industries;
- proposing corresponding standards and system component certification for integrated solar heat processes; and
- the development of financial templates to simplify financing procedures.

However, the main barriers to the further deployment of solar heating and cooling systems in agricultural and food processing industries are high investment costs, lack of subsidies and governmental financial support, cheap or subsidized fossil fuels, lack of public awareness, lack of technical information transfer, scale issues, lack of system instructional documents, lack of relevant training, and lack of dissemination of solar thermal energy systems proposed for agricultural systems.

References

[1] Hossain M, et al. Review on solar water heater collector and thermal energy performance of circulating pipe. Renew Sustain Energy Rev 2011;15(8):3801−12.

[2] Jamar A, et al. A review of water heating system for solar energy applications. Int Commun Heat Mass Transf 2016;76:178−87.

[3] Jaisankar S, et al. A comprehensive review on solar water heaters. Renew Sustain Energy Rev 2011;15(6):3045−50.

[4] Vengadesan E, Senthil R. A review on recent development of thermal performance enhancement methods of flat plate solar water heater. Sol Energy 2020;206:935−61.

[5] Dubey S, Tiwari G. Thermal modeling of a combined system of photovoltaic thermal (PV/T) solar water heater. Sol energy 2008;82(7):602−12.

[6] Ogueke N, Anyanwu E, Ekechukwu O. A review of solar water heating systems. J Renew Sustain energy 2009;1(4):043106.

[7] Sadhishkumar S, Balusamy T. Performance improvement in solar water heating systems—a review. Renew Sustain Energy Rev 2014;37:191−8.

[8] Shukla R, et al. Recent advances in the solar water heating systems: a review. Renew Sustain Energy Rev 2013;19:173−90.

[9] Poppi S. Solar heat pump systems for heating applications: analysis of system performance and possible solutions for improving system performance. KTH Royal Institute of Technology; 2017.

[10] Chikaire J, et al. Solar energy applications for agriculture. J Agric Veterinary Sci 2010;2:58−62.

[11] Farjana SH, et al. Solar process heat in industrial systems−a global review. Renew Sustain Energy Rev 2018;82:2270−86.

[12] Liu, Y.-M., et al. Solar water heating for livestock industry in Taiwan. In International solar energy society, ISES solar world congress 2015, SWC 2015. International Solar Energy Society; 2015.

[13] Mekhilef S, et al. The application of solar technologies for sustainable development of agricultural sector. Renew Sustain Energy Rev 2013;18:583−94.

[14] Mekhilef S, Saidur R, Safari A. A review on solar energy use in industries. Renew Sustain Energy Rev 2011;15(4):1777−90.

[15] Panchal H, Patel R, Parmar K. Application of solar energy for milk pasteurisation: a comprehensive review for sustainable development. Int J Ambient Energy 2020;41 (1):117−20.

[16] Premkumar S, et al. Solar industrial process heating associated with thermal energy storage for feed water heating. Middle East J Sci Res 2014;20(11):1686−8.

[17] Sharma AK, et al. Solar industrial process heating: a review. Renew Sustain Energy Rev 2017;78:124−37.

[18] Torshizi MV, Mighani AH. The application of solar energy in agricultural systems. J Renew Energy Sustain Dev 2017;3(2):234−40.

[19] Vannoni C, Battisti R, Drigo S. Potential for solar heat in industrial processes. IEA SHC Task 2008;33:174.

[20] Zhang B, et al. Experimental study of the burning-cave hot water soil heating system in solar greenhouse. Renew Energy 2016;87:1113−20.

[21] Rezvani S, et al. Techno-economic and reliability assessment of solar water heaters in Australia based on Monte Carlo analysis. Renew energy 2017;105:774−85.

[22] Fudholi A, Sopian K. Review on solar collector for agricultural produce. Int J Power Electron Drive Syst 2018;9(1):414.

[23] Fudholi A, Sopian K. A review of solar air flat plate collector for drying application. Renew Sustain Energy Rev 2019;102:333−45.

[24] Fudholi A, et al. Review of solar dryers for agricultural and marine products. Renew Sustain energy Rev 2010;14(1):1−30.

[25] Mustayen A, Mekhilef S, Saidur R. Performance study of different solar dryers: a review. Renew Sustain Energy Rev 2014;34:463−70.

[26] Phadke PC, Walke PV, Kriplani VM. A review on indirect solar dryers. ARPN J Eng Appl Sci 2015;10(8):3360−71.

[27] Kabeel AE, et al. Solar air heaters: design configurations, improvement methods and applications−A detailed review. Renew Sustain Energy Rev 2017;70:1189−206.

[28] Arunkumar H, Karanth KV, Kumar S. Review on the design modifications of a solar air heater for improvement in the thermal performance. Sustain Energy Technol Assess 2020;39:100685.

[29] Chan H-Y, et al. Comparison of thermal performances between low porosity perforate plate and flat plate solar air collector. Journal of Physics: Conference Series. IOP Publishing; 2018.

[30] Vadiee A, Martin V. Energy management in horticultural applications through the closed greenhouse concept, state of the art. Renew Sustain Energy Rev 2012;16(7):5087−100.

[31] Alinejad T, Yaghoubi M, Vadiee A. Thermo-environomic assessment of an integrated greenhouse with an adjustable solar photovoltaic blind system. Renew Energy 2020;156:1−13.

[32] Allouhi A, et al. Solar driven cooling systems: an updated review. Renew Sustain Energy Rev 2015;44:159−81.

[33] Daut I, et al. Solar powered air conditioning system. Energy Procedia 2013;36:444−53.

[34] Desideri U, Proietti S, Sdringola P. Solar-powered cooling systems: technical and economic analysis on industrial refrigeration and air-conditioning applications. Appl Energy 2009;86(9):1376−86.

[35] Ge T, Dai Y, Wang R. Review on solar powered rotary desiccant wheel cooling system. Renew Sustain Energy Rev 2014;39:476−97.

[36] Ghaith FA. Performance of solar powered cooling system using parabolic trough collector in UAE. Sustain Energy Technol Assess 2017;23:21−32.

[37] Islam MA, Saha BB. TEWI Assessment of Conventional and solar powered cooling systems. Solar Energy. Springer; 2020. p. 147−77.

[38] Sadi M, Arabkoohsar A, Joshi AK. Techno-economic optimization and improvement of combined solar-powered cooling system for storage of agricultural products. Sustain Energy Technol Assess 2021;45:101057.

[39] Atmaca I, Yigit A. Simulation of solar-powered absorption cooling system. Renew Energy 2003;28(8):1277−93.

[40] Shirazi A, et al. A systematic parametric study and feasibility assessment of solar-assisted single-effect, double-effect, and triple-effect absorption chillers for heating and cooling applications. Energy Convers Manag 2016;114:258−77.

[41] Vadiee A. Energy management in large scale solar buildings: the closed greenhouse concept. KTH Royal Institute of Technology; 2013.

[42] Lychnos G, Davies PA. Modelling and experimental verification of a solar-powered liquid desiccant cooling system for greenhouse food production in hot climates. Energy 2012;40(1):116−30.

[43] Vadiee A, Yaghoubi M. Enviro-economic assessment of energy conservation methods in commercial greenhouses in Iran. Outlook Agriculture 2016;45(1):47−53.

[44] Vadiee A, Yaghoubi M. Exergy analysis of the solar blind system integrated with a commercial solar greenhouse. Int J Renew Energy Res 2016;6(3):1189−99.

[45] Vadiee A, et al. Energy analysis of solar blind system concept using energy system modelling. Sol Energy 2016;139:297−308.

[46] Dur WWMS. Global market development and trends in 2019 I detailed market figures 2018 IEA. Solar Heating & Cooling Programme; 2020.

[47] Duffie JA, Beckman WA. Solar engineering of thermal processes. John Wiley & Sons; 2013.

[48] Belessiotis V, Delyannis E. Solar drying. Sol energy 2011;85(8):1665−91.

[49] Udomkun P, et al. Review of solar dryers for agricultural products in Asia and Africa: an innovation landscape approach. J Environ Manag 2020;268:110730.

[50] García-Valladares O, et al. Solar thermal drying plant for agricultural products. Part 1: direct air heating system. Renew Energy 2020;148:1302−20.

[51] Briam R, Walker ME, Masanet E. A comparison of product-based energy intensity metrics for cheese and whey processing. J Food Eng 2015;151:25−33.

[52] Jermann C, et al. Mapping trends in novel and emerging food processing technologies around the world. Innovative Food Sci & Emerg Technol 2015;31:14−27.

[53] Ladha-Sabur A, et al. Mapping energy consumption in food manufacturing. Trends Food Sci & Technol 2019;86:270−80.

[54] Morawicki R, Hager, T. Energy and greenhouse gases footprint of food processing; 2014.

[55] Wang L. Energy efficiency technologies for sustainable food processing. Energy efficiency 2014;7(5):791−810.

[56] Tosato, R.K.G.S.G., Solar heat for industrial process-technology brief. IRENA-energy technology systems analysis programme (ETSAP), 2015.

[57] Muller H. Solar process heat in the food industry−methodological analysis and design of a sustainable process heat supply system in a brewery and a dairy; 2016.

[58] Botzios-Valaskakis MKA. Solar systems applications in the dairy industry. The Centre for Renewable Energy Sources and Saving (CRES).

[59] Allouhi A, et al. Design optimization of a multi-temperature solar thermal heating system for an industrial process. Appl Energy 2017;206:382−92.

[60] L., R.-C.K.a.S., Solar thermal technologies: clean fit for food and beverage industries. WWF South Africa; 2018.

[61] Godley R. Solar water heating and dairy farming − potential in the peak district national park: review of technology & issues. T4 Sustainability Limited; 2008.

[62] Liu Y-M, et al. Solar thermal application for the livestock industry in Taiwan. Case Stud Therm Eng 2015;6:251−7.

[63] Hess S, et al. Initial study on solar process heat for sout African sugar mills. In South African sugar technologists' association congress; 2016.

[64] Cuce E, Cuce PM. A comprehensive review on solar cookers. Appl Energy 2013;102:1399−421.

[65] Kundapur A. Review of solar cooker designs. TIDE 1998;8(1):1−37.

[66] Muthusivagami R, Velraj R, Sethumadhavan R. Solar cookers with and without thermal storage—a review. Renew Sustain energy Rev 2010;14(2):691−701.

[67] Yettou F, et al. Solar cooker realizations in actual usean overview. Renew Sustain Energy Rev 2014;37:288−306.

[68] Mahavar S, et al. Design development and performance studies of a novel single family solar cooker. Renew energy 2012;47:67−76.

[69] Kahsay MB, et al. Theoretical and experimental comparison of box solar cookers with and without internal reflector. Energy Procedia 2014;57:1613−22.

[70] Kumar N, et al. Design and development of efficient multipurpose domestic solar cookers/dryers. Renew Energy 2008;33(10):2207−11.

[71] Zamani H, Moghiman M, Kianifar A. Optimization of the parabolic mirror position in a solar cooker using the response surface method (RSM). Renew Energy 2015;81:753−9.

[72] Harmim A, et al. Mathematical modeling of a box-type solar cooker employing an asymmetric compound parabolic concentrator. Sol Energy 2012;86(6):1673−82.

[73] Harmim A, et al. Design and experimental testing of an innovative building-integrated box type solar cooker. Sol Energy 2013;98:422−33.

[74] Saxena A, Agarwal N. Performance characteristics of a new hybrid solar cooker with air duct. Sol Energy 2018;159:628−37.

[75] Nahar N, Gupta J, Sharma P. Performance and testing of two models of solar cooker for animal feed. Renew energy 1996;7(1):47−50.

[76] Nahar N, Gupta J, Sharma P. A novel solar cooker for animal feed. Energy Convers Manag 1996;37(1):77−80.

[77] Nahar N, Sharma P, Chaudhary G. Processing of agricultural products in solar cooker for income generation. In: International solar food processing conference; 2009.

[78] Öztürk HH. Experimental determination of energy and exergy efficiency of the solar parabolic-cooker. Sol energy 2004;77(1):67−71.

[79] Al-Soud MS, et al. A parabolic solar cooker with automatic two axes sun tracking system. Appl Energy 2010;87(2):463−70.

[80] Biermann E, Grupp My, Palmer R. Solar cooker acceptance in South Africa: results of a comparative field-test. Sol energy 1999;66(6):401−7.

[81] Kalbande S, et al. Design theory and performance analysis of paraboloidal solar cooker. Appl Sol Energy 2008;44(2):103−12.

[82] Edmonds I. Low cost realisation of a high temperature solar cooker. Renew Energy 2018;121:94−101.

[83] Kaushik S, Gupta M. Energy and exergy efficiency comparison of community-size and domestic-size paraboloidal solar cooker performance. Energy Sustain Dev 2008;12(3):60−4.

[84] Farooqui SZ. A vacuum tube based improved solar cooker. Sustain Energy Technol Assess 2013;3:33−9.

[85] Farooqui SZ. Impact of load variation on the energy and exergy efficiencies of a single vacuum tube based solar cooker. Renew energy 2015;77:152−8.

[86] Aramesh M, et al. A review of recent advances in solar cooking technology. Renew Energy 2019;140:419−35.

[87] Panwar N, Kothari S, Kaushik S. Techno-economic evaluation of masonry type animal feed solar cooker in rural areas of an Indian state Rajasthan. Energy policy 2013;52:583−6.

[88] Epp JMaB. India: temple possesses world's largest solar steam cooking system. <https://www.solarthermalworld.org/news/india-temple-possesses-worlds-largest-solar-steam-cooking-system>; 2009 [accessed 26.06.21].

[89] Vadiee A. Energy analysis of the closed greenhouse concept: towards a sustainable energy pathway. KTH Royal Institute of Technology; 2011.

[90]   Taki M, Rohani A, Rahmati-Joneidabad M. Solar thermal simulation and applications in greenhouse. Inf Process Agriculture 2018;5(1):83−113.

[91]   Vadiee A, Martin V. Energy management strategies for commercial greenhouses. Appl Energy 2014;114:880−8.

[92]   Vadiee A, Martin V. Energy analysis and thermoeconomic assessment of the closed greenhouse−the largest commercial solar building. Appl Energy 2013;102:1256−66.

[93]   Voogt J, Van Weel P. Climate control based on stomatal behavior in a semi-closed greenhouse system'Aircokas'. Int Workshop Greenh EnvirControl Crop Prod Semi-Arid Reg 2008;797.

[94]   van t Ooster A, et al. Development of concepts for a zero-fossil-energy greenhouse. in International symposium on high technology for greenhouse system management: Greensys2007 801; 2007.

[95]   Marcelis L, et al. Climate and yield in a closed greenhouse. In International symposium on high technology for greenhouse system management: Greensys2007 801; 2007.

[96]   Opdam J, et al. Closed greenhouse: a starting point for sustainable entrepreneurship in horticulture. In: International conference on sustainable greenhouse systems-Greensys2004 691; 2004.

[97]   De Zwart H. Overall energy analysis of (semi) closed greenhouses. In International symposium on high technology for greenhouse system management: Greensys2007 801; 2007.

[98]   Zaragoza G, Buchholz M. Closed greenhouses for semi-arid climates: critical discussion following the results of the Watergy prototype. Int Workshop Greenh EnvirControl Crop Prod Semi-Arid Reg 2008;797.

[99]   Hoes, H., et al. The GESKAS project, closed greenhouse as energy source and optimal growing environment. In: International symposium on high technology for greenhouse system management: Greensys2007 801; 2007.

[100]   Genovese, A., et al. Photovoltaic as sustainable energy for greenhouse and closed plant production system. In: International workshop on greenhouse environmental control and crop production in semi-arid regions 797; 2008.

[101]   Vox G, et al. Solar absorption cooling system for greenhouse climate control: technical evaluation. Acta Hortic 2014;1037:533−8.

[102]   Santamouris M, et al. Passive solar agricultural greenhouses: a worldwide classification and evaluation of technologies and systems used for heating purposes. Sol Energy 1994;53(5):411−26.

[103]   Monghasemi N, Vadiee A. A review of solar chimney integrated systems for space heating and cooling application. Renew Sustain Energy Rev 2018;81:2714−30.

[104]   Faisal M, et al. Design and development of a photovoltaic power system for tropical greenhouse cooling. Am J Appl Sci 2007;4(6):386−9.

[105]   Carlini M, Honorati T, Castellucci S. Photovoltaic greenhouses: comparison of optical and thermal behaviour for energy savings. Math Probl Eng 2012;2012.

[106]   Ganguly A, Misra D, Ghosh S. Modeling and analysis of solar photovoltaic-electrolyzer-fuel cell hybrid power system integrated with a floriculture greenhouse. Energy Build 2010;42(11):2036−43.

[107]   Al-Ibrahim A, Al-Abbadi N, Al-Helal I. PV greenhouse system: system description, performance and lesson learned. In: International symposium on greenhouses, environmental controls and in-house mechanization for crop production in the tropics 710; 2004.

[108]   Yano A, et al. Development of a greenhouse side-ventilation controller driven by photovoltaic energy. Biosyst Eng 2007;96(4):633−41.

[109] Heldman DR, Lund DB, Sabliov C. Handbook of food engineering. CRC press; 2018.

[110] Eltawil MA, Samuuel D. Vapour compression cooling system powered by solar PV array for potato storage; 2007.

[111] Best R, et al. Solar cooling in the food industry in Mexico: a case study. Appl Therm Eng 2013;50(2):1447–52.

[112] Edwin M, Sekhar SJ. Techno-economic studies on hybrid energy based cooling system for milk preservation in isolated regions. Energy Convers Manag 2014;86:1023–30.

Chapter 8

Solar desalination technology to supply water for agricultural applications

Shiva Gorjian[1,2], Mushtaque Ahmed[3], Omid Fakhraei[1], Sina Eterafi[1] and Laxmikant D. Jathar[4]

[1]*Biosystems Engineering Department, Faculty of Agriculture, Tarbiat Modares University (TMU), Tehran, Iran,* [2]*Renewable Energy Department, Faculty of Interdisciplinary Science and Technology, Tarbiat Modares University (TMU), Tehran, Iran,* [3]*Department of Soils, Water, and Agricultural Engineering, College of Agricultural and Marine Sciences, Sultan Qaboos University, Al-Khoudh, Muscat, Oman,* [4]*Department of Mechanical Engineering, Imperial College of Engineering and Research, Pune, India*

8.1 Introduction

Water is the most important constituent of the earth's hydrosphere which plays an important role in nearly all aspects of human life and the ecosystem [1,2]. The earth's hydrosphere contains 1.386 billion km^3 with a vast area of about 71% which has been covered by oceans. It is worthy to be noted that nearly 96.54% of the whole water bodies on Earth are saline water with only 2.53% freshwater [3]. From this value, 0.3% is directly available in lakes and rivers as surface water, 68.7% is in the form of ice and permanent snow cover frozen in polar ice caps and glaciers, and 30.1% is available as groundwater [4,5]. Water scarcity occurs when there are insufficient water resources in both terms of quantity and quality to meet the standard water demand. In recent decades, unequal distribution of water on earth along with a considerable increase in the global freshwater demand has resulted in the world's water scarcity. Statistics show that about 700 million people in 43 countries around the world do not have access to fresh water, and by 2025, 1.8 billion people will be living in the regions with absolute water scarcity [6].

Water shortages may be caused by increased population, overuse of water, expansion of agriculture, and climate change. Due to the existence of a close relationship between the climate and the hydrological cycle, climate change has shown significant impacts on water resources [7,8].

Solar Energy Advancements in Agriculture and Food Production Systems.
DOI: https://doi.org/10.1016/B978-0-323-89866-9.00002-X

In this regard, increased temperatures will increment evaporation, causing the precipitation to rise, albeit there are regional variations in rainfall. In this situation, frequent droughts and floods in different regions at different times as well as dramatic variations in snowfall and snowmelt, especially in mountainous areas, are more expected [9].

8.1.1 Water resources and usage in agriculture

The agricultural sector is responsible to meet the worldwide demand for food and other agricultural products driven by the population growth projected to reach 9.8 billion by 2050 from 7.9 billion in 2021. The food production is almost entirely supported by water, and therefore 11% growth in water withdrawal for agriculture and an additional land area of 32 million ha are expected to be demanded by 2050 to provide global food security[1] [10,11]. As shown in Fig. 8.1, the agriculture sector accounts for approximately 70% of global water usage followed by energy supply, industry, and domestic sectors [13,14]. Industrialization, urbanization, contamination, as well as climate change through variations in annual rainfall patterns, melting snowpack, and incidence of periodic floods and droughts, have put more pressure on the agriculture sector with more competition for water resources. In this regard, supplying water for irrigation is essential to achieve food security particularly in arid and semiarid regions where water resources are crucial for economic growth [15−18].

Two main water supplying methods are available to farmers and ranchers to cultivate crops as *rainfed* and *irrigation* [19,20]. Rainfed is the natural use of water for crop production provided through direct rainfall, covering 80%

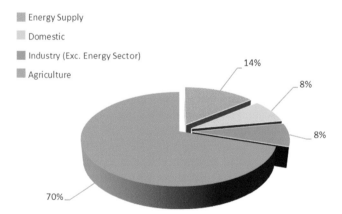

FIGURE 8.1 Global water demand by sector in 2021. *Data from Valuing Water. 2021;191; AQUASTAT - FAO's Global Information System on Water and Agriculture. Food Agric Organ United Nations; http://www.fao.org/aquastat/en/overview/methodology/water-use; 2021 [11,12].*

---

1. Food security is a measure of the availability of food and individuals' ability to access it.

of croplands in the world, whereas irrigation covers about 20% of cultivated lands, accounting for 40% of food production which is expected to rise over 45% by 2030. Employing the irrigation method, a controlled amount of water is applied to the soil through different systems including pumps, pipes, and sprays, and is mainly utilized in regions where irregular rainfall or drought is expected. On the other hand, in rainfed farming, the contamination of food products is less likely compared to irrigation-based farming but this method is more sensitive to water shortages due to a reduction in rainfall. In addition to crop production, water in agriculture is utilized for poultry, livestock, and inland fishery production [21,22].

Several irrigation methods vary in the way the water is provided for plants are as follows [17,23]:

- *Surface irrigation*, in this method, water through surface irrigation systems (furrow, flood, or level basin) flows across the surface of farmlands and permeates into the soil by gravity or the land's slope.
- *Drip irrigation*,[2] this method is so-called *localized* or *trickle irrigation* uses a system in which water is distributed in a pre-planned pattern through a network of pipes under low pressure in which a small water discharge is applied to the plant or its nearby region.
- *Sprinkler irrigation*, in this method which is so-called *overhead irrigation*, the water is piped to one or more center points inside the field and distributed by overhead high-pressure sprinklers or guns mounted on permanently installed risers.
- *Sub-irrigation*,[3] in this method, the water table is artificially raised to wet the soil from below the root zone of the plants. Often these systems are integrated with drainage infrastructures installed in permanent grasslands of lowlands or river valleys.

Improper planning and management intensify irrigated agriculture, resulting in biodiversity decline and other environmental problems in the agroecosystems [24,25]. Under such circumstances, conventional water sources cannot solely supply the current or future irrigation demands, and therefore new solutions are required to keep the agricultural production sustainable [10,15]. Irrigation water is mostly supplied through groundwater (springs or wells) and surface water (rivers, lakes, or reservoirs) [26]. To mitigate water scarcity in the agriculture sector, several applicable methods are available including conservation of water resources, repairing of infrastructures, and improving the catchment and distribution systems. But these methods can only amend the current use of water resources, while in most cases, increasing water supply beyond the available amount from the hydrological cycle is required [24,27]. In this context, an applicable technical approach to make agricultural production sustainable and effectively eliminate the climatological and hydrological

---

2. So-called microirrigation system.
3. This method is also used in commercial greenhouses (usually for potted plants).

limitations is increasing the water supply through seawater and brackish water desalination as well as wastewater reclamation and reuse [26,28].

8.1.2 Salinity of irrigation water and its impact on crop production

Soil is considered a nonrenewable resource, and therefore it cannot be recovered shortly after it is lost. After soil erosion, soil salinity is known as the main cause of land degradation, disrupting the ecosystem, biodiversity loss, crop yields declination, abandonment of already productive farmlands, and contamination of freshwater [29,30]. All agricultural soils and irrigation waters contain mineral salts but the type and amount of salts differ upon the make-up of both the soil and irrigation water [29,31]. Currently, about 20% of global irrigated lands (45 million ha), producing one-third of the global food, are salt-affected and almost 10 million ha of the world's agricultural lands are annually destroyed by salt accumulation mostly because of low-quality irrigation water, poor drainage, and massive irrigation in intensive farming [32]. There can be several reasons for soil salinity but irrigation water quality has been historically the primary cause. The lack of sufficient freshwater resources for irrigation has caused the moderately saline water to be used as a supplemental irrigation water source, causing the addition of a considerable quantity of salts into the soil over a long period. Salinity is predominantly a hazard in irrigated lands and areas with saline soils, but normally it is not an issue in rainfed agriculture [21,33]. Fig. 8.2 shows the crop yield as a function of soil salinity. The crop growth and crop yield are decreased due to both salty irrigation water and soil salinity mainly because of the osmotic effect[4] that depresses the external water potential, while some specific ions may have adverse chemical effects [31,32].

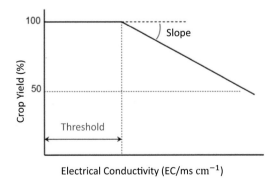

FIGURE 8.2 Dependency of crop yield with soil or irrigation water salinity [34].

---

4. High salt concentration of soil solution that reduces the ability of plants to acquire water (so-called water-deficit effect).

Although most plants respond to salinity as a function of the total osmotic potential of soil water, some herbaceous plants and most of the woody species are susceptible to specific ion toxicities. The salinity also affects photosynthesis by reducing the availability of carbon dioxide (CO_2) as a result of diffusion limitations and decreasing the contents of photosynthetic pigments [10,32]. The salinity thresholds of vegetable crops determined by irrigation water and soil salinity are presented in Table 8.1. As shown in this table, the lowest threshold level of irrigation water with no impact on the crop growth is 0.7 dS/m which is less than the threshold level for soil salinity (1 dS/m). Additionally, it is found that most vegetable crops have a salinity threshold of ≤ 2.5 dS/m.

TABLE 8.1 Determined salt tolerance of vegetable crops by the salinity of soil (EC_e) and irrigation water (EC_W) [32].

	Soil		Irrigation Water	
Vegetable	Threshold (dS/m) EC_e	Slope (% per dS/m)	Threshold (dS/m) EC_W	Rating
Asparagus	4.1	2.0	2.7	T
Bean	1.0	19.0	0.7	S
Broccoli	2.8	9.2	1.9	MS
Carrot	1.0	14.0	0.7	S
Cauliflower	—	—	1.9	MS
Celery	1.8	6.2	1.2	MS
Eggplant	1.1	6.9	0.7	MS
Lettuce	2.0	13.0	0.9	MS
Muskmelon	1.0	1.0	—	MS
Okra	1.2	—	—	S
Onion	1.2	16.0	0.8	S
Pea	1.5	14.6	—	MS
Pepper	1.5	14.0	1.0	MS
Potato	1.7	12.0	1.1	MS
Purslane	6.3	9.6	—	MT
Red beet	4.0	—	2.7	MT
Spinach	2.0	7.6	1.3	S
Strawberry	1.0	33.0	0.7	S
Tomato	2.5	9.9	1.7	MS

S, Sensitive; MS, moderately sensitive; MT, moderately tolerant; T, tolerant.

8.2 Desalination and agriculture

In recent years, brackish water and seawater desalination to supply irrigation water for crop production has been dramatically increased, while the seawater desalination with 50% higher costs compared to brackish water desalination is mainly limited to arid-climate coastal regions, high-return agriculture, and small islands with no conventional water sources [10,13]. The current demand on the agricultural sector seems to be increased in the future, imposing challenges to developing countries [35−37]. In 2019, around 21,123 total desalination plants were installed in 150 countries, devoting a total global cumulative desalination capacity of about 126.57 million m³/d [38]. Fig. 8.3 shows the top 10 countries in terms of desalination capacity. As shown in this figure, Saudi Arabia has currently the largest global desalination capacity by supplying 60% of total water demand through desalination [39]. Several countries around the world are using desalinated water to meet the water demand for agricultural operations. In this case, Spain with a total desalination capacity of 1.4 million m³/day, uses about 22% of the desalinated water for fertigation, while Kuwait, with a desalination capacity of over 1 million m³/day, uses about 13% of this amount for fertigation [13,38]. Other countries using desalinated water for agricultural operations are Italy (desalination capacity; 64,700 m³/day −1.5% for agriculture), Bahrain (desalination capacity 620,000 m³/day−0.4% for agriculture), Qatar (0.1% for agriculture), USA (1.3% for agriculture) and Israel. Saudi Arabia as the largest global producer of desalinated water uses only 0.5% of its desalination capacity for agricultural purposes [38,40]. The Gulf Cooperation Council (GCC) countries including Bahrain, Saudi Arabia, Kuwait, Oman, Qatar, and United Arab Emirates (UAE) are classified as arid countries with low and erratic rainfall rates, limited groundwater resources, and high evapotranspiration rates. Therefore, seeking alternative water resources

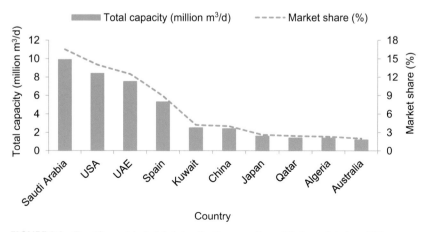

FIGURE 8.3 Top 10 countries' global desalination capacity and their market share [39].

and the employment of innovative strategies to manage water to make agriculture in GCC countries sustainable is crucial [41].

Choosing the most appropriate desalination technology for agriculture depends on two main factors of the net economic returns of agricultural products and environmental costs [42]. The desalinated water should meet the quality standard for irrigation water which has been established by Food and Agriculture Organization (FAO) primarily based on salinity [43]. Through desalination, some mineral nutrients essential for plant growth may be lost which can be overcome by adding complementary minerals using blended water. The most important parameters of the treated water for irrigation are salinity, sodium (Na) content, trace elements, excess chloride (Cl^-), and nutrients [10,13].

8.2.1 Conventional desalination technologies

In a typical desalination process, salt and mineral components are taken away from saline water, making it drinkable as well as consumable in agricultural applications. Several desalination technologies are available to extract freshwater from saline water which are broadly classified as thermal evaporation and membrane-based technologies [44−46]. In thermal evaporation methods, heat is utilized to vaporize water and then the water vapor is condensed to produce freshwater. The most common types of thermal desalination methods are; multistage flash (MSF), multieffect distillation (MED), and vapor compression (VC). On the other side, membrane-based technologies work based on the concentration gradient, electric potential, and mechanical pressure across semipermeable membranes which assists in separating the salt from the water stream [47−49]. The membrane technologies include reverse osmosis (RO) [50], forward osmosis (FO) [51], electro-dialysis (ED) [52], membrane distillation (MD) [53], nanofiltration (NF) [54], microfiltration (MF), ultrafiltration (UF), and membrane bioreactor (MBR) [49]. The RO is the most mature desalination process which is globally used, devoting 61% of the worldwide share, followed by MSF (26%), MED (8%), ED (3%), and other desalination technologies [39]. The increase in the global capacity of different desalination technologies is shown in Fig. 8.4.

Among different desalination methods, membrane techniques are employed in different countries to supply irrigation water with a leading share for seawater RO desalination, mainly because of its minimal energy expense in comparison with other technologies. However, the cost of brackish water desalination is typically a third of seawater desalination [13,56]. Despite the recent advancements, the cost of desalination for irrigated agriculture is still high. However, desalinated water can be used to provide water for intensive horticulture of high-value cash crops such as vegetables and flowers, especially those cultivated in greenhouses located in coastal areas [57,58].

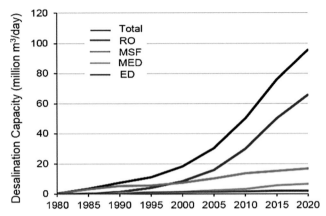

FIGURE 8.4   Global operational desalination capacity by technology [55].

8.2.2   Seawater greenhouses

A seawater greenhouse (SWGH) is a technology based on air humidification–dehumidification (HDH) that occurs in a greenhouse, creating a desirable growing environment for crop production by reducing the amount of required irrigation water and simultaneously providing new resources of freshwater from saline water [59]. The SWGHs have several benefits over traditional structures in desert climates and the proximity of saltwater resources [60]. The process that occurs in a typical SWGH is a recreation of the "hydrologic cycle" when water evaporates from saline water and regains as freshwater by condensation [18]. Under tropical and temperate conditions, a HDH water desalination system can produce 11.6 and 20.4 m^3/year-ha of freshwater with a power consumption of 1.6 and 1.9 kWh/m^3, respectively [60]. To produce freshwater and cool the air, a SWGH uses sunlight, saline water, and air, creating favorable temperate conditions for the cultivated crops inside. The working principle of a typical SWGH is presented in Fig. 8.5. As shown in this figure, pumping seawater into the greenhouse creates a cool and humid environment. The ambient warm dry air passes through the first evaporator wetted by seawater where the air is drawn into the greenhouse [61].

The seawater is evaporatively cooled in the first evaporator where it is collected and pumped to the condenser. Then, the coolant flows back to the first evaporator and the cold-water circuit is completed. In this case, the temperature of the air passing through the cropping area is increased by gaining solar heat, and its humidity is raised by evapotranspiration [18,62]. Since the air moisture-holding capacity increases by temperature, the second evaporator placed at the end of the cropping area enriches the air up to saturation point. Then, the saturated air passes through the condenser and is cooled using cold deep seawater which removes salt and impurities [63,64].

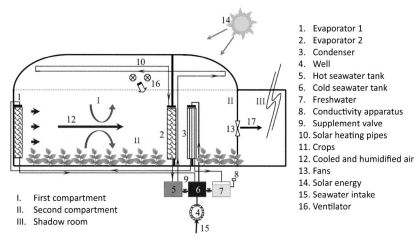

1. Evaporator 1
2. Evaporator 2
3. Condenser
4. Well
5. Hot seawater tank
6. Cold seawater tank
7. Freshwater
8. Conductivity apparatus
9. Supplement valve
10. Solar heating pipes
11. Crops
12. Cooled and humidified air
13. Fans
14. Solar energy
15. Seawater intake
16. Ventilator

I.   First compartment
II. Second compartment
III. Shadow room

FIGURE 8.5   Working principle of a typical seawater greenhouse [59].

8.2.3   Solar-powered desalination technologies (an overview)

The use of renewable energy sources (RESs) to power desalination units is an exceptionally promising choice particularly in remote and dry areas where the utilization of traditional energy is expensive or impossible. A renewable energy-driven desalination system is a novel approach to desalinate water in a cost-effective and eco-friendly way [65]. Globally, in the last few years, more than 130 renewable-powered desalination plants have come into operation. Among different RESs, solar, wind, geothermal, wave, and tidal energy are the major renewable sources that can be harnessed to drive desalination plants [66]. Despite this, hydropower and biomass are not appropriate technologies for integration with desalination plants since they need water resources that are not available in water-scarce countries [65]. Solar-powered desalination plants are mainly classified as direct and indirect systems with many subgroups as presented in Fig. 8.6. In the first type, solar energy is directly absorbed by the saline water and vaporizes it, while in indirect systems, the solar energy is absorbed by solar collecting equipment and then is transferred with higher quality to the saline water [68,69]. Despite their simpler design, due to low operating temperatures, the productivity of direct solar desalination systems is proportionate to the surface area of the device, and therefore they cannot produce a large quantity of water per day [70,71].

According to a recent review presented by Chauhan et al. [72] on direct solar desalination techniques, considering various designs configurations and recent advancements, direct solar desalination systems cannot provide yields of more than 10 L/m^2/d with few exceptional cases. On the other hand, indirect solar desalination systems are composed of two distinct

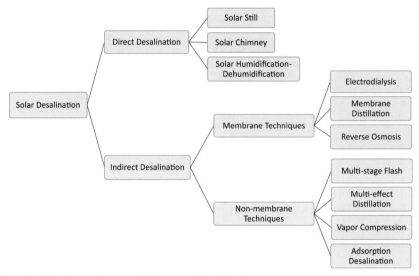

FIGURE 8.6   Classification of solar-powered desalination techniques [67].

subsystems of solar energy collection and proven desalination plants. In this regard, membrane desalination plants such as RO and ED can be directly powered by the electricity generated by photovoltaic (PV) systems or concentrated solar power (CSP) plants. In contrast, solar thermal collectors can be integrated with thermal desalination processes such as MSF, MED, HDH, and MD to supply their required thermal power. However, the integration of CSP plants with desalination processes is only proposed for large-scale desalination plants and SWGs where CSP either export electricity to drive RO, mechanical VC (MVC), and ED, or deliver the exhaust steam to power MED or MSF desalination processes [73,74].

To better understand the integration between the desalination processes and renewable energy sources, it is critical to investigate the heat and electricity demands of different desalination techniques as presented in Table 8.2 for both thermal and membrane desalination plants with specified water production capacity. The total cost of water production using desalination consists of three main components: (1) energy cost, (2) operational and maintenance cost, and (3) capital investment with the highest share of 50% for the energy cost [75,76]. As mentioned above, to operate solar desalination systems, two forms of energy as solar heat and solar electricity are available according to the type of the desalination process. The energy and capital investment costs with a share of over 20% for energy cost, and over 40% for capital cost are two dominant factors influencing the expense of desalinated water [65]. The cost of desalination using small- to medium-scale solar desalination plants is quite high compared to the conventional systems, and only large-scale solar desalination systems can compete with

TABLE 8.2 The energy demand of conventional desalination technologies [65].

Desalination technique	Capacity (m³/day)	Electric energy consumption (kWh/m³)	Thermal energy consumption (kJ/kg)	Total equivalent energy consumption (kWh/m³)
MSF	50,000–70,000	4–6	190–390	13.5–25.5
MED-TVC	10,000–35,000	1.5–2.5	145–390	11–28
MED	5000–15,000	1.5–2.5	230–390	6.5–11
MVC	100–2500	7–12	—	7–12
RO	24,000	3–7	—	3–7
ED	145,000	2.6–5.5	—	2.6–5.5

TABLE 8.3 Water cost of various solar desalination plants in different scales (in 2018) [73].

Desalination plant	Very small	Small-scale	Medium-scale	Large-scale
Solar still	USD 6.0−65.0/m³	—	—	—
Solar HDH	USD 4.4/m³	USD 2.9−22.1/m³	—	—
Solar MD	USD 12.0−18.0/m³	—	—	—
Solar MSF	—	—	—	USD 1.4−1.6/m³
Solar MED	—	USD 18.0−22.0/m³	USD 4.1−8.0/m³	USD 0.9−1.3/m³
PV-RO	USD 15.6/m³	USD 6.5−12.8/m³	USD 0.8−8.4/m³	—
PV-ED	—	USD 0.2−16.0/m³	USD 5.7−12.1/m³	—
Solar thermal RO	—	—	—	USD 2.2/m³
CSP + RO/ MED/MSF	—	—	—	USD 0.9−1.2/m³ (MED)

those powered by fossil fuels with an estimated water cost ranges from USD 0.5 to 1.5/m³ [73,77]. Table 8.3 represents the water cost of different solar-powered desalination plants considering their capacity.

8.3 Solar-powered desalination for agricultural applications

The agricultural development transition towards sustainable intensification is a strategic pathway to efficiently use natural resources including water. The concept of sustainable intensification means producing more from the same land considering conservation of resources, mitigation of negative environmental impacts, and enhancement of natural capital and the flow of ecosystem services [11]. In this regard, the use of solar desalination technology to supply the water demand of agricultural operations is considered a sustainable solution. In a study by Atkinson [78], a solar-powered RO desalination system with the firm's membrane technology was fabricated to provide freshwater for a village in Kenya for drinking and irrigation purposes. In this

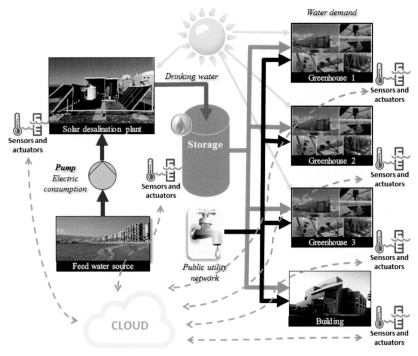

FIGURE 8.7   Schematic diagram of the smart grid irrigation framework developed in Ref. [79].

system, eight Lewabrane[5] membrane elements with a total surface area of more than 50 m^2 were used to supply clean water for the population of the village. The system was directly powered by a 10-kW solar PV system without using battery storage. The PV-RO desalination plant could provide water up to 20 m^3/d for drinking and irrigation, and 12 m^3 for fisheries in the village and its neighboring regions. To manage an intelligent network including solar membrane distillation (SMD) facilities for irrigation of several greenhouses, Muñoz et al. [79] employed an Internet of Things (IoT) that cloud architecture according to the FIWARE[6] standard to operate a smart-grid framework including solar desalination appliances and several greenhouses that require irrigation water to minimize the operating costs (see Fig. 8.7). In this case, greenhouses and SMD facilities were connected through proportional control systems operated based on IoT to obtain sustainable

---

5. A technology based on press material issued by the LANXESS Deutschland GmbH.
6. An open source initiative that defines a universal set of standards for context data management to facilitate the development of smart solutions for different domains such as *smart cities*, *smart industry*, *smart agri-food*, and *smart energy*.

irrigated crops. The results revealed that almost 75% of the total operating costs can be saved using the proposed approach.

As an appealing solution for agricultural irrigation in rural and remote areas, Dehesa-Carrasco et al. [80] investigated the performance of a PV-powered NF (PV-NF) desalination system. In this study, the effect of solar radiation and influent concentration on energy consumption, permeate production, recovery rate, and quality of the permeate product was explored. The results indicated that the PV-NF system can produce 2.16−4.8 m³/d of freshwater with the maximum energy consumption of 1.55 kWh/m³ (at the TDS of 2539 mg/L). Also, the cost of the permeate water unit was calculated as USD 1.05−0.47/m³. They concluded that the quality of the obtained permeate can satisfy the standard norm of irrigation. Considering the water scarcity in the south-eastern of Spain and generally in countries of the Mediterranean sea, Gil et al. [81] investigated a predictive model for the efficient use of a divided energy system including a SMD facility and crop cultivation in a greenhouse as the most popular form of farming in the mentioned area. In this study, every MD module embedded in the desalination facility was fed according to the greenhouse's water demand and the MD plant's thermal energy demand (see Fig. 8.8). Simulation results indicated that the proposed distributed approach can optimally manage industrial-scale plants. Moreover, the comparison between the automatic and the manual (a nonoptimal) operation showed that the thermal performance of operations applying the proposed technique can be improved by 5%. They claimed that the proposed system brings savings around 50 MWh/season in thermal energy for an 8-ha cultivated land.

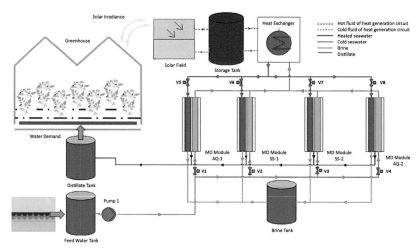

FIGURE 8.8 The schematic diagram of the proposed solar membrane distillation unit integrated with a greenhouse [81].

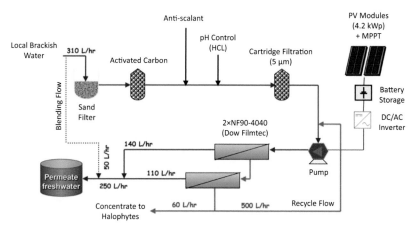

FIGURE 8.9 The design scheme of the developed photovoltaic-powered nanofiltration pilot desalination plant in Ref. [82].

To provide irrigation water for high-value crops cultivated in arid environments such as Jordan, Palestinian Authority, and Israel, Ghermandi et al. [82] developed and evaluated a PV-NF desalination system called the AGRISOL project to decrease groundwater abstraction rates, increase crop yields, and enhance farmers' welfare by enlarging their product portfolio (see Fig. 8.9). In this study, experiments were conducted to refine the economic and technical viability for two pilot solar desalination plants installed in mentioned regions. The results indicated that a market potential for the proposed innovation exists both in Israel and Jordan. In addition, experiments performed on strawberries in the region demonstrated the potential of this technology in making the cultivation of salt-sensitive cash crops possible. They concluded that desalination can be considered as a worthwhile strategy towards more sustainable water management in regional dry-land agriculture. To supply agricultural water in California, Weiner et al. [83] developed a CSP hybrid RO-MED system. They reported that this hybrid system can operate more efficiently than the current stand-alone MED system and has recovery ratios more than the current stand-alone RO system, reducing energy and disposal costs. Using a CSP plant, both work and renewable heat can be supplied without emissions, allowing for a remote operation. They claimed that using the developed system, the levelized cost of water is reduced by over 41% compared to a stand-alone MED system. To minimize the levelized cost of water, employment of five MED effects and operating both RO stages at an average flux of 18.2 L/m^2/h was proposed. Zarzoum et al. [84] carried out a testable study on a SMD unit equipped with a direct contact membrane (DCMD) as shown in Fig. 8.10. For the heating of the feedwater hot stream, a heat exchanger that could operate using the heat in the range of 60°C−80°C

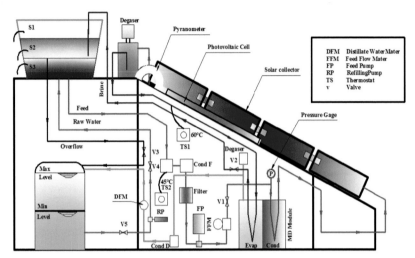

FIGURE 8.10   A schematic diagram of the developed solar membrane distillation plant in Ref. [84].

was employed with no heat exchanger for cooling the feedwater cold stream. The results indicated a good correlation between global solar energy and the system's yield productivity. They concluded that the system is very promising with more suitability for use in dry regions in Arabian countries and those located in North Africa.

Stuber et al. [85] designed a solar-powered desalination system for high-recovery treatment of subsurface agricultural drainage water as well as other brackish groundwater sources as a reuse strategy. In this case, an open-cycle vapor-absorption heat pump was integrated with a MED system and a large parabolic trough concentrator (PTC). The function of distillation without the heat pump indicated the thermal energy consumption of 261.87 kWh$_{th}$/m^3, and a decrease by over 49% with the heat pump to reach 133.2 kWh$_{th}$/m^3. They concluded that this decrease is equal to the reduction in the solar array's area required to supply equal freshwater without the heat pump. From the results, thermal energy performance of 34.9 kWh$_{th}$/m^3 was obtained for an optimized design with a 10-effect MED plant operating at 85% recovery. Hipólito-Valencia et al. [86] proposed a new superstructure formulated as a multiobjective model, where an equipped steam Rankin cycle was fed by fossil fuels and solar energy to supply power demands of the desalination process and agricultural activity. They indicated the highest amount of saving in the groundwater and external electrical power consumption as about 66.12% and 86.2% respectively. In addition, it was claimed the total cost is mainly affected by the desalination plant and annual consumption of electricity and seawater, while its possibility mostly depends on the desalination location, and available sea and specific aquifer conditions such as privileges and availability in each country.

To have a sustainable resilient greenhouse, Akrami et al. [87] developed a zero-liquid-discharge (ZLD) system by using solar still desalination technique, HDH, and rainwater harvesting (see Fig. 8.11). In this case, experiments were performed to appraise the efficiency of the developed procedure under climate conditions of the UK and Egypt. The results revealed that the proposed system is a successful stand-alone greenhouse model that can supply its water requirements. From the experimental results obtained under the climate of the UK, solar still could produce a maximum amount of potable water as 58 mL/d for a distillation area of 0.72 m^2, while in Egypt, maximum water of 1090 mL/d was produced for each solar still. Additionally, the HDH could generate 7 L of distilled water which was added to the harvested rainwater, and therefore, maximum water of 7 L/d could be produced under climate conditions of the UK. They claimed that developed greenhouses can be implemented as stand-alone systems in countries with different climatic conditions to supply the agricultural water needs.

Roca et al. [88] simulated a case study in which a solar MED system was used to produce freshwater for irrigation of agricultural lands. They employed an appropriate control system in a solar MED plant to supply the irrigation water demand of a greenhouse installed in Spain (see Fig. 8.12). The challenge was the proper operation of the desalination plant to supply the daily water demand of the crops cultivated inside the greenhouse. In this case, a hierarchical controller was proposed to reduce the cost of the produced water by the solar desalination plant, maintaining a distillate volume for greenhouse irrigation. They found that the proposed control scheme could assure the water demand, decrease solar electricity costs, and store more thermal energy in the storage system. Promising results showed that dynamic models developed for both water demand and production systems can be the

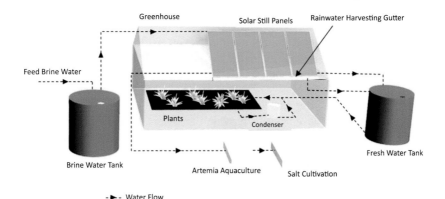

FIGURE 8.11 A schematic representation of the developed solar membrane distillation unit in Ref. [87].

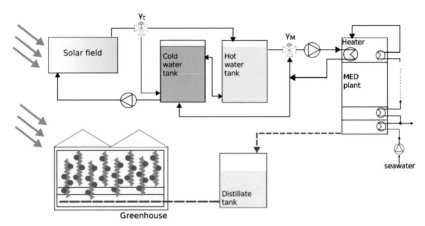

FIGURE 8.12 Schematic of a solar-powered multieffect distillation desalination plant coupled with a greenhouse [88].

keys to achieve more competitive distillate prices when model prediction controllers are applied.

In another study by Rabhy et al. [89], a fully transparent solar distiller was designed to supply irrigation water for an agricultural greenhouse. In this study, a numerical simulation based on equations of transient mass and energy balance using computational fluid dynamics (CFDs) was developed. In this regard, heat transfer, parameters of flow and phase change of water, and humid air with a free surface inside the distiller were simulated and the glass and basin temperature variations were specified as CFD boundary conditions. The simulation results obtained from the lumped and the coupled lumped-CFD models were compared with the experimental data to evaluate the transparent solar still under the Alexandria climate conditions in Egypt. The numerical results showed an increase of approximately 22% in the daily yield due to improving the insulation of the basin. They concluded that equipping the greenhouse roof with transparent distillers can supply 37.5% of water for irrigation while reducing the power consumption of the cooling system by 60% in the greenhouse.

In another study by Mashaly et al. [68], a solar distillation system to treat three types of feedwater including seawater, groundwater, and agricultural-drainage water was developed to provide the water demand of greenhouses. The schematic view of the solar still is presented in Fig. 8.13. In this study, two methods of the Fernandez (F) and the adapted Penman-Monteith (A-PM) were applied to anticipate the crop-water requirements (CWR) of greenhouses. The results revealed that the A-PM method can supply the greenhouse CWR of about 2 m^2 and therefore, the produced water by 1 m^2 of the solar still is sufficient. They claimed if the capacity of water production exceeds the CWR of the greenhouse, the described solar-desalination system can meet the CWR.

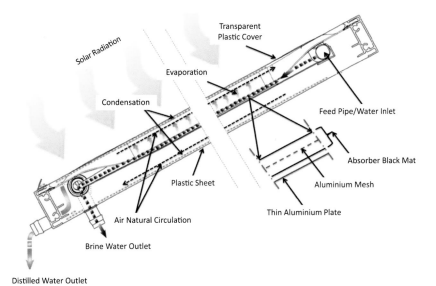

FIGURE 8.13 Cross-sectional view of the developed solar still panel in Ref. [68].

Mahmood and Al-Ansari [90] developed and analyzed a novel solar-powered self-sustainable greenhouse utilizing HDH phenomena to desalinate saline groundwater and improve the food security of regions with arid climates. The proposed greenhouse was composed of a greenhouse unit, solar PTCs, thermal energy storage (TES), organic Rankine cycle (ORC), and an absorption cooling system (see Fig. 8.14). In this study, a thermodynamic model based on mass, entropy, energy, and exergy of all ingredients of the system was developed and the performance of the system was evaluated to find the relationship between different parameters. The results of the parametric analysis revealed that the proposed system can produce 17.5−27.3 m^3/d freshwater, 4.3 MW cooling power, and 1.03 MW electricity with an output gained ratio of nearly 2.10−3.3. They claimed that the suggested system can provide the annual greenhouse requirements sustainably. For sustainable food production in arid lands and highlighting the utilization of solar energy in greenhouses to supply freshwater derived from seawater or brackish water with less environmental impact, Shekarchi and Shahnia [91] evaluated various solar-powered desalination plants by exploring their specifications, advantages, and limitations. Moreover, they introduced and discussed different types of greenhouses considering their total water demands as a function of the type of cultivated crops, the technology of greenhouse, type of irrigation, installation location, and climate conditions. Also, the integration of solar desalination plants with greenhouses was investigated. The results indicated that for choosing the most proper solar-driven desalination technology the parameters associated with the greenhouses' location such as

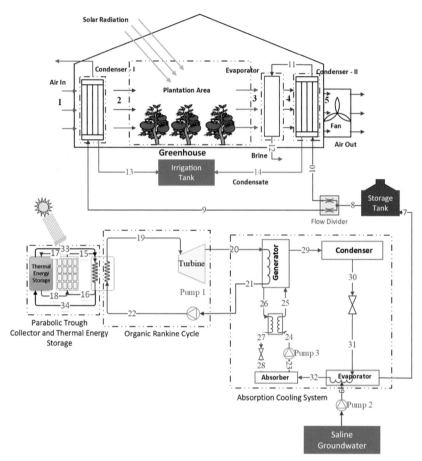

FIGURE 8.14 Schematic representation of the developed self-sustained agriculture greenhouse in Ref. [90].

available solar radiation, environmental temperature and humidity, and sea-water salinity along with specifications of greenhouses including the type of cultivated crops, CWR, type of the greenhouse, type of the irrigation system, and soil specifications should be considered. They claimed that by consider-ing the water production rate and cost, direct solar distillation systems and PV-ED desalination plants are most suitable for use in small-scale green-houses, while solar thermal-powered RO, MED, and MSF desalination plants are more preferred for use in medium- and large-scale applications.

Among various solar desalination technologies, the amount of water pro-duced by solar distillation systems is not enough to supply the irrigation water demand for crop production, especially for those crops cultivated in open fields, while their integration with agricultural greenhouses is a more efficient concept. In contrast, PV-RO plants can fit greenhouses with any capacity from the view of energy consumption and cost of water production, while PV-RO and PV-ED systems are highly recommended for use in small-scale greenhouses when brackish water and seawater are the only feedwater resources. For medium- to large-scale greenhouses, the use of solar thermal-powered RO and MED desalination plants is more preferred. According to the literature, the most widespread desalination technology to provide irrigation water in vast croplands is PV-RO.

8.4 Case studies (implemented projects around the world)

Several countries around the world are employing solar desalination technologies to provide the required water for agricultural purposes. Most of these countries have faced physical water shortages due to the drought in the past decades while having access to high levels of solar radiation as well as large bodies of salt water. In this section, some installed solar desalination plants around the world to provide water for agricultural applications are presented and discussed.

8.4.1 Commercial solar-powered desalination plants

- *PV-RO desalination plants installed in Santiago*: Osmosun is a French company active in the field of solar water desalination plants and annually carries out several projects in different scales and capacities around the world to supply drinking water in areas with no access to freshwater. In 2020, the company implemented a freshwater supply project in Moia Moia, Santiago Island with a financial package carried out by FIDA (FAO) (see Fig. 8.15). In this project, a PV-RO desalination plant with a production capacity of 50 m^3/d was installed to provide the required water for cultivation in a 6-ha farmland. They reported that the solar desalination plant operates only with solar energy without using battery storage [92].
- *PV-RO desalination plant installed in Namibia*: Solar Water Solutions is a Finnish company that activates in the field of desalination. Since 2015,

FIGURE 8.15 Photovoltaic-reverse osmsis desalination plant installed at the field of crops, Moia Moia, Santiago island [92].

FIGURE 8.16 Namibia's solar-powered desalination project to supply irrigation water [93].

this company with the cooperation of leading Dutch institutional investors, Climate Fund Managers (CFM), as well as private companies and
local governments have installed and implemented several solar desalination plants in various locations. In 2019, the company, in a joint project
called "Carbon Garden project" with the cooperation of the University of
Turku and the University of Namibia in Henties Bay, launched a PV-RO
solar desalination project with the freshwater production capacity of
3.5 m³/h (see Fig. 8.16). The feed water resource for this system in its
initial phase was seawater, but for the next phases, groundwater wells of
Namibia[7] were also considered. In this project, 4 ha of land were allocated for planting and afforestation. This project aims to absorb CO_2
emission through afforestation, timber trading of trees after their growth,
and the replacement and replanting of trees [93,94].
- *Solar-powered MED desalination plant installed in California*: In 2013,
  the American company of WaterFX launched implementing a CSP system and a desalination plant to supply water for agriculture. This project
  is located in the Panoche area of Central Valley of California. One-third
  of the nation's food is grown in the Central California Valley, and therefore the freshwater shortage is a major problem of the agriculture sector
  in the region.[8] In the initial phase of this project, a CSP system using
  PTCs with a total area of 656 m² and thermal power production capacity
  of 400 kW, along with a three-stage MED desalination system with a
  freshwater production capacity of 53 m³/d were installed (see Fig. 8.17).
  The WaterFX company noted that the system consumes one-fifth of the
  electricity required by conventional RO plants installed in San Diego and
  Carlsbad (cities in California), leading to a 50%−60% decrease in water
  production costs. In the second operational phase, the decision was made
  to launch the project by 2020 on a larger scale with a 10-stage MED

---

7. Namibia is a dry country with scarcity of freshwater and most of saline groundwater.
8. The water source for irrigation in this area is drained saline water of agricultural lands.

FIGURE 8.17 Solar-powered desalination plant installed in Panoche, California [95]: (A) Solar parabolic trough concentrator and (B) multieffect distillation plant.

desalination plant and a production capacity of about 7500 m^3 of freshwater per day [85,95].

- *PV-RO desalination plant installed in Baja California*: San Quintín[9] is a region with a temperate climate and has a favorable condition for cultivation all year round. But from 1985 to 2015, due to continuous drought, the area under cultivation was decreased from 28,000 ha to 7800 ha. In 2012, Berrymex which is a leading company in the agricultural industry with the main products of strawberries, raspberries, blueberries, and blackberries, decided to use an independent irrigation network for its lands using desalinated seawater that later became the first water desalination project in Mexico for agricultural applications. In this project, Berrymex in collaboration with World Water and NEWEN Energías Alternas launched a solar water desalination project which took seven years to be completed and come into operation. The Berrymex project employed a RO desalination plant with a production capacity of 810 m^3/h integrated with a PV system with a power capacity of 500 kW to supply the power demand of the RO desalination system [96] (see Fig. 8.18).

- *PV-RO desalination plant for reirrigation installed in Spain*: In 2021, a desalination and drainage treatment project for hydroponic greenhouses was launched in Spain.[10] Almost 15% of agriculture in the south and southeast of Spain is in the form of hydroponic greenhouse cultivation. Therefore, the drained effluent of these greenhouses which is a huge volume of water containing pollutants (nitrate and other fertilizers) has to be managed. In this regard, the University of Cartagena in collaboration with the University of Almeria and the support of the European Union for the realization of new technologies implemented a PV-RO desalination plant in the Agricultural Research Center of the University of Almeria to supply the irrigation water from the drained water of a hydroponic tomato greenhouse with an area of 1454 m^2 [98] (see Fig. 8.19).

---

9. San Quintín is a coastal town located in the west coast of the Mexican state of Baja California.

10. Spain is the country employs the most desalination plants for the agricultural sector in the world.

FIGURE 8.18 Baja California's photovoltaic-reverse osmosis desalination project to supply irrigation water [97].

FIGURE 8.19 The reirrigation photovoltaic-reverse osmosis desalination plant installed in Spain: (A) Installed photovoltaic system, (B) reverse osmosis desalination plant, and (C) inside view of the greenhouse [98,99].

It was reported that the use of this system results in successful reirrigation of the greenhouse with the drainage water and positive impacts on crop growth. In this regard, it was suggested to expand this project to a larger scale for all hydroponic greenhouses installed not only in Spain but also in Portugal, Italy, Malta, and Greece [99].

8.4.2 Commercial solar seawater greenhouses

- *Solar seawater greenhouse installed in Qatar*: In 2011, the Norwegian company of Sahara Forest launched a solar-powered multipurpose project in Qatar including SWGHs to provide both suitable conditions for year-round cultivation of high-value vegetables and water for outside vegetation and evaporative hedges, cultivation of algae (for biofuel production and feed for livestock and aquatic animals), and dry salt production. The project was completed with a fund equal to USD 5.3 million received by Yara ASA[11] and Qafco[12] companies. In this project, a CSP plant consisting of a large-scale PTC was installed to supply the heat demand of an employed MSF desalination plant with the capacity of $10 \, m^3/d$ to supply water requirements of the plants cultivated inside the greenhouses as well as the outside, while the waste heat was rejected to warm the greenhouses in

---

11. One of the largest fertilizer companies in the world.
12. The largest single-site producer of ammonia and urea worldwide.

FIGURE 8.20   Sahara Forest solar desalination project installed in Qatar [100]: (A) Picture of the complex, and (B) photo of the implemented project.

winter and to regenerate the desiccant employed to dehumidify the air. Additionally, a PV system with a power production capacity of 40 kW was installed to supply the electricity demand of the project (see Fig. 8.20).

This project was built on 1 ha of land with a cultivation area of 600 m^2. It is expected that the cultivation area of 20 ha can meet the total import requirements for agricultural products including cucumber, tomato, pepper, and eggplant in Qatar. The production capacity of this complex is 75 kg/m^2/year which is competitive with European industrial greenhouses [101]. In this project, the seawater with a salinity of 3.5% is used for the cultivation of marine plants and algae in a capacity of 50 m^3 that can be used as raw material for biofuels production, fodder in livestock, and fish feed. The effluent from this stage with a salinity of 5%−7% is used in the cooling cycle of the greenhouse, the CSP plant, and the desalination plant. To create suitable conditions of humidity and air temperature for cultivation outside the greenhouses, the water leaving the CSP system and the greenhouse cooling system with 15%−20% salinity collides with the vertical hedges outside the greenhouses. In this way, like the evaporator, the optimal environment for the growth of food material, fodder for grazing livestock, and even native plant species is provided. Finally, the effluent from the hedges with salinity above 30%, which tends to precipitate, enters in evaporation pools to produce dry salt [100,101].

- *Solar seawater greenhouse installed in Jordan*: This project was implemented in 2016 on the coast of the Red Sea and the Port of Aqaba Special Economic Zone by the Norwegian company Sahara Forest with the support of the European Union and the Jordanian government. The purpose of this project was to provide freshwater for agricultural applications from an alternative source in Jordan. Currently, twelve groundwater aquifers in this country are used as main sources of freshwater which are running out because of the extraction rate twice their natural charge rate. Decisions by the executor company have been made to extend the project from a research scale to a commercial type with a total cultivated area of 60,000 m^2 after the mitigation of the COVID-19 pandemic [102]. Similar to Qatar's project, it is expected that water demands of greenhouse cultivation, farmland cultivation, and dry salt production to be provided on a

FIGURE 8.21 The solar-powered SWGH implemented in Jordan, Aqaba [103], (A) Overview of the project, (B) Inside view of the greenhouse.

FIGURE 8.22 Photos of the solar seawater greenhouse project implemented in Somaliland, Berbera [108].

larger scale. In this project, a PV-RO desalination plant is used to irrigate plants, while the cooling of greenhouses in hot weather is performed using seawater and HDH desalination process so-called salt water-cooled greenhouses (see Fig. 8.21) [104]. It has been reported that 130 tons of vegetables such as green pepper and buckwheat are annually produced along with 10 m^3 of freshwater in this complex [105,106].

- *Solar seawater greenhouse installed in Somaliland*: In 2017, the British Seawater Greenhouse company, in collaboration with the PENHA, the Pastoral & Environmental Network in the Horn of Africa, and Aston University built a SWGH in Somalia, 17 km far from Berbera, on the shores of the Gulf of Aden. This project aimed to provide a solution to solve food security problems and widespread disruptions in the region. Additionally, the HDH cooling technology was employed inside the greenhouse using seawater, where a shade net system[13] was installed on the walls, naturally assisting to cool the greenhouse [107] (see Fig. 8.22). The whole project was approximately 1 ha, including two greenhouse units of 1200 and 500 m^2, and an area that was considered for cultivation outside the greenhouse. The employed seawater desalination system was a RO plant integrated with a 10-kW solar PV system. The crops grown in this project are lettuce, tomatoes, cucumbers, carrots, onions, and beans [108].
- *Solar seawater greenhouse installed in Australia*: In 2010, the British company of Seawater Greenhouse implemented a solar SWGH project in the

---

13. A low-cost, rugged and modular facility used under the harshest environments on Earth.

port of Augusta, South Australia. This project was the first commercial-scale project of this company which was implemented expending USD 2 million. This project aimed to mitigate water tensions and drought as well as food security threats in this region. The initial phase of the project was launched with a greenhouse with an area of 2000 m^2. In this project, the greenhouse employs a HDH cooling process using seawater. Moreover, the irrigation water supply is provided by RO and HDH desalination systems and the cooling process. A PV system was integrated to supply the power demand of the desalination system as well as the entire project (see Fig. 8.23). The annual tomato production capacity of 100,000 kg was reported in the first year of operation. It was also decided to use the extra produced freshwater to irrigate citrus trees outside the greenhouse [109,110].

- *Solar seawater greenhouse installed in Australia*: The SWGH project in the port of Augusta was launched in its second phase of operation under the new ownership of Sundrop on 49 ha with 20 ha of greenhouse cultivated area. The project was launched in 2014 and completed in 2016 by expending 200 million AUDUSD, and co-fundings of the international investment Corporation Kohlberg Kravis Robert (KKR) as 100 million AUDUSD, Australia's Clean Energy Finance Corporation as 40 million AUDUSD, and Australian Government as 6 million AUDUSD. This project, known as the first large-scale project in the world, employs a solar tower system with 23,000 mirrors, a TES system, and a MED desalination plant as shown in Fig. 8.24. The cooling system of greenhouses uses seawater as the feed of a

FIGURE 8.23 Solar-powered SWGH installed in Port Augusta, Australia [109]: (A) Overview of the project and (B) Inside view of the greenhouse.

FIGURE 8.24 The solar-powered SWGH integrated with a solar tower installed in Port Augusta, Australia [111]: (A) Overview of the project and (B) inside view of the greenhouse.

HDH desalination process where a part of the produced freshwater is obtained from the greenhouses' cooling system. The solar tower provides the heat demands of the desalination plant, the greenhouse, and the steam cycle to generate electricity. The solar tower produces 39 MW of heat per day which leads to the production of 10 m^3/d of freshwater and the total electricity requirement of the project, resulting in 2000 m^3 of saved diesel per year. The greenhouse production has been reported as 17,000 tonnes per year, accounting for 13% of the Australian market [111–113].

8.4.3 Research-scale solar seawater greenhouses

Iran is currently constructing two solar SWGH units in Sistan-Baluchestan and on the Makran coast funding by the technology development council of water, drought, erosion, and environment at science and technology vice presidency. Each solar SWGH unit has an area of 400 m^2 as shown in Fig. 8.25 and is expected to become operational in 2021. This project aims to enlarge agriculture and create employment through the use of high-quality solar energy in southern regions of the country. The project has been entirely done by knowledge-based companies in the country [115]. Israel is the most advanced country in the field of desalination technologies and has implemented several desalination projects around the world to supply drinking and agricultural water due to droughts. Because of being energy-intensive, some companies such as IDE Technologies, and research centers such as Ben-Gurion University increased their attention to solar thermal and PV systems to supply power demands of desalination plants. In 2012, to supply water for intensive agriculture use in Jordan, the Palestinian Authority and Israel, a solar-powered desalination project known as AGRISOL was implemented in Israel (Hatzeva) and Jordan (Karama). In this project, two pilot PV-NF desalination plants with a water production capacity of 5 m^3/d were installed. The potential market of the desalinated irrigation water was explored through surveys to find the farmers' viewpoints and their potential to switch to this alternative water resource for crop production. It was claimed that solar desalination can be considered a sustainable solution in regional arid land agriculture [82].

8.5 Economics and environmental impacts

The process of water cost assessment depends on several factors including the desalination technology, water production benefits during a long period, quality of feed water, accessible options for waste disposal, the annual market value of the production, and the costs for maintenance, marketing, labor, transportation, and employees during the plant life [116]. The costs of desalination using RO technology, as the most mature membrane-based desalination technology, are steadily declining, mainly because of the dramatic increase in plant capacity, better process design, and more beneficial

FIGURE 8.25 Installed homegrown solar SWGH in Sistan-Baluchestan, Iran [114].

materials and membranes. Considering seawater desalination, the water production cost of RO plants is less than MED and MSF since the RO has high productivity in recovery tools. But for brackish water desalination, the most economical desalination technologies are RO and ED [60]. For small-scale solar-powered MD systems, the cost of the desalinated water is more than that of commercial RO plants mainly because of their low productivity and high capital costs [116,117].

In terms of environmental impacts, high-temperature rejected brine streams can cause the death of marine organisms and coral reefs, and negative polluting impacts on the seabed by entering the sea. Hence, a highly polluted saltwater source limits the amount of available saline water for desalination [60]. However, the effect of rejected brine on marine life conditions is still unknown to a large extend, and therefore further studies are required in this field [118]. Due to both risks of depleting conventional energy resources and increasing greenhouse gas (GHG) emissions, the fossil fuel-powered desalination systems are no longer profitable to overcome the global water crisis [3]. In a review study conducted by Kumar et al. [36], the main benefits and limitations associated with desalination technologies in agricultural applications were discussed considering favorable quality parameters required for agricultural water and available water resources. According to this study, the results from several bench- and pilot-scale experiments indicated that integration of solar energy with RO plants coupled with low energy demand technologies such as FO, NF, and MF can significantly improve cost savings. For use in irrigated agriculture, the desalinated freshwater is still too expensive compared to conventional water resources.

In agricultural operations, the net cost of desalinated water is the sum of costs for labor, energy, chemicals, payback, membrane replacement, operation, management, and maintenance as presented in Table 8.4. For seawater RO desalination, the operational costs as a function of the plant size, intake, the distance between intake and RO plant, and product pumping are around €0.35−0.5/m^3 (without payback) and the cost of water production ranges

TABLE 8.4 RO desalination costs using brackish water and seawater as feed [36].

Type of cost	Seawater (USD/acre/foot)[a]	Brackish water
Energy	327−401	119−178
Labor	27−120	30−104
Chemical materials	27−80	30−45
Replace of membrane	1−54	22−33
Chemical cleaning	1−3	2−4
Maintenance	27−48	18−27
Operation and management	461−728	223−401
Payback costs	223−327	104−134
Total costs	669−1,055	312−535

[a] $1\ acre/foot = 1233.48\ m^3$.

from 5% to 25% of overall costs for the crop production. The trends of recent desalination indicate that a more rapid decrease in RO costs compared to thermal technologies due to innovations and implementation of large-scale desalination plants to minimize labor costs and utilization of mechanical equipment with high efficiency will result in a remarkable reduction in costs.

In a study by Awaad et al. [60], the state of the art in economic aspects and environmental impacts of desalination processes were investigated and discussed. In this study, the growth of crops including tomatoes, peppers, cucumbers, strawberries, lettuces, flowers, and herbs in greenhouses using desalinated water was evaluated and controlled through the management of climate variables as well as the quantity and quality of irrigated water along with fertilizers. The results indicated that if farmers follow the technical recommendations to produce a high crop yield, the process of water desalination can have economic benefits and will support agricultural development concerning the production and export of agricultural products such as fruit and vegetables. They asserted that although irrigation of extensive crops such as wheat, corn, and rice with desalinated water is not economically efficient, it is still an affordable method to be used for the production of high-value crops.

Morad et al. [119] developed and evaluated two solar-powered desalination systems with distinct operational conditions; one using a condenser integrated with a solar flat-plate collector (FPC) and vacuum pump, and the other was an ordinary solar desalination system without a vacuum pump. In this study, the performance of both systems as a function of the level of water tank's flow rate and water salinity variation was examined and aspects of temperatures, the efficiency of the condenser, production of water, and costs were evaluated. The results of experiments revealed that for all feedwater salinity levels, the developed system increases the water productivity more than the ordinary system due to the existence of the vacuum pump. Also, the maximum freshwater production of 10.94 and 7.27 L/d considering the price values of USD 0.031 and 0.030/L were obtained at groundwater tank's flow rates of 0.80 and 0.40 L/h using developed and ordinary systems, respectively. It was claimed that by increasing the flow rate of water in the developed system, the water productivity is increased while the costs are decreased.

An economic evaluation for different desalination scenarios to reach a practical solution to the water scarcity issue for irrigation and drinking in the central region and southern area of Iran was performed by Zehtabiyan-Rezaie et al. [120]. In this study, an ordinary solar still and a still combined with a solar collector were considered at an interest rate of 4%. The results revealed that production of freshwater in a solar desalination farm comprised of solar distillation units is an economic and favorite scenario for the south region with a cost of USD $2.25/m^3$, while in contrary, transportation of saline water to central areas is neither economical nor environmentally possible. They claimed that the best alternative solution is transporting the produced freshwater from the south to the center with a cost of USD $4.12/m^3$.

Daghari and Zarrogh [121] investigated the performance of some desalination processes including MED and MSF integrated with a CSP plant to produce freshwater for irrigation. For cereals as the main type of agricultural products imported by Tunisia, the yield is less than 1.65 tons/ha, while in Egypt which is located in semiarid and arid bioclimatic zones, it is 6.61 tons/ha. It is mainly because Egyptian farmers irrigate cereals with the water that comes from the Nile river, while in Tunisia, the insufficient availability of irrigation water in both terms quality and quantity most often leads to low agricultural yields.

To assess the economic possibility of solar-powered desalination and water pumping in agriculture, Jones et al. [122] evaluated a variable speed PV pumping and desalination systems without battery storage through hourly simulation using three types of power supplies including PV, diesel engine (DE), and grid electricity (GE), four types of membranes, four inverter configurations, and two recovery rates of RO system. From the results, it was found that according to assumed electrical costs, PV is more profitable than diesel-powered systems considering assumed electrical costs but less affordable than grid-powered systems. Additionally, it was found that the economy of the modeled PV desalination system is adversely affected by oversizing requirements to supply water demand in agricultural operations. It was claimed that PV water pumping and desalination is profitable only for high-yielding crops and low water requirements. Further, economic results indicated a clear advantage for the performance of PV pumping and desalination system without energy storage compared to the previous system using energy storage which is mainly due to the larger system size, low-energy demand membranes, and the system which is being optimized. Alzoheiry and Hemeda [123] proposed a method to predict the performance of a solar-powered desalination plant for drip irrigation to grow bell pepper in a greenhouse and evaluate its economic performance. The analysis indicated that for bell pepper cultivated in the greenhouse, the predicted seasonal crop evapotranspiration is 370.8 mm while the measured value is 324.6 mm with an overall overestimation of 15%. They asserted that although the extra cost added to the final product is high (USD 1.09/kg), this unit can open the possibility for irrigation when desalination and mixing processes are not available.

In a study by Dehesa-Carrasco et al. [80], the performance of a low-pressure desalination system (PV-NF) for use in rural areas considering four cases with different inlet influent concentrations was evaluated and the unit cost of permeated water was estimated. The results showed that the PV-NF system can produce $2.16-4.8$ m^3/d, with a permeate water's unit cost of USD $1.05-0.47$/m^3, making it possible to irrigate between 0.5 and 1 ha of crops. A comparison between the water cost produced by solar-powered desalination plants and conventional types indicates that the unit price of solar-powered desalinated seawater is USD $3.45-9.9$/m^3 which is considerably higher than the costs associated with conventional fossil-powered desalinated water as USD $0.38-2.97$/m^3. In this regard, Table 8.5 confirms that solar-powered desalination plants are more expensive and currently less profitable compared to conventional systems, but contribute less CO_2 emissions as the main causes of global warming. Since renewable energy sources such as solar or wind operate with no water, therefore they will not pollute water resources. However, it is expected that solar desalination costs to be decreased in the future, especially for PV-powered desalination systems due to the declination trend in PV module costs [124]. Hence, there would be tremendous potential for the integration of solar energy with desalination

TABLE 8.5 Cost of water production using different energy sources and feedwaters [117].

Source of feedwater	Source of energy	Cost (USD/m^3)
Seawater	Conventional energy	0.38−2.97
	PV electricity	3.45−9.9
	Wind power	1.1−5.5
Brackish water	Conventional energy	0.23−1.17
	PV electricity	4.95−11.35
	Geothermal energy	2.2

plants, making the price of the desalinated water more affordable especially in locations with high potentials of solar energy [117].

8.6   Conclusion and prospects

Water scarcity is a global issue and still a challenging topic in several regions around the world specifically in locations with arid and semiarid climate specifications. For many years, desalination has been used to supply freshwater especially in areas with access to saltwater. The largest saltwater resources on earth are seas and oceans, but brackish groundwaters are also available. Agriculture as a water-intensive sector with a share of 70% of global water withdrawals is responsible to provide food for the growing population. In many regions, the only available water resource for irrigation is saltwater. The continuous utilization of saltwater for irrigation can cause severe problems both for soil and cultivated crops. In this regard, desalination can be considered a feasible solution to meet the water demand of the agriculture sector. Despite many advancements in the field of desalination, the cost of using desalinated water for irrigation is still high. In terms of greenhouses, the use of desalinated water is more affordable when high-value crops are cultivated inside. However, brine disposal remains the major environmental concern with desalination processes.

Most of the desalination plants installed to supply water for agriculture are powered by fossil fuels and therefore, water desalination using these plants is no longer sustainable mainly due to the caused environmental impacts. One of the abundant renewable energy resources with the highest compatibility for integration with desalination plants is solar energy. In this case, the most common solar desalination plants utilized to provide water in agriculture are PV-RO but other technologies such as PV-ED and PV-NF have also been used especially for crop production in greenhouses. To provide water for large-scale greenhouse cultivation systems, the integration of CSP plants with thermal desalination systems is also considered an appropriate solution. Due to their low productivity and high dependency on climate parameters, solar distillation systems have less been used to supply irrigation water, but their use in small-scale greenhouses for the cultivation of high-value crops has been reported in some cases. Solar SWGHs are also mature technologies that are used both for the cultivation of crops and producing freshwater and therefore, they can be considered as self-sustained facilities with most applicability in coastal areas due to high accessibility to seawater.

According to discussed topics in this chapter, although solar-powered desalination plants can offer several benefits over conventional types (especially in terms of environment conservation), their costs are the main barrier in front of their worldwide deployment. However, technical improvements in solar systems along with cost reductions can assist in paving the way for extensive use of these sustainable facilities, making them more competitive with conventional desalination plants to supply water in the agriculture sector. As an alternative solution, the use of solar systems in combination with other RESs in the form of hybrid systems can considerably increase reliability and decrease total costs.

Acknowledgment

The authors would like to thank Tarbiat Modares University (TMU) (http://www.modares.ac.ir) for the received financial support (grant number IG/39705) for the "Renewable Energies Research Group."

References

[1] Salzmann CG. Advances in the experimental exploration of water's phase diagram. J Chem Phys 2019;150:060901. Available from: https://doi.org/10.1063/1.5085163.
[2] Gorjian S, Ghobadian B, Tavakkoli Hashjin T, Banakar A. Experimental performance evaluation of a stand-alone point-focus parabolic solar still. Desalination 2014;352:1−17. Available from: https://doi.org/10.1016/j.desal.2014.08.005.
[3] Gorjian S, Ghobadian B. Solar desalination: a sustainable solution to water crisis in Iran. Renew Sustain Energy Rev 2015;48:571−84. Available from: https://doi.org/10.1016/j.rser.2015.04.009.
[4] Water distribution on Earth. https://en.wikipedia.org/wiki/Water_distribution_on_Earth; 2021 [accessed 02.04.21].
[5] Hirschmann M, Kohlstedt D. Water in Earth's mantle. Phys Today 2012;65:40−5. Available from: https://doi.org/10.1063/PT.3.1476.
[6] Houngbo GF. Nature-based solutions for water; 2018.
[7] van Vliet MTH, Jones ER, Flörke M, Franssen WHP, Hanasaki N, Wada Y, et al. Global water scarcity including surface water quality and expansions of clean water technologies. Env Res Lett 2021;16:024020. Available from: https://doi.org/10.1088/1748-9326/abbfc3.
[8] Galgano FA. Water in the Middle East. Adv Mil Geosci 2019;296:73−89. Available from: https://doi.org/10.1007/978-3-319-90975-2_5.
[9] Gorjian S, Ghobadian B, Ebadi H, Ketabchi F, Khanmohammadi S. Applications of solar PV systems in desalination technologies. Photovolt Sol Energy Convers 2020;237−74. Available from: https://doi.org/10.1016/B978-0-12-819610-6.00008-9 Elsevier.
[10] Martínez-Alvarez V, González-Ortega MJ, Martin-Gorriz B, Soto-García M, Maestre-Valero JF. Seawater desalination for crop irrigation—current status and perspectives. Emerging Technologies for Sustainable Desalination Handbook. Elsevier; 2018. p. 461−92. Available from: https://doi.org/10.1016/B978-0-12-815818-0.00014-X.
[11] Valuing Water. vol. 191. 2021.
[12] AQUASTAT - FAO's Global Information System on Water and Agriculture. Food Agric Organ United Nations. http://www.fao.org/aquastat/en/overview/methodology/water-use; 2021.
[13] Suwaileh W, Johnson D, Hilal N. Membrane desalination and water re-use for agriculture: state of the art and future outlook. Desalination 2020;491:114559. Available from: https://doi.org/10.1016/j.desal.2020.114559.
[14] Gorjian S, Singh R, Shukla A, Mazhar AR. On-farm applications of solar PV systems. In: Gorjian S, Shukla ABT-PSEC, editors. Photovoltaic solar energy conversion. Elsevier; 2020. p. 147−90. Available from: https://doi.org/10.1016/B978-0-12-819610-6.00006-5.

[15] Singh A. Conjunctive use of water resources for sustainable irrigated agriculture. J Hydrol 2014;519:1688−97. Available from: https://doi.org/10.1016/j.jhydrol.2014.09.049.

[16] Gruère G, Ashley C, Cadilhon J-J. Reforming water policies in agriculture: Lessons from past reforms. Available from: https://doi.org/10.1787/1826beee-en; 2018.

[17] Molden D. Water for food water for life. Routledge; 2013. Available from: https://doi.org/ 10.4324/9781849773799.

[18] Al-Ismaili AM, Jayasuriya H. Seawater greenhouse in Oman: a sustainable technique for freshwater conservation and production. Renew Sustain Energy Rev 2016;54:653−64. Available from: https://doi.org/10.1016/j.rser.2015.10.016.

[19] Types of Agricultural Water Use, Other Uses of Water, Healthy Water, CDC. https:// www.cdc.gov/healthywater/other/agricultural/types.html; n.d. [accessed 04.04.21].

[20] Bentzen JS, Kaarsen N, Wingender AM. Irrigation and autocracy. J Eur Econ Assoc 2016;15:1−53. Available from: https://doi.org/10.1111/jeea.12173.

[21] Malakar A, Snow DD, Ray C. Irrigation water quality—a contemporary perspective. Water 2019;11:1482. Available from: https://doi.org/10.3390/w11071482.

[22] Zwarteveen M. Drip irrigation for agriculture. Routledge; 2017. Available from: https:// doi.org/10.4324/9781315537146.

[23] Lankford B, Closas A, Dalton J, López Gunn E, Hess T, Knox JW, et al. A scale-based framework to understand the promises, pitfalls and paradoxes of irrigation efficiency to meet major water challenges. Glob Env Chang 2020;65:102182. Available from: https:// doi.org/10.1016/j.gloenvcha.2020.102182.

[24] Shaffer DL, Yip NY, Gilron J, Elimelech M. Seawater desalination for agriculture by integrated forward and reverse osmosis: improved product water quality for potentially less energy. J Memb Sci 2012;415−416:1−8. Available from: https://doi.org/10.1016/j. memsci.2012.05.016.

[25] Feroze Ahmed M, Ahmed T. Status of remediation of arsenic contamination of groundwater in Bangladesh. Comprehensive Water Quality and Purification, vol. 1. Elsevier; 2014. p. 104−21. Available from: https://doi.org/10.1016/B978-0-12-382182-9.00002-5.

[26] Niu G, Cabrera RI. Growth and physiological responses of landscape plants to saline water irrigation: a review. HortScience 2010;45:1605−9. Available from: https://doi.org/ 10.21273/HORTSCI.45.11.1605.

[27] Martínez-Alvarez V, González-Ortega MJ, Martin-Gorriz B, Soto-García M, Maestre-Valero JF. The use of desalinated seawater for crop irrigation in the Segura River Basin (south-eastern Spain). Desalination 2017;422:153−64. Available from: https://doi.org/ 10.1016/j.desal.2017.08.022.

[28] Seelen LMS, Flaim G, Jennings E, De Senerpont Domis LN. Saving water for the future: Public awareness of water usage and water quality. J Env Manage 2019;242:246−57. Available from: https://doi.org/10.1016/j.jenvman.2019.04.047.

[29] Shahid SA, Zaman M, Heng L. Introduction to soil salinity, sodicity and diagnostics techniques. Guideline for Salinity Assessment, Mitigation and Adaptation Using Nuclear and Related Techniques. Cham: Springer International Publishing; 2018. p. 1−42. Available from: https://doi.org/10.1007/978-3-319-96190-3_1.

[30] Zörb C, Geilfus C-M, Dietz K-J. Salinity and crop yield. Plant Biol 2019;21:31−8. Available from: https://doi.org/10.1111/plb.12884.

[31] Kurunc A, Aslan GE, Karaca C, Tezcan A, Turgut K, Karhan M, et al. Effects of salt source and irrigation water salinity on growth, yield and quality parameters of Stevia rebaudiana Bertoni. Sci Hortic (Amst) 2020;270:109458. Available from: https://doi.org/ 10.1016/j.scienta.2020.109458.

[32] Machado R, Serralheiro R. Soil salinity: effect on vegetable crop growth. management practices to prevent and mitigate soil salinization. Horticulturae 2017;3:30. Available from: https://doi.org/10.3390/horticulturae3020030.

[33] Feng G, Zhang Z, Zhang Z. Evaluating the sustainable use of saline water irrigation on soil water-salt content and grain yield under subsurface drainage condition. Sustainability 2019;11:6431. Available from: https://doi.org/10.3390/su11226431.

[34] Maas EV, Hoffman GJ. Crop salt tolerance—current assessment. J Irrig Drain Div 1977;103:115−34. Available from: https://doi.org/10.1061/JRCEA4.0001137.

[35] Sun J, Li YP, Suo C, Liu YR. Impacts of irrigation efficiency on agricultural water-land nexus system management under multiple uncertainties—a case study in Amu Darya River basin, Central Asia. Agric Water Manag 2019;216:76−88. Available from: https://doi.org/10.1016/j.agwat.2019.01.025.

[36] Kumar R, Ahmed M, Bhadrachari G, Thomas JP. Desalination for agriculture: water quality and plant chemistry, technologies and challenges. Water Supply 2018;18:1505−17. Available from: https://doi.org/10.2166/ws.2017.229.

[37] Zhang Y, Sivakumar M, Yang S, Enever K, Ramezanianpour M. Application of solar energy in water treatment processes: a review. Desalination 2018;428:116−45. Available from: https://doi.org/10.1016/j.desal.2017.11.020.

[38] Aende A, Gardy J, Hassanpour A. Seawater desalination: a review of forward osmosis technique, its challenges, and future prospects. Processes 2020;8:901. Available from: https://doi.org/10.3390/pr8080901.

[39] Darre NC, Toor GS. Desalination of water: a review. Curr Pollut Rep 2018;4:104−11. Available from: https://doi.org/10.1007/s40726-018-0085-9.

[40] Cong VH. Desalination of brackish water for agriculture: challenges and future perspectives for seawater intrusion areas in Vietnam. J Water Supply Res Technol - AQUA 2018;67:211−17. Available from: https://doi.org/10.2166/aqua.2018.094.

[41] Al-Jabri S, Ahmed M. Use of renewable energy for desalination in urban agriculture in the GCC countries: possibilities and challenges. J Agric Mar Sci [JAMS] 2018;22:48. Available from: https://doi.org/10.24200/jams.vol22iss1pp48-57.

[42] Al Jabri SA, Zekri S, Zarzo D, Ahmed M. Comparative analysis of economic and institutional aspects of desalination for agriculture in the Sultanate of Oman and Spain. Desalin WATER Treat 2019;156:1−6. Available from: https://doi.org/10.5004/dwt.2019.24066.

[43] Jeong H, Kim H, Jang T. Irrigation water quality standards for indirect wastewater reuse in agriculture: a contribution toward sustainable wastewater reuse in South Korea. Water 2016;8:169. Available from: https://doi.org/10.3390/w8040169.

[44] Panagopoulos A, Haralambous KJ, Loizidou M. Desalination brine disposal methods and treatment technologies - a review. Sci Total Environ 2019;693:133545. Available from: https://doi.org/10.1016/j.scitotenv.2019.07.351.

[45] Ahmadi E, McLellan B, Mohammadi-Ivatloo B, Tezuka T. The role of renewable energy resources in sustainability of water desalination as a potential fresh-water source: an updated review. Sustainability 2020;12:5233. Available from: https://doi.org/10.3390/su12135233.

[46] Shanmugan S, Essa FA, Gorjian S, Kabeel AE, Sathyamurthy R, Muthu Manokar A. Experimental study on single slope single basin solar still using TiO_2 nano layer for natural clean water invention. J Energy Storage 2020;30:101522. Available from: https://doi.org/10.1016/j.est.2020.101522.

[47] Elsaid K, Kamil M, Sayed ET, Abdelkareem MA, Wilberforce T, Olabi A. Environmental impact of desalination technologies: a review. Sci Total Environ 2020;748:141528. Available from: https://doi.org/10.1016/j.scitotenv.2020.141528.

[48] Eke J, Yusuf A, Giwa A, Sodiq A. The global status of desalination: an assessment of current desalination technologies, plants and capacity. Desalination 2020;495:114633. Available from: https://doi.org/10.1016/j.desal.2020.114633.

[49] Feria-Díaz JJ, López-Méndez MC, Rodríguez-Miranda JP, Sandoval-Herazo LC, Correa-Mahecha F. Commercial thermal technologies for desalination of water from renewable energies: a state of the art review. Processes 2021;9:262. Available from: https://doi.org/10.3390/pr9020262.

[50] Qasim M, Badrelzaman M, Darwish NN, Darwish NA, Hilal N. Reverse osmosis desalination: a state-of-the-art review. Desalination 2019;459:59−104. Available from: https://doi.org/10.1016/j.desal.2019.02.008.

[51] Awad AM, Jalab R, Minier-Matar J, Adham S, Nasser MS, Judd SJ. The status of forward osmosis technology implementation. Desalination 2019;461:10−21. Available from: https://doi.org/10.1016/j.desal.2019.03.013.

[52] Abou-Shady A. Recycling of polluted wastewater for agriculture purpose using electrodialysis: perspective for large scale application. Chem Eng J 2017;323:1−18. Available from: https://doi.org/10.1016/j.cej.2017.04.083.

[53] Ye Y, Yu S, Hou L, Liu B, Xia Q, Liu G, et al. Microbubble aeration enhances performance of vacuum membrane distillation desalination by alleviating membrane scaling. Water Res 2019;149:588−95. Available from: https://doi.org/10.1016/j.watres.2018.11.048.

[54] Niewersch C, Battaglia Bloch AL, Yüce S, Melin T, Wessling M. Nanofiltration for the recovery of phosphorus—development of a mass transport model. Desalination 2014;346:70−8. Available from: https://doi.org/10.1016/j.desal.2014.05.011.

[55] Jones E, Qadir M, van Vliet MTH, Smakhtin V, Kang S. The state of desalination and brine production: a global outlook. Sci Total Environ 2019;657:1343−56. Available from: https://doi.org/10.1016/j.scitotenv.2018.12.076.

[56] Burn S, Hoang M, Zarzo D, Olewniak F, Campos E, Bolto B, et al. Desalination techniques—a review of the opportunities for desalination in agriculture. Desalination 2015;364:2−16. Available from: https://doi.org/10.1016/j.desal.2015.01.041.

[57] Dévora-Isiordia GE, Martínez-Macías M, del R, Correa-Murrieta MA, Álvarez-Sánchez J, Fimbres-Weihs GA. Using desalination to improve agricultural yields: success cases in Mexico. Desalination and Water Treatment, vol. 32. InTech; 2018. p. 137−44. Available from: https://doi.org/10.5772/intechopen.76847.

[58] Beltrán J, Koo-Oshima S. Water desalination for agricultural applications. Proc. FAO Expert Consult. Water Desalin. Agric. Appl. 2006, p. 60.

[59] Mahmoudi H, Spahis N, Abdul-Wahab SA, Sablani SS, Goosen MFA. Improving the performance of a seawater greenhouse desalination system by assessment of simulation models for different condensers. Renew Sustain Energy Rev 2010;14:2182−8. Available from: https://doi.org/10.1016/j.rser.2010.03.024.

[60] Awaad HA, Mansour E, Akrami M, Fath HES, Javadi AA, Negm A. Availability and feasibility of water desalination as a non-conventional resource for agricultural irrigation in the MENA region: a review. Sustainability 2020;12:7592. Available from: https://doi.org/10.3390/su12187592.

[61] Mahmoudi H, Abdul-Wahab SA, Goosen MFA, Sablani SS, Perret J, Ouagued A, et al. Weather data and analysis of hybrid photovoltaic−wind power generation systems adapted to a seawater greenhouse desalination unit designed for arid coastal countries. Desalination 2008;222:119−27. Available from: https://doi.org/10.1016/j.desal.2007.01.135.

[62] El-Awady MH, El-Ghetany HH, Latif MA. Experimental investigation of an integrated solar green house for water desalination, plantation and wastewater treatment in remote

arid Egyptian communities. Energy Procedia 2014;50:520−7. Available from: https://doi. org/10.1016/j.egypro.2014.06.063.

[63] Ghaffour N, Reddy VK, Abu-Arabi M. Technology development and application of solar energy in desalination: MEDRC contribution. Renew Sustain Energy Rev 2011;15:4410−15. Available from: https://doi.org/10.1016/j.rser.2011.06.017.

[64] Al-Ismaili AM, Jayasuriya H, Al-Mulla Y, Kotagama H. Empirical model for the condenser of the seawater greenhouse. Chem Eng Commun 2018;205:1252−60. Available from: https://doi.org/10.1080/00986445.2018.1443081.

[65] Alkaisi A, Mossad R, Sharifian-Barforoush A. A review of the water desalination systems integrated with renewable energy. Energy Procedia 2017;110:268−74. Available from: https://doi.org/10.1016/j.egypro.2017.03.138.

[66] Caldera U, Breyer C. Assessing the potential for renewable energy powered desalination for the global irrigation sector. Sci Total Environ 2019;694:133598. Available from: https://doi.org/10.1016/j.scitotenv.2019.133598.

[67] Bait O. Direct and indirect solar−powered desalination processes loaded with nanoparticles: A review. Sustain Energy Technol Assess 2020;37:100597. Available from: https://doi.org/10.1016/j.seta.2019.100597.

[68] Mashaly AF, Alazba AA, Al-Awaadh AM, Mattar MA. Area determination of solar desalination system for irrigating crops in greenhouses using different quality feed water. Agric Water Manag 2015;154:1−10. Available from: https://doi.org/10.1016/j.agwat.2015.02.009.

[69] Gandhi AM, Shanmugan S, Gorjian S, Pruncu CI, Sivakumar S, Elsheikh AH, et al. Performance enhancement of stepped basin solar still based on OSELM with traversal tree for higher energy adaptive control. Desalination 2021;502:114926. Available from: https://doi.org/10.1016/j.desal.2020.114926.

[70] Jathar LD, Ganesan S, Shahapurkar K, Soudagar MEM, Mujtaba MA, Anqi AE, et al. Effect of various factors and diverse approaches to enhance the performance of solar stills: a comprehensive review. J Therm Anal Calorim 2021;1−32. Available from: https://doi. org/10.1007/s10973-021-10826-y.

[71] Ketabchi F, Gorjian S, Sabzehparvar S, Shadram Z, Ghoreishi MS, Rahimzadeh H. Experimental performance evaluation of a modified solar still integrated with a cooling system and external flat-plate reflectors. Sol Energy 2019;187:137−46. Available from: https://doi.org/10.1016/j.solener.2019.05.032.

[72] Chauhan VK, Shukla SK, Tirkey JV, Singh Rathore PK. A comprehensive review of direct solar desalination techniques and its advancements. J Clean Prod 2021;284:124719. Available from: https://doi.org/10.1016/j.jclepro.2020.124719.

[73] Zhang Y, Sivakumar M, Yang S, Enever K, Ramezanianpour M. Application of solar energy in water treatment processes: a review. Desalination 2018;428:116−45. Available from: https://doi.org/10.1016/j.desal.2017.11.020.

[74] Eterafi S, Gorjian S, Amidpour M. Thermodynamic design and parametric performance assessment of a novel cogeneration solar organic Rankine cycle system with stable output. Energy Convers Manag 2021;243:114333. Available from: https://doi.org/10.1016/j. enconman.2021.114333.

[75] Yadav MC. Water desalination system using solar heat: a review ARenew Sustain Energy Rev 2017;67:1308−30. Available from: https://doi.org/10.1016/j.rser.2016.08.058.

[76] Sharon H, Prabha C, Vijay R, Niyas AM, Gorjian S. Assessing suitability of commercial fibre reinforced plastic solar still for sustainable potable water production in rural India through detailed energy-exergy-economic analyses and environmental impacts. J Environ Manage 2021;295:113034. Available from: https://doi.org/10.1016/j.jenvman.2021.113034.

[77] El-Bialy E, Shalaby SM, Kabeel AE, Fathy AM. Cost analysis for several solar desalination systems. Desalination 2016;384:12−30. Available from: https://doi.org/10.1016/j.desal.2016.01.028.

[78] Atkinson S. Solar-powered desalination system provides villagers with access to clean water for drinking and use in farming. Membr Technol 2020;2020:7−8. Available from: https://doi.org/10.1016/s0958-2118(20)30125-7.

[79] Muñoz M, Gil J, Roca L, Rodríguez F, Berenguel M. An IoT architecture for water resource management in agroindustrial environments: a case study in Almería (Spain). Sensors 2020;20:596. Available from: https://doi.org/10.3390/s20030596.

[80] Dehesa-Carrasco U, Ramírez-Luna JJ, Calderón-Mólgora C, Villalobos-Hernández RS, Flores-Prieto JJ. Experimental evaluation of a low pressure desalination system (NF-PV), without battery support, for application in sustainable agriculture in rural areas. Water Supply 2017;17:579−87. Available from: https://doi.org/10.2166/ws.2016.147.

[81] Gil JD, Álvarez JD, Roca L, Sánchez-Molina JA, Berenguel M, Rodríguez F. Optimal thermal energy management of a distributed energy system comprising a solar membrane distillation plant and a greenhouse. Energy Convers Manag 2019;198:111791. Available from: https://doi.org/10.1016/j.enconman.2019.111791.

[82] Ghermandi A, Naoum S, Alawneh F, Offenbach R, Tripler E, Safi J, et al. Solar-powered desalination of brackish water with nanofiltration membranes for intensive agricultural use in Jordan, the Palestinian Authority and Israel. Desalin WATER Treat 2017;76:332−8. Available from: https://doi.org/10.5004/dwt.2017.20624.

[83] Weiner AM, Blum DH, Lienhard JH, Ghoniem AF, Weiner AM. Design of a hybrid RO-MED solar desalination system for treating agricultural drainage water in California; 2015.

[84] Zarzoum K, Zhani K, Ben Bacha H, Koschikowski J. Experimental parametric study of membrane distillation unit using solar energy. Sol Energy 2019;188:1274−82. Available from: https://doi.org/10.1016/j.solener.2019.07.025.

[85] Stuber MD, Sullivan C, Kirk SA, Farrand JA, Schillaci PV, Fojtasek BD, et al. Pilot demonstration of concentrated solar-powered desalination of subsurface agricultural drainage water and other brackish groundwater sources. Desalination 2015;355:186−96. Available from: https://doi.org/10.1016/j.desal.2014.10.037.

[86] Hipólito-Valencia BJ, Mosqueda-Jiménez FW, Barajas-Fernández J, Ponce-Ortega JM. Incorporating a seawater desalination scheme in the optimal water use in agricultural activities. Agric Water Manag 2021;244:106552. Available from: https://doi.org/10.1016/j.agwat.2020.106552.

[87] Akrami M, Salah A, Dibaj M, Porcheron M, Javadi A, Farmani R, et al. A zero-liquid discharge model for a transient solar-powered desalination system for greenhouse. Water 2020;12:1440. Available from: https://doi.org/10.3390/w12051440.

[88] Roca L, Sánchez J, Rodríguez F, Bonilla J, de la Calle A, Berenguel M. Predictive control applied to a solar desalination plant connected to a greenhouse with daily variation of irrigation water demand. Energies 2016;9:194. Available from: https://doi.org/10.3390/en9030194.

[89] Rabhy OO, Adam IG, Elsayed Youssef M, Rashad AB, Hassan GE. Numerical and experimental analyses of a transparent solar distiller for an agricultural greenhouse. Appl Energy 2019;253:113564. Available from: https://doi.org/10.1016/j.apenergy.2019.113564.

[90] Mahmood F, Al-Ansari TA. Design and thermodynamic analysis of a solar powered greenhouse for arid climates. Desalination 2021;497:114769. Available from: https://doi.org/10.1016/j.desal.2020.114769.

[91] Shekarchi N, Shahnia F. A comprehensive review of solar-driven desalination technologies for off-grid greenhouses. Int J Energy Res 2019;43:1357−86. Available from: https://doi.org/10.1002/er.4268.

[92] Cape Verde, OSMOSUN® 6 BW, agriculture found at MOIA MOIA. Mascara Renewable Water 2021. https://www.osmosunwater.solutions/en/project/cape-verde-osmosun-6-bw-agriculture-found-at-moia-moia/; 2021.

[93] In Namibia's carbon sink project, irrigation water is produced by solar power. Solar Water Solution 2019. https://solarwatersolutions.fi/fi/article/namibian-hiilinieluhankkees-sa-kasteluvesi-tuotetaan-aurinkovoimalla/; 2019.

[94] Väänänen K. University of Turku Plants Trees in Namibia as Carbon Sinks. Univ Turku 2019. https://www.utu.fi/en/news/news/university-of-turku-plants-trees-in-namibia-as-carbon-sinks; 2019.

[95] Kurup P, Turchi C. Initial investigation into the potential of CSP industrial process heat for the Southwest United States. Golden, CO (United States). https://doi.org/10.2172/1227710; 2015.

[96] BerryMex inaugura la primera planta desalinizadora de agua de mar para uso agrícola en el continente americano. 3ersector. http://www.3ersector.mx/index.php/noticias-2018/84-desarrollo/5702-inaugura-berrymex-la-primera-planta-desalinizadora-de-agua-de-mar-para-uso-agricola-en-el-continente-americano; 2019.

[97] BerryMex inaugura la primera planta desalinizadora de agua de mar para uso agrícola en el continente. Energiasalternas; 2019.

[98] Proyecto_Innovacion Deseacrop Detalle Proyecto. Sacyr. https://www.sacyr.com/en/-/proyecto_innovacion-deja-tu-huella; 2021.

[99] Calzada M, Campos E, Buendia R, Concesiones SA. Proyecto LIFE Deseacrop. Interempresas; 2021.

[100] Qatar - Sahara Forest Project. https://www.saharaforestproject.com/qatar/; 2012.

[101] Alessandro Flammini, Manas Puri LPOD. Environment and Natural Resources Management Working Paper in the Context of the Sustainable Energy for All Initiative Walking the Nexus Talk: Assessing the Water-Energy-Food Nexus. Office of Knowledge Exchange, Research and Extension Food and Agriculture Organization of the United Nations Viale delle Terme di Caracalla; 2013.

[102] Whitman E. A land without water: the scramble to stop Jordan from running dry. Nature 2019;573:20−3. Available from: https://doi.org/10.1038/d41586-019-02600-w.

[103] Naumann E. Aqaba Official Opening. Flickr. https://www.flickr.com/photos/saharaforest-project/albums/72157684942354852/with/36893615766/; 2018.

[104] Saltwater-cooled greenhouses. Sahara Forest Project. https://www.saharaforestproject.com/saltwater-cooled-greenhouses/; 2013.

[105] Jordan. Sahara Forest Project. https://www.saharaforestproject.com/jordan/; 2018.

[106] Faulkner C. Water in the desert: Aqaba farming project nurtures hope in Jordan's arid south. The National 2020. https://www.thenationalnews.com/world/mena/water-in-the-desert-aqaba-farming-project-nurtures-hope-in-jordan-s-arid-south-1.1118875#4; 2020.

[107] Seawater Greenhouse Somaliland. https://www.sgsomaliland.com/; 2021.

[108] Somalia solar seawater greenhouse. https://seawatergreenhouse.com/construction-blog/2017/8/17/seawater-greenhouse-somaliland-construction-is-underway; 2021.

[109] Australia—Seawater Greenhouse. https://seawatergreenhouse.com/australia-1; 2021.

[110] Seawater Greenhouses Produce Tomatoes in the Desert. https://blogs.ei.columbia.edu/2011/02/18/seawater-greenhouses-produce-tomatoes-in-the-desert/; 2021.

[111] Australia's Sundrop farms are designed for the desert. WIRED UK. https://www.wired.co.uk/article/sundrop-farms-australian-desert; 2021.

[112] Our Technology - The Sundrop System - Sundrop Farms. Sundrop. https://www.sundrop-farms.com/our-technology/; 2021.

[113] Desert Sun & Seawater Farming. https://mimiculture.com/posts/002-desert-sun-and-seawater-farming.php; 2021.

[114] Two Seawater Greenhouses Planned in Southern Iran. Financial Tribune. https://financialtribune.com/articles/sci-tech/98355/two-seawater-greenhouses-planned-in-southern-iran; 2021 [accessed 05.06.21].

[115] Seawater greenhouse using renewable energy to be established. Tehran Times. https://www.tehrantimes.com/news/435867/Seawater-greenhouse-using-renewable-energy-to-be-established; 2019.

[116] Albloushi A, Giwa A, Mukherjee D, Calabro V, Cassano A, Chakraborty S, et al. Renewable energy-powered membrane systems for water desalination. Current Trends and Future Developments on (Bio-) Membranes. Elsevier; 2019. p. 153−77. Available from: https://doi.org/10.1016/B978-0-12-813545-7.00007-6.

[117] Ullah I, Rasul M. Recent developments in solar thermal desalination technologies: a review. Energies 2018;12:119. Available from: https://doi.org/10.3390/en12010119.i.

[118] Gorjian S, Jamshidian FJ, Hosseinqolilou B. Feasible solar applications for brines disposal in desalination plants; 2019, p. 25−48. Available from: https://doi.org/10.1007/978-981-13-6887-5_2.

[119] Morad MM, El-Maghawry HAM, Wasfy KI. A developed solar-powered desalination system for enhancing fresh water productivity. Sol Energy 2017;146:20−9. Available from: https://doi.org/10.1016/j.solener.2017.02.002.

[120] Zehtabiyan-Rezaie N, Alvandifar N, Saffaraval F, Makkiabadi M, Rahmati N, Saffar-Avval M. A solar-powered solution for water shortage problem in arid and semi-arid regions in coastal countries. Sustain Energy Technol Assess 2019;35:1−11. Available from: https://doi.org/10.1016/j.seta.2019.05.015.

[121] Daghari I, Zarroug MREL. Concepts review of solar desalination technologies for irrigation. J N Sci 2020;71:4319−26.

[122] Jones MA, Odeh I, Haddad M, Mohammad AH, Quinn JC. Economic analysis of photovoltaic (PV) powered water pumping and desalination without energy storage for agriculture. Desalination 2016;387:35−45.

[123] Alzoheiry A, Hemeda S. Techno-economical evaluation of a solar desalination system for irrigation purposes (case study on greenhouse bell pepper). J Soil Sci Agric Eng 2019;10:139−46. Available from: https://doi.org/10.21608/jssae.2019.36693.

[124] Benda V, Černá L. PV cells and modules − state of the art, limits and trends. Heliyon 2020;6:e05666. Available from: https://doi.org/10.1016/j.heliyon.2020.e05666.

Chapter 9

Solar applications for drying of agricultural and marine products

Ankit Srivastava, Abhishek Anand, Amritanshu Shukla and
Atul Sharma
*Non-Conventional Energy Laboratory, Rajiv Gandhi Institute of Petroleum Technology, Jais,
Amethi, India*

9.1 Principles of drying technology

Drying is the most ancient postharvest handling method for agricultural commodities that can increase storage life, improve product quality, minimize food losses during storage, and reduce transportation costs because the water content is removed through the drying process [1,2]. Drying farm products in regulated temperature and humidity conditions assist in reasonably quick drying while retaining moisture and ensuring superior quality [3,4]. Controlled temperature drying is mainly practiced in industrial drying systems that use huge amounts of fossil fuels and electricity. Drying is an efficient postprocessing method for crops to increase their storage life, enhance commodity quality, and minimize shipping costs because the weight of the commodity is reduced after the drying process [2].

The conversion of liquid water to water vapor is a primary process of drying energy usage (2258 kJ/kg at 101.3 kPa). Water in the crops and marine products may exist in a variety of forms, including free moisture and bound form which directly affects the drying rate. Table 9.1 lists the psychometry terms and their definitions that are commonly used. The moisture content of crops and marine products is stated on a dry basis (d.b.) or wet basis (w.b.) where the w.b. (X_w) is the ratio of the mass of moisture content to the total mass of the product [6]:

$$X_w = \frac{m_w}{m_w + m_d}, \text{kg per kg of the mixture} \tag{9.1}$$

and, on a d.b. (X_d), it is represented as the ratio of the mass of dry content to the mass of moisture content in the product [6]:

$$X_d = \frac{m_d}{m_w} \tag{9.2}$$

Solar Energy Advancements in Agriculture and Food Production Systems.
DOI: https://doi.org/10.1016/B978-0-323-89866-9.00003-1

TABLE 9.1 Basic terms of psychometry and drying [5].

Psychrometry and drying parameters	Definition
Bound moisture	Bound moisture is a dense liquid at a certain temperature that exerts less pressure of vapor than the pure liquid.
Equilibrium moisture content	The limiting moisture to dry a given material under particular air temperature and humidity conditions.
Constant-rate period	Constant-rate period is the drying time during which the water removal rate per unit drying surface is constant.
Critical moisture content	The average moisture content, at which the drying rate begins to decline.
Dry-weight basis	The moisture content of wet solids is expressed as the ratio of the mass of dry content in kg to the mass of wet content in kg.
Initial moisture distribution	At the beginning of drying, the initial distribution of moisture relates to the distribution of moisture in a solid.
Falling-rate period	Falling rate is a drying time during which instant drying rate declines continuously.
Hygroscopic material	Hygroscopic material is a material that can have bound moisture content.
Funicular state	Funicular state is the condition where a porous body gets dried when capillary suction allows air to sweep through the pores.
Pendular state	It is the condition of a fluid in a porous material when there is no longer a continuous layer of fluid within and between discrete particles, preventing capillary movement.
Capillary flow	Capillary flow is the flow of liquid between intercellular spaces and over the surface of a solid caused by molecular attraction, between the liquid and the solid surface.
Internal diffusion	Internal diffusion can be described as moving liquid or vapor through a solid due to a difference in concentration.
Wet-weight basis	Wet-weight basis expresses moisture in a substance as a percentage of wet solid weight.
Free-moisture content	The liquid that is soluble at a given temperature is known as free-moisture material, and humidity will contain both bound and unbound moisture.

(*Continued*)

TABLE 9.1 (Continued)

Psychrometry and drying parameters	Definition
Moisture gradient	Moisture gradient refers to the water distribution in a solid in the drying phase at a given moment.
Moisture content	A solid's moisture content is typically represented as moisture quantity per unit weight of dry or wet solid.
Unaccomplished moisture change	The ratio of free humidity present at all times to that initially present.
Unbound moisture	Unbound moisture in a hygroscopic material is moisture that exceeds the moisture content of the equilibrium equivalent to the moisture content of the saturation.
Nonhygroscopic material	Material that cannot contain bound moisture is nonhygroscopy material.

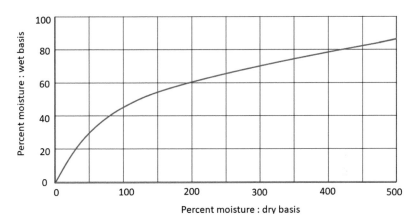

FIGURE 9.1   Percentage moisture relationship between wet basis (w.b.) and dry basis (d.b.) [5].

where m_w is the mass of moisture content in kg and m_d is the mass of dry content in kg.

Fig. 9.1 shows the moisture percentage relationship between w.b. and d.b. In general, for statistical analysis, the moisture content is expressed on a d.b., and agricultural moisture content is generally expressed on a w.b.

9.2 Drying psychrometry

Psychrometry is an important term in drying phenomena since it relates to the mixture of air-vapor properties that regulate the drying rate. The temperature and rate of liquid vaporization are determined by the amount of vapor in the environment that provides adequate heat for drying. Drying occurs at saturation temperature when there is free moisture or a liquid base, just as free water at the atmospheric pressure of 101.325 kPa vaporizes at 100°C at 100% of the steam atmosphere. On the contrary, the temperature at which vaporization occurs depends on the vapor content in the surrounding air when advanced vapor is purged from the dryer setting using a second inert gas [7]. In purging steam, the solvent is often heated to a temperature equal to or greater than its vapor pressure, while condensation occurs in the opposite. In most drying processes, the moisture is dissipated, and the air is used as the purge stream.

Fig. 9.2 depicts the psychrometric chart introducing the thermodynamic properties of moist air at constant pressure, which is sometimes equated to an elevation from the sea level. In addition, the following can be explained:

1. The saturation temperature line or the wet-bulb on the psychrometric chart shows the maximum weight of water vapor to be generated by 1 kg of dry air in the saturation humidity of the dry bulb. The partial pressure of water in the air at this temperature is the pressure of water vapor. The saturation humidity (Hs) is the maximum amount of water vapor that air can contain at a certain temperature, without phase separation and given by Eq. (9.3):

$$Hs = \frac{18}{28.9} \frac{p_s}{(P - p_s)} \qquad (9.3)$$

2. The relative humidity (H_R) is described by Eq. (9.4):

$$H_R = 100.\frac{P}{p_s} \qquad (9.4)$$

where P (bar) is the partial pressure of water vapor in air and p_s (bar) is the saturation pressure at that temperature. The curves named "volume m^3/kg dry air" give humid volumes. The difference between the real amount of dry air and the volume of humid air at a specific temperature is the volume of water vapor.

3. Enthalpy data is given based on kJ/kg of dry air. Enthalpy-at saturation data are accurate only at the saturation temperature and humidity of the air.

4. There are no lines for humid heats in Fig. 9.2. It is the constant-pressure specific heat of moist air, per unit mass of the dry air and can be calculated by

$$C_s = 1.0 + 1.87H(\text{J/kgK}) \qquad (9.5)$$

where H is the specific humidity in kg water vapor per kg dry air in the mixture.

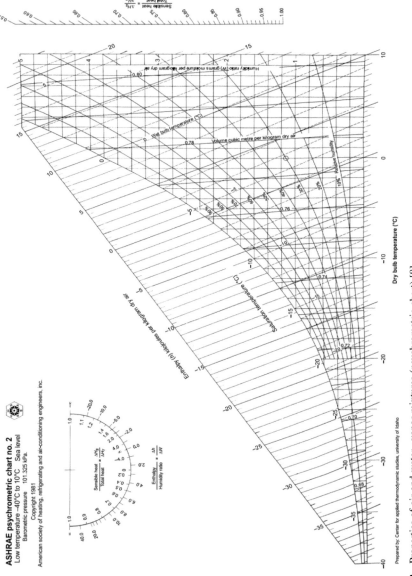

FIGURE 9.2 Properties of air and water–vapor mixtures (psychrometric chart) [8].

5. The wet-bulb-temperature lines which depend on the relationship, also represent the adiabatic saturation lines for air and water vapor only.

$$Hs - H = \left(\frac{c_s}{\lambda}\right).(T - Ts) \qquad (9.6)$$

where λ is the latent heat of evaporation of water in kJ/kg, T is the actual temperature, and Ts is the saturation temperature both in K.

9.3   Solar drying technology

Solar drying is among the most appealing and successful solar energy technology applications, especially in tropical and subtropical locations. Traditional open sun drying, carried out on a wide scale, is high in rural areas in most developing nations as a result of inappropriate drying/under drying, fungal growth, insect invasions, birds, and rodents that occur due to open sun drying. Properly designed solar dryers can provide an important and appropriate option for drying many agricultural products in the world [9]. Even so, the major issue in solar drying systems is the irregular availability of solar radiation as the primary source of energy that can be minimized by storing excess energy during peak time and consuming it during night hours or when solar radiation is not sufficient [10]. To overcome the limitation of open sun drying, solar dryers have been developed. These devices provide value addition in drying products. Further enhances the income of farmers working in the drying product business. These devices eliminate moisture content, which aids in the preservation of food products for a long period. They also protect food items from contamination, ensuring that food products retain their quality after drying [11]. The objective of a solar dryer is to deliver heat of higher quality to the crop. In conventional dryers, a significant amount of heat is wasted and solar radiation is not adequately utilized to dry food products. Solar dryers assist in decreasing heat loss by enhancing the use of solar radiation to dry food products [12].

A broad classification of solar dryers is shown in Fig. 9.3. As presented in this figure, solar dryers are classified as direct or indirect with additional subclasses of each type. Solar dryers are categorized mainly by their mode of heating and the way that solar radiation is used. In this case, they are mainly classified as passive and active dryers. The term "passive solar dryer" refers to a dryer in which air is naturally circulated, whereas an "active solar dryer" is a dryer in which air is circulated by force using a fan and blower [14].

9.3.1   Direct solar dryers

In a direct solar dryer, the crop is directly placed under the exposure of sunlight. This is a popular dryer design in which the materials to be dried are

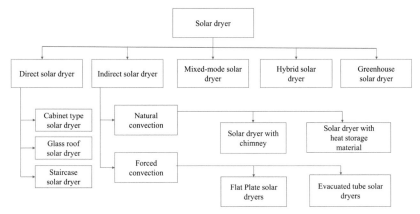

FIGURE 9.3   Classification of solar dryers considering drying modes [13].

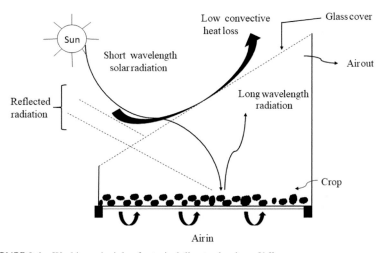

FIGURE 9.4   Working principle of a typical direct solar dryer [16].

enclosed in a chamber with a glass cover. Heat is generated primarily on the commodity surface as well as within the drying chamber by absorbing solar radiation. The product is kept in the drying chamber for 20–30 days based on the product's initial moisture content [15]. Fig. 9.4 shows the working principle of a typical direct solar dryer. This is a passive dryer in which air is naturally circulated within the drying chamber. The air approaches the drying chamber from below and exits from above. Here, humidity is eliminated from the top of the drying chamber which is similar to the open-sun drying method. The only distinction is that the food product is placed in a drying chamber or cabinet which is covered by a glass cover [14]. From the total

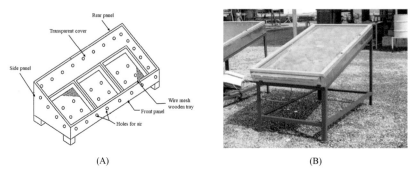

(A)   (B)

FIGURE 9.5 (A) Main parts of a cabinet-type solar dryer [17], (B) photo of a cabinet-type solar dryer [19].

solar radiation incidents on the dryer, a part is reflected over the glass cover, a part is absorbed by the glass, and the remaining portion of solar radiation is transported by conduction within the drying chamber. Solar radiation is absorbed by the product and the drying chamber, resulting in heated air circulation. The transparent glass cover reduces direct convective losses to the atmosphere and has a significant effect on the temperature increase of the product and the drying chamber or cabinet [18].

A simple design of a direct solar dryer is a cabinet-type dryer. It is a compact wood box with a maximum temperature of 80°C within the dryer. The cabinet dryer has some drawbacks that one of them is a long drying time due to natural airflow convection which results in a low coefficient of heat transfer and poor efficiency [19]. A schematic view and photo of a cabinet dryer indicating its different parts are shown in Fig. 9.5. Other common designs of direct solar dryers are glass-roof and staircase solar dryers. In a glass-roof solar dryer, the glass roof is located over the drying chamber [5], and in the staircase solar dryer, the drying chamber is arranged as like stairs and placed under the direct solar radiation which is covered with a glass sheet [20].

9.3.2 Indirect solar dryers

The indirect solar drying approach has been so far more successful than solar drying in its direct form because some quality issues arise in direct dryers such as the color and aroma of the final dried product affected due to direct exposure to the sun [21]. In this type, the air is heated by absorbing solar radiation through a flat plate or a concentrated solar collector and circulated within the drying chamber. The hot air is blown into the drying chamber where the crops or marine products are stored. The moisture in the product will be then evaporated by convection and diffusion [22]. In indirect

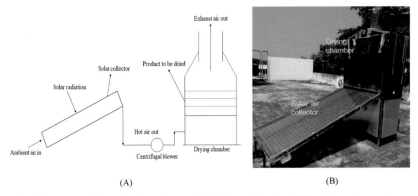

(A)   (B)

FIGURE 9.6   (A) Schematic view of a typical indirect solar dryer [25], (B) experimental setup of an indirect solar dryer [26].

dryers, the drying rate is faster and varies with the dryer design, initial moisture content of food items, air velocity within the drying chamber, ambient temperature, and humidity [23]. The efficiency of indirect solar dryers ranges from 2% to 48%, while this value is between 0% and 34% for direct solar dryers [24]. Fig. 9.6 shows a schematic view and photo of an indirect solar dryer.

Flat-plate collectors (FPCs) are the most common type of collectors integrated with dryers, but evacuated tube collectors (ETCs), as well as concentrators, have also been utilized. An evacuated tube solar dryer (ETSD) was designed and built by Malakar et al. [27] to dry garlic cloves. The drying performance parameters of drying rate, collector efficiency, and dryer efficiency of garlic cloves were evaluated during experimental tests. The experiments were carried out at three air velocities of 1, 2, and 3 m/s. During the drying process, the developed solar dryer indicated better performance at an airflow velocity of 2 m/s. In the drying chamber, the maximum reached temperature was 86.7°C and at the airflow velocity of 2 m/s, the highest collector and dryer efficiency was calculated as 42.56% and 56%, respectively.

Concentrators can be integrated with solar dryers to perform fast drying for the crops requiring drying temperatures between 60°C and 80°C. In a study by Varghese et al. [28], a parabolic dish collector (PDC) with an embedded insulator constructed of plywood and wrapped with aluminum foil was developed and integrated with a drier consisting of four trays to dry peanuts. An exhaust fan positioned at the top of the dryer retained the flow of hot air through the trays. According to the results, the PDC could provide a maximum temperature of 79°C to dry 1.5 kg of peanut in 5 h. Additionally, the heated air from the PDC could reduce peanut's moisture content by 21% more than the open sun drying.

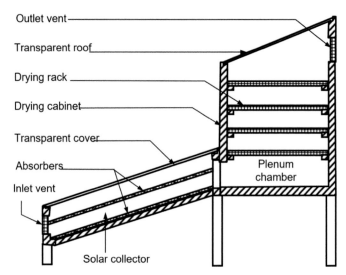

FIGURE 9.7   Schematic view of a mixed-mode natural convection solar dryer [29].

9.3.3   Mixed-mode solar dryers

A mixed-mode solar dryer combines the benefits of direct and indirect solar dryers. In mixed-mode solar dryers, the energy available for the drying process is provided by the combined effect of solar radiation incident directly on the drying material and the preheated air in the solar air heating collectors. In a mixed-mode solar dryer, the drying crop is placed in direct exposure to solar radiation and heated air by the solar collector is circulated within the drying chamber. For the design of this kind of dryer, several considerations must be taken into accounts such as thermal performance, cost-effectiveness, lifespan, reliability, repair, and installation [16]. Fig. 9.7 shows a schematic view of a mixed-mode natural convection solar dryer. These dryers can be operated either in natural or forced convection. In natural convection, the air circulates within the drying chamber naturally, and in forced convection, the air is circulated by fans or blowers [30].

9.3.4   Hybrid solar dryers

In hybrid solar dryers, an auxiliary energy source is integrated to not only improve the performance of the solar dryer but also enhances the quality of dried products due to continuous drying operation [31]. The common components of a hybrid solar dryer are as follows: (1) solar collector/reflector, (2) heat exchanger, (3) heat storage unit, and (4) drying chamber. To maintain optimum drying conditions, other energy sources such as biogas, wind,

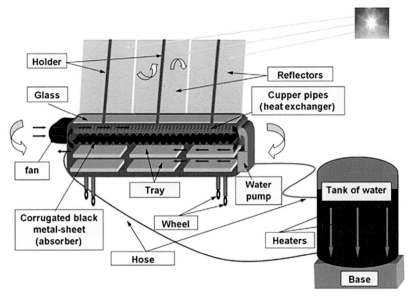

FIGURE 9.8 Schematic of the hybrid solar dryer developed in [33].

photovoltaic (PV), and electrical heaters are integrated. As a consequence, solar energy along with other energy sources are utilized to provide the heat requirement of the drying process [32].

Fig. 9.8 shows a schematic view of a hybrid solar dryer developed by Amer et al. [33]. Several other designs are possible for hybrid solar dryers as hybrid solar-biomass dryers [34,35], hybrid photovoltaic-thermal (PVT) solar dryers [36−38], hybrid geothermal-solar dryers [39], hybrid gas-fired solar dryers [40], and hybrid solar-electric dryers [33,41].

9.3.5   Solar greenhouse dryers

A solar greenhouse drying method is based on the concept of the greenhouse effect in which short-wave solar radiation is trapped inside and long-wave radiation is emitted back to the atmosphere. Solar greenhouse dryers (SGHDs) are used for seed harvesting, poultry, aquaculture, soil solarization, and field drying [42]. Greenhouses may be in passive mode (natural convection) or active mode (forced convection). The chimney acts as a respirator in passive SGHDs to increases the airflow rate inside [43], while in an active SGHD, the exhaust fan is provided for the humid airflow outside the dryer [44]. SGHDs are mostly utilized in large-scale drying processes to dry herbs, vegetables, fruits, and seafood. They can provide at least a 50% faster drying rate than conventional dryers [45]. A schematic diagram of the SGHD

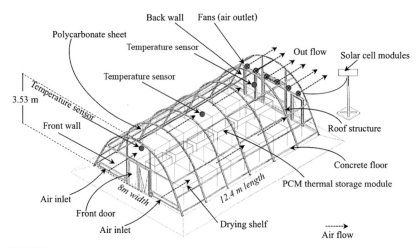

FIGURE 9.9 Schematic representation of the developed SGHD in [46].

developed by Pankaew et al. [46] is shown in Fig. 9.9. SGHDs can keep the temperature inside the dryer up to 70°C. The UV radiation led to the loss of vitamins and important nutritional content in dried products. But SGHDs filters UV light well, retaining nutritional values of the final dried products in addition to their original color, aroma, and taste [47]. SGHDs can also be integrated with thermal energy storage (TES) units to expand the drying process when solar radiation is not available. The continuous drying process also reduces the reabsorption of moisture from the environment and therefore reduces the drying time [46,48,49].

9.4 Optimal conditions for drying agriculture and marine products in solar dryers

Using solar dryers, the agricultural product can be stored for six to twelve months with preserving food quality and nutrient value in the dried product [50]. Various solar dryer designs are available which can be selected based on the type of the drying product since different products have different drying rates and therefore a proper solar dryer should be chosen to achieve the highest efficiency. The optimal conditions in solar drying depend upon various parameters such as the product to be dried, solar dryer design, environmental temperature and pressure, and ambient humidity [51]. Tables 9.2–9.5 indicate drying characteristics of different agricultural products including time and the final moisture content to get a high-quality dried product. Table 9.6 also represents the initial and final moisture content of several agricultural products along with their maximum allowable temperature at which the product may be dried which is crucial for safe drying.

TABLE 9.2 Drying characteristics of fruits [52].

Fruits	Drying time (h)	Preparation form	Reduction in moisture content (%)
Cherries	11	Complete cherries	40
Apples	8	10-mm thick slice of apples	40
Plantains	8	8-mm thick slices	50
Pears	10	Quartered	25
Apricots	12	Halve	30
Peaches	10	Thin slices	20
Prunes	12	Halve	35
Brambles/ Raspberries	12	Completely	25
Elders	12	Completely	30
Strawberries	14	Halve	25
Whortleberries	10	Completely	30

TABLE 9.3 Drying characteristics of vegetables [53].

Vegetables	Drying time (h)	Preparation form	Reduction in moisture content (%)
Beans	5	Small slices	60
Onion-vegetables	5	Small Pieces	50
Cauliflower	6	Blanch in small pieces	50
Pod pepper	6	10 mm strips	30
Tomatoes	8	Little slices	25
Carrots	6	Small piece	25

TABLE 9.4 Drying characteristics of some grains [54].

Grains	Drying time (h)	Preparation	Reduction in moisture content (%)
Maize	6	Whole grains	80
Grains	5	Without shell	90
Seeds, kernels, nuts	4	Without shell	90

TABLE 9.5 Drying characteristics of herbs [55].

Herbs	Drying time
Solid, brittle herbal types	At a natural open-air condition: 2−3 wk
Smoothly leaves herbal-types	In a moderate temperature dryer: 4−6 h

TABLE 9.6 Drying characteristics of some agricultural and marine products [56].

Agricultural product	Maximum allowable temperature for drying (°C)	Moisture content (%)	
		Initial	Final
Chili	65	80	5
Potato	75	75	13
Brinjal	60	95	6
Spinach	N.A.	80	10
Onion ring	55	80	10
Grape	70	80	15−20
Banana	70	80	15
Guavas	65	80	7
Copra	65	80	20
Pineapple	65	80	10
Fish raw	30	75	15
Fish water	50	75	15

(Continued)

TABLE 9.6 (Continued)

Agricultural product	Maximum allowable temperature for drying (°C)	Moisture content (%)	
		Initial	Final
Prunes	55	85	15
Nutmeg	65	80	20
Sorrel	65	80	20
Coffee	N.A.	50	11
Coffee beans	N.A.	55	12

9.5 Integration of solar dryers with photovoltaic and photovoltaic-thermal modules

PV modules are generally integrated with solar dryers to run fan(s)/blower(s) for forced air circulation within the drying chamber [2]. PVT modules that combine the generation of solar electricity and heat in a single component can also improve the performance of dryers. PVT technologies have been explored extensively over the years, with variations in design, working fluids, and other performance-effective variables being investigated. Many factors affect the thermal and electrical output of PVT modules including ambient temperature, solar radiation, wind speed, working fluid temperature and flow rate [57].

Hidalgo et al. [58] constructed a PV-assisted direct solar dryer and evaluated its performance under natural and forced convection modes. In this design, the PV module supplied the electricity demand of eight blowers, allowing the inside air to be recirculated. During the drying of green onions, moisture and colorimetric parameters were measured. Drying kinetics showed a constant rate followed by decreasing rate periods for both working modes of natural and forced convection. The effective diffusivity values for drying under natural and forced convection runs were obtained as 5.15×10^{-9} m^2/s and 1.15×10^{-8} m^2/s, respectively. The average efficiencies and specific energy consumption of the solar dryer were calculated as 34.2% and 18.3 kWh/kg for the natural convection drying process, and 38.3% and 16.4 kWh/kg for the forced convection drying process. Under two distinct operating circumstances, there was little color difference between fresh and dried green onions.

Goud et al. [59] developed an indirect type solar dryer in which the airflow was controlled by an inlet fan operated by a PV module. The experimental setup consisted of a trapezoidal duct to which fans fixed at the inlet of the duct, PV modules, solar air collector, drying chamber, and chimney as

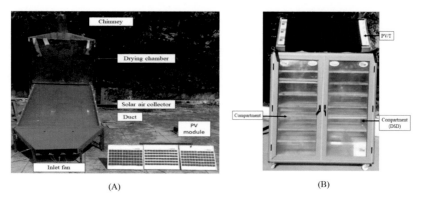

(A)   (B)

FIGURE 9.10   (A) Indirect solar dryer with photovoltaic module [59], (B) solar dryer with photovoltaic-thermal air collector [60].

shown in Fig. 9.10A. Investigations were performed during the drying of green chili and okra. Performance parameters and drying kinetics of the indirect solar dryer were estimated. The efficiency values of solar air collector and drying process were obtained as 74.13% and 9.15% during the drying of green chili, while these values were obtained as 78.30% and 26.06% during the drying of okra. It was concluded that the system is much improved compared to the natural convection system, and for the running of the inlet fan, the PV module can generate the appropriate amount of electricity.

Fterich et al. [60] fabricated a mixed solar dryer with induced convection for drying tomatoes, under climate conditions of Tunisia. The dryer was composed of a PVT air collector and drying compartment as shown in Fig. 9.10B. In this design, the airflow enters the tubular aluminum channels below the PV module and at the same time spreads into the upper gap. As a result, heat exchange occurs on both sides of the PV module, facilitating the thermal energy transportation to the drying room and cooling solar cells. The tomatoes were placed in two trays and dried and the obtained results were compared to a sample of naturally dried tomatoes. According to the results, the product moisture content was reduced from 91% to 22% in tray 1 and 28.9% in tray 2. It was also observed that the drying temperature, as well as the quality of the dried product, are improved, allowing farmers to decrease their crop losses by storing them for longer periods. It was also claimed that in remote areas, the developed solar dryer arrangement can provide sufficient electricity.

Tiwari et al. [61] investigated the integration of PVT air collectors with drying systems. Additionally, a thermal model describing the integration of PVT air collector with greenhouse drying system was presented. Since drying systems are frequently needed in distant places where grid connectivity is unavailable, the primary goal of fabricating a greenhouse drying system integrated with PVT air collector was to make it self-sustained. The results indicated that the PVT air collector integrated with the drying system provides better temperature control over other drying systems. Additionally, the average thermal efficiency, electric efficiency, and total thermal efficiency of the PVT air collector were calculated as 26.68%, 11.26%, and 56.30%, respectively, under the air mass flow rate of 0.01 kg/s. According to the literature, it can be observed that the integration of PV and PVT modules with solar dryers will improve their performance as PV modules can provide the electricity demand of the solar dryers, while PVT modules can also fulfill the heat demand for drying processes.

9.6   Integration of dryers with solar thermal collectors

Different criteria may be used to classify solar thermal collectors, but in the most common type, they are categorized as concentrating and nonconcentrating collectors. Two distinct surface areas may be described using this classification as the collector area, which intercepts the incident solar radiation, and the absorber area, which effectively absorbs the solar heat [62]. A concentrating solar device, specifically, gathers radiation on an intercepting area and concentrates it onto a relatively small absorber by using reflecting surfaces (the intercepting area is wider than the absorber one). In this regard, these collectors are ideal for medium- and high-temperature applications [63].

Rabha and Muthukumar [64] discussed the efficiency of a solar dryer with forced convection and a paraffin wax-based shell and tube latent heat storage device. Two double-pass solar air heaters, a paraffin wax-based heat storage module, a blower, and a drying chamber were combined to form the solar dryer as shown in Fig. 9.11. The dryer was evaluated by drying 20 kg of red chili at temperatures ranging from 36°C to 60°C. After four days, the moisture content of the red chili was reduced from 73% to 9.7% during 10 h

FIGURE 9.11   Crop dryer integrated with solar air heater and paraffin wax heat exchanger [64].

of operation in a day. From the results, the first solar heater's total energy and exergy performance values were found to be 32.5% and 0.9%, respectively. While the values for the second solar heater connected in series with the first one were reported as 14% and 0.8%, respectively. Moreover, the drying chamber's exergy efficiency was found to be between 25% and 98%, with an average value of 52%.

Essalhi et al. [65] investigated the thermal performance of two different solar air collector designs for an indirect type solar dryer. The absorber in the first design was made of a polyethylene alveolar plate with a $100\ mm^2$ surface area. The collector in the second design was constructed of two aluminum plates corrugated and placed in opposition to form 90 mm parallel cylinders. The results of the study revealed that the corrugated absorber collector provides better thermal performance compared to the polyethylene plate collector. This is because as the airflow rate increases, the efficiency of the collector with a corrugated aluminum absorber improves. The average efficiency was also calculated as 77% under forced convection.

The thermal performance of a solar dryer integrated with an ETC was studied by Singh et al. [66]. The main components of the set-up are, ETC, shell tube heat exchanger, and drying chamber as shown in Fig. 9.12. In this configuration, the ETC was linked to the shell and tube heat exchanger using chlorinated polyvinyl chloride (CPVC) pipes and fittings linked to the drying chamber via a duct. An axial flow exhaust fan was installed within the duct to direct the air from the heat exchanger into the chamber. With the aid of the frame, the ETC was positioned south facing and inclined at an inclination of about 15 degrees to the horizon. The collection area of the ETC in this

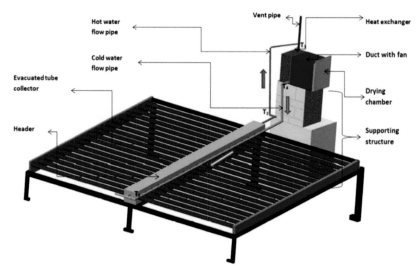

FIGURE 9.12 Schematic of the dryer integrated with an ETC [66].

study was 2.22 m^2. From the results, the maximum temperature difference between the air inside the drying chamber and the ambient was determined as 35.4°C, and the maximum set-up efficiency was calculated as 55%.

Daghigh et al. [67] designed and built a dryer integrated with a PVT collector and an ETC. The goal of this system was to experimentally evaluate the solar dryer's performance under weather conditions in Sanandaj, Iran, and investigate the effect of using different collectors integrated with the dryer. To evaluated the system, a blower was used to direct the air into the ETC or the heat exchanger installed behind the PV module, where it was heated and circulated to the dryer. In this study, the solar dryer was evaluated in two modes: a solar dryer coupled with a PVT collector and a solar dryer coupled with an ETC. A schematic drawing of the experimental setup is shown in Fig. 9.13. The experimental results indicated that the ETC collector performs better than the PVT collector in terms of both the output temperature and efficiency with the average efficiency values of 13.7% for the dryer with PVT and 28.2% for the dryer with ETC.

Ssemwanga et al. [68] constructed an improved hybrid indirect solar dryer with a metallic solar concentrator and drying cabinet. A conventional active-mode solar PV and electric dryer along with an external thermal backup system were also installed. The fruit drying performance of the hybrid indirect solar dryer, solar PV and electric dryers, and the conventional open sun drying technique were examined during the drying of mangoes and pineapples and the obtained results were compared. The drying times for the solar PV and electric dryers, the hybrid indirect solar dryer, and the open sun

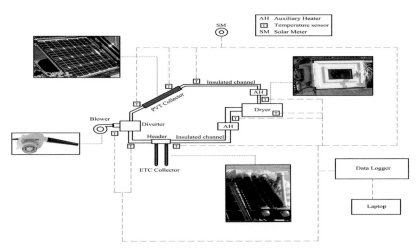

FIGURE 9.13 Schematic view of the experimental setup of the hybrid indirect solar dryer [67].

drying were recorded as 10, 18, and 30 h, respectively. It was found from the experiments that the drying efficiency of the improved hybrid indirect solar dryer is close to the solar PV and electric dryers and is 18% higher than the open sun drying.

It has been observed that the appropriate design of solar thermal collectors and using effective materials in their structures can increase the performance of the solar dryer that they integrated with solar dryer. Solar thermal collectors are also responsible to provide the optimum use of solar energy.

9.7 Integration of solar dryers with energy storage systems

Solar radiation is intermittent due to insufficient/no availability of solar radiation during the cloudy or dusty days as well as night times. Solar heat availability can be improved by storing excess thermal energy when it exceeds demand and using the same stored energy when it is required. Solar heat surpluses can be supplied from the stored heat using thermal, electrical, chemical, and mechanical methods. The storage of solar heat in TES mediums in solar-powered dryers will improve the system's usability as well as reliability. Storing solar thermal energy may be performed in several ways. Classification of TES methods is presented in Fig. 9.14, including the storage in the form of thermal (sensible and latent heat), chemical, or a combination of these [69,70].

In sensible heat storage, thermal energy is stored by increasing liquid or solid temperature, by using heat control and changing the substance temperature during the charging and discharging processes. The amount of stored heat is determined by the rise in the temperature of the material and the quantity of sensible heat storage material [71]. In general, water is a robust

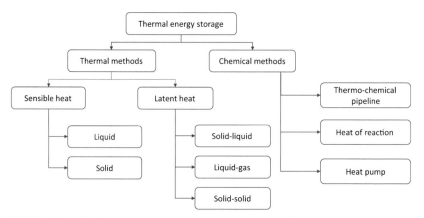

FIGURE 9.14 Classification of thermal energy storage systems [70].

sensible TES (STES) substance due to its high specific heat and low cost [72,73]. The sum of deposited energy is determined by the material's temperature change and is expressed as [74]:

$$E = m \int_{T_1}^{T_2} C_p dT = mC_p(T_2 - T_1) \tag{9.7}$$

where m is the mass of sensible heat storage material in kg and C_p is the specific heat of material in J/kg-K at constant pressure. T_1 and T_2 signify the lower and upper-temperature levels in K. The difference $(T_2 - T_1)$ is denoted as the temperature rise.

The phase transformation of a substance is used to store latent heat with the most common method of solid–liquid phase transition accomplished by melting and solidifying a substance. Heat is added to the substance as it melts, retaining vast amounts of heat at a steady temperature until the material solidifies, at which the heat is released. The phase change materials (PCMs) as latent TES (LTES) materials are extensively investigated and used for a wide variety of applications [75]. PCMs absorb and release heat depending on their properties during melting and solidification. Using STES, the temperature of storing material rises and is stabilized during the transition process, allowing a large amount of energy to be stored in the form of temperature rise. When PCM is heated, it undergoes a phase change (typically melting), retaining energy in the form of latent heat. In this case, the quantity of stored energy (E) in J depends on the mass of material (m) in kg and latent fusion heat (λ) in J/kg [76]:

$$E = m\lambda \tag{9.8}$$

The stored and released energy in breaking and reforming molecular bonds in a fully reversible chemical reaction is the basis for thermo-chemical heat storage methods. The amount of heat contained in this case is determined by the amount of storage material, the endothermic heat of reaction, and the degree of conversion [77]:

$$Q = m\Delta h_r \tag{9.9}$$

where Δh_r is the energy for the reaction in kJ, and m is the amount of involved substance in kg. Mortezapour et al. [78] suggested a new solar crop dryer design that used a reversed absorber plate type collector and natural airflow thermal storage. The schematic view and photo of the solar dryer set-up are shown in Fig. 9.15. To dry onions, the performance was evaluated using a $(1 \times 1 \text{ m}^2)$ crop dryer area with a packed bed and an airflow tube. The parametric investigation involved the impacts of airflow rate and relative humidity (RH) of drying air on the drying process. In this study, an absorber plate with 30 degrees inclination, a built thermal storage channel, and an

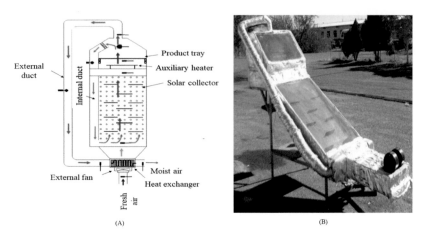

FIGURE 9.15 Developed solar dryer with TES [78]: (A) Schematic view, (B) the experimental set-up.

airflow channel with 0.12 m diameter were employed. During the investigation, the drying rate was measured per hour and it was observed that initially drying rate is faster which later on is decreased. Moreover, it was reported that the highest color quality is obtained when the smallest air flow rate and highest RH of drying air are employed.

Kesavan et al. [79] evaluated the performance of a triple-pass solar dryer to investigate the drying characteristics of potato slices. The developed dryer was made up of a solar collector with three passes, a centrifugal blower, an absorber plate, a sand bed (STES), metallic wire mesh, and a drying chamber. The experimental setup of the triple-pass solar dryer is shown in Fig. 9.16. The experiments were carried out at the maximum air mass flow rate of 0.062 kg/s. The highest collector outlet air temperature of 62°C was measured at 1.00 pm, with solar irradiation of 998 W/m². Experimental results showed that in 4.5 h, the moisture content of potato slices is decreased from 76% (w.b.) to 13% (w.b.). The efficiency of this solar dryer also ranged from 12% to 66%, with an average value of 45%.

Swami et al. [80] evaluated the performance of a solar fish dryer integrated with PCM. The experimental setup consisted of a drying chamber, chimney, and air blower. The schematic view of the experimental setup is shown in Fig. 9.17A. The capacity of the drying chamber was 5 kg for drying fish with the dimensions of 745 × 525 × 540 mm. A FPC with dimensions of 1165 × 480 × 150 mm was employed to absorb solar radiation. In experiments, two separate PCM units with a total mass of 9 kg, one using paraffin wax C31−33 and the other using paraffin wax C23−24. From the experimental results, it was inferred that an optimized airflow rate of 5 m/s, a mass flow rate of 0.314 kg/s, and a heating chamber depth of 10 cm are suitable to provide the solar fish dryer's required operating temperature.

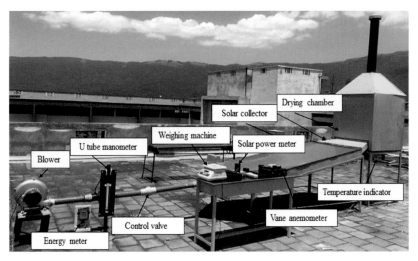

FIGURE 9.16 Experimental setup of the triple-pass solar dryer developed in [79].

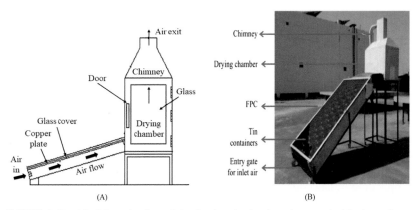

FIGURE 9.17 (A) Schematic view of the developed solar dryer integrated with phase change material [80], (B) passive solar dryer integrated with phase change material containers [81].

Using PCM, the fish drying period was reduced by nearly 70% and it was proved that PCM usage can be an outstanding technique for heat accumulation in this dryer. Due to melting point temperature and being suited for operating temperatures of fish drying, the PCM-1 was proved to be more efficient than PCM-2.

In another study, Babar et al. [81] designed a passive solar FPC dryer with a capacity of 7.5 kg and a drying chamber's cross-sectional area of 0.74 m × 0.58 m. A solar collector with an aspect ratio of 2, a collector depth of 0.25 m, and hence the volume of 0.5 m^3 was employed in the experimental setup (Fig. 9.17B). Mushroom buttons (Agaricus bis porus) were dried in the FPC solar dryer and under open sun drying. To ensure adequate heat transfer between the air and the PCM, PCM containers were placed horizontally within the bottom of the absorber plate. PCM containers were cylindrical Tins with a diameter of 10.5 cm and a height of 17.5 cm, each contained 750 mL of PCM. From the results, the overall difference between the average dryer inlet and the air temperature was obtained as 24.6°C, when the average daily solar radiation was 882.35 W/m^2. In FPC solar dryer, the humidity ratio of mushroom buttons reached zero after 21 h, while in open sun drying, the same humidity ratio was achieved during 33 h. It was concluded that the FPC solar dryer takes 36.36% less time than open sun drying for drying mushroom buttons.

Vigneshkumar et al. [82] investigated the drying of sliced potatoes in an indirect solar dryer using PCM. In this work, an indirect type single-pass forced convection solar dryer integrated with paraffin as PCM was fabricated to enhance the performance during off sunshine hours. For this purpose, various parameters were measured in two cases as the solar dryer with and without PCM to dry sliced potatoes. During experiments, the flow rate of air within the drying chamber was maintained as 0.065 kg/s. The results indicated that the presence of PCM within the solar collector considerably increases the temperature of the air inside the drying room in two hours after the sunshine. They claimed that using paraffin, the percentage weight of removed moisture from potato slices was increased by 5.1% per day.

9.8 Emerging solar drying technologies

Many advances in technologies related to the drying of agricultural products have occurred in recent years. Many kinds of research have shown that comprehensive food drying approaches can be used to increase the efficiency of solar dryers and also to enhance product quality by controlling the drying temperature [40,83]. Selected energy-saving strategies, as well as new hybrid drying technologies, are discussed in this section.

Hybrid drying technologies blend two or more different drying processes that can have a synergistic effect, resulting in reduced energy demand and drying time while retaining the highest quality of the final dried product in terms of flavor, nutrients, color, fragrance, and texture [84]. The use of integrated combination and hybrid drying techniques is found to consume 20%—40% less specific energy. Hybrid drying technologies are promoted as a way to overcome the drawbacks of traditional drying methods, reduce product deterioration, and still deliver a product with the desired residual moisture content. New hybrid drying systems concentrate on energy savings and maintain the quality of dried products [85]. Hybrid solar dryer with biomass backup burner [86], hybrid solar-wind dryer [87], and hybrid solar-geothermal FPC dryer [39] are solar-assisted hybrid technologies that are developed to enhance the performance characteristics of solar dryers.

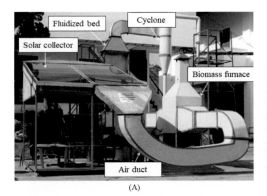

(A)   (B)

FIGURE 9.18 (A) Solar-assisted fluidized bed dryer integrated with biomass furnace [88], (B) developed prototype solar dryer with a wind generator [87].

Yahya et al. [88] constructed a hybrid fluidized bed drying system integrated with a biomass furnace to investigate the drying kinetics of paddy. The system included a fluidized bed, a solar collector, and a biomass furnace as shown in Fig. 9.18A. The average air-drying temperature was between 61°C and 78°C. With a mass flow rate of 0.125 kg/s, the moisture content of the paddy was reduced from 20% (w.b.) to 14% (w.b.). For average air temperature drying at 61°C and 78°C, the thermal efficiencies were calculated as 13.45% and 16.28%, respectively. Ndukwu et al. [87] designed and manufactured an active mix mode solar dryer with an axial fan operated by a wind air generation, as well as a passive mix mode solar dryer that was evaluated with and without glycerol as heat storage for drying pretreated potatoes (Fig. 9.18B). The results indicated that the solar dryer combined with a wind-powered axial fan and thermal storage has the shortest drying time (14 h), whereas the passive solar dryer without thermal storage has the longest drying time (25 h). Furthermore, the invented solar dryers can reduce open sun drying time by 9−16 h. The solar dryer combined with a wind-powered axial fan demonstrated superior drying efficiency (31.5%), but a passive solar dryer with thermal storage demonstrated worse energy efficiency (nearly 25%).

Ananno et al. [39] presented a comprehensive design and numerical study of a hybrid geothermal-PCM flat-plate solar collector (PA-FPSC) as shown in Fig. 9.19, which was a self-sustaining renewable power food drying system. In this study, the suitability of the hybrid geothermal-solar dryer for developing nations with a considerable underground geothermal temperature difference was explored. The performed numerical study suggested that this near-zero hybrid energy technology will be very efficient. Using the recommended conceptual design and the related mathematical model, the functional aspects of PA-FPSC were evaluated. When the flow speed was 0.02 kg/s, numerical models showed a 20.5% higher efficiency than a normal

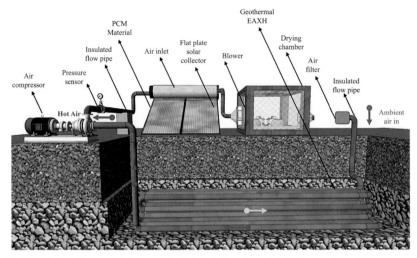

FIGURE 9.19 Conceptual design of the PA-FPSC geothermal hybrid solar dryer [39].

solar FPC in the PA-FPSC geothermal hybrid. The results indicated that the temperature of the PA-FPSC geothermal hybrid ranges between 18.79°C and 16.55°C which is higher than the ambient temperature after 5 h of sunset, and 15.9°C−13.4°C higher than the ambient temperature after 12 h of sunset.

This section shows that emerging solar drying techniques can be utilized to develop various kinds of hybrid solar dryers utilizing solar energy in combination with other renewable sources. Proper design of hybrid solar dryers enhances the drying parameters and properly controlled temperature within the drying chamber enhances the quality of dried products.

9.9 Economics of solar dryers

Economic analysis of various solar drying systems can be conducted using the cost analysis, sinking fund factor method, and cash flow diagrams by calculating the payback period[1] (PP) of them. For proper economic analysis, the carbon credit obtained by the use of clean energy alternatives must be measured. It is also important to consider the effect of renewable energies on the environment. Economic analysis can also show how much costs are annually saved and how much costs are saved over time when a solar dryer is utilized. Carbon mitigation, small PP, and a large life span of solar dryers indicate the long-term benefits of their usage [89].

ELkhadraoui et al. [90] suggested a novel mixed-mode SGHD for drying red pepper and grapes. The economic study revealed a PP of 1.6 years while

---

1. This is a time which it takes to recover the expense of investment.

the system's lifetime was calculated as 20 years. Kurt et al. [91] investigated the economic efficiency of solar slurry drying using the greenhouse solar drying process. Results indicated that although the SGHD has a higher capital cost than a conventional solar dryer, it is more cost-effective due to lower energy requirements. Nabnean et al. [92] proposed a new design to dry osmotic cherry tomatoes. The dryer consists of a drying cabinet, a heat exchange, a solar water collection of $16\,m^2$, and a hot water storage unit. The cabinet measures 1.0 m width, 3.0 m length, and 1.4 m high, with an output of 100 kg for dehydrated osmotic cherry tomatoes. Each batch was dried up with 100 kg of osmotically dehydrated cherry tomatoes. The new solar dryer greatly reduced drying time compared to natural sun drying. The effectiveness of the solar collector was 21% to 69% and the dryer's PP was calculated as 1.37 years.

Dejchanchaiwong et al. [93] studied two mixed-mode and indirect solar dryers. It was observed that the drying time of the mixed-mode solar dryer is about 56% shorter than the indirect solar dryer, but the PP of the mixed-mode solar dryer is about 1 year more than the indirect solar dryer. Tiwari and Tiwari [94] made a mixed-mode PVT dryer and compared it to a direct solar drying system. From the results, it was found that the mixed-mode PVT dryer provides better quality for the final dried product with the PP for the mixed-mode PVT dryer as 3.8 years. Chaudhari et al. [84] performed an economic analysis and feasibility study of a hybrid solar dryer to dry ginger. The designed hybrid dryer consisted of a drying chamber, a heater, a flushing air, a temperature controller, and a reflector to enhance the solar radiation to the required level. From the results, the dryer's PP was calculated as 6 months and the benefit−cost ratio (BCR) was obtained as 2.3. It was concluded that the calculated PP is smaller than the expected lifetime of a solar drying system. Haque et al. [95] developed and tested an indirect solar dryer for domestic use in rural areas. The indirect dryer was dismantlable, compact, and long-lasting. At the peak time, the temperature in the drying chamber was measured about 15°C higher than the ambient. From the results, the PP of the designed dryer was calculated as a 0.56 year.

Sandali et al. [96] performed the thermal and economic analysis of a direct solar dryer. The thermal analysis was carried out using numerical modeling, while the economic assessment was carried out by estimating the solar dryer's life cycle cost (LCC) and life cycle benefit (LCB). In this study, different heat supply techniques including the use of heat exchangers, porous media, and PCMs were investigated. From the results, the heat supply approach of using a heat exchanger with geothermal water was chosen as the most preferred one due to the short calculated PP of 0.9 years. Atalay and Cankurtaran [76] evaluated the performance, exergo-economics, and environmental impacts of a large-scale indirect type solar dryer with an energy storage medium. The energy PP was estimated to be 6.82 years and the CO_2 reduction for the system's projected lifespan was estimated to be 99.60

tonnes. Malakar et al. [27] designed and fabricated an ETC-based solar dryer with a heat pipe for the drying of garlic cloves (10 kg) to be dried from the moisture content of 69% (w.b.) to 8% (w.b.) for the experimental performance assessment. At 2 m/s air velocity, the ETSD could dry the garlic clove in 8 h. The established ESTD was found to have a PP of 1.3 years. Oppong Akowuah et al. [97] explored the financial and economic feasibility of a solar-biomass hybrid dryer for maize drying. To determine the financial feasibility of the dryer operation and the profitability of the investment, a cost–benefit analysis was conducted using net present value (NPV), benefit–cost ratio (BCR), internal rate of return (IRR), and PP. The results indicated that the total required capital investment to set up the drying system is US$5263 with an annual operating cost of US$1166. Additionally, the investment in the dryer was assessed to be viable with an NPV of $4876 and an IRR of 38% at a capital cost of 24% using a 10-year economic utilization period. It was also found that at a drying cost of US$2.11/bag, the initial capital investment may be recouped in PP of 2.7 years with a BCR of 1.48. Zachariah et al. [48] conducted an environmental and economic study on a PV-assisted mixed-mode solar dryer. To improve drying performance, the author's suggested solar dryer incorporates a TES unit based on paraffin wax and exhaust air recirculation. The results indicated that the suggested dryer with TES has an energy PP of 1.91 years and a PP of 0.8 years, which is significantly less than the dryer life, making it both environmentally and economically sustainable.

The economic studies show that various designs of solar dryers have shorter PP in the range of 5 months to five years while having a large service life of 15 years to 30 years. The PP is the most effective parameter used for the economic comparison of different solar dryers. Table 9.7 enlists different

TABLE 9.7 Different types of dryers and their payback period.

Type of dryer	Payback period (PP)	Reference
Mixed-mode SGHD	1.6 years	[90]
Mixed-mode PVT dryer	3.8 years	[94]
Hybrid solar dryer	6 months	[84]
Indirect solar dryer	0.56 year	[95]
Direct solar dryer	0.9 year	[96]
Large-scale indirect type solar dryer	6.82 years	[76]
ETC-based solar dryer	1.3 years	[27]
Solar-biomass hybrid dryer	2.7 years	[97]
PV-assisted mixed-mode solar dryer	0.8 years	[48]

types of solar dryers considering their PP. From the data presented in the table, the PP for a direct solar dryer is shorter than that of an indirect solar dryer, while indirect solar dryers have a smaller PP than hybrid and mixed-mode dryers.

9.10 Implemented projects

The All-India Women's Conference (AIWC), a nonprofit organization with 150,000 participants with more than 500 branches throughout India, began researching solar dryers in four different locations. This study aimed to dry enough fruit and vegetables to provide a livelihood for one individual. The dried fruit and vegetables were of high quality and could be used in place of fruit and vegetables dried using electricity or gas. This project aimed to bring value to the fruit and vegetable processing industry at the microlevel by using solar dryers. The solar dryer project used by self-helping group (SHG) is shown in Fig. 9.20. This project assures SHG participants (mostly women) who can run this modern equipment with zero energy expense and manage it themselves with minimal assistance. The most significant advantage of the enterprise is that it does not require a full-time commitment and therefore can be undertaken by day workers, housewives, or other women engaged in any other active service. Aside from that, it prevents a massive amount of rural produce from spoiling. The use of solar dryers is concluded to be ideal for wealth formation among poor people and to play a significant role in mitigating global warming. This could be easily repeated in any country that grows a lot of fruits and vegetables, as well as medicinal plants and fish, all of which could be solar-dried to make value-added and hygienic products [98].

Nambangan Perak is a fishing community in Kenjeran Surabaya that still uses the traditional method of drying fish. A solar dryer case study was conducted in the Fishery Community of Kenjeran Surabaya. The traditional method requires stable weather and a large drying area and is highly dependent on the sun which only shines for around 6 h per day. Furthermore, the

FIGURE 9.20 Photos of the solar dryer project used by self-helping groups (SHGs).

village area is densely packed with conventional fish tray dryers, narrowing the village roads as a result. To address this problem, a dryer technology was proposed to support the fishing community in the drying process and improve the quality of dried fish. This technology can be used in inclement conditions, rain or shine, and at any time of day or night. Solar collectors are built into this dryer, allowing for absorbing heat from the sun and PV panels to supply electricity demand. The design of the solar dryer, as well as fish before and after drying, are shown in Fig. 9.21. It has been reported that the dryer and its solar collector can reduce the drying time by up to 4 h compared to traditional processes and since the final moisture content of fish is less than 10%, the dryer can be used three times a day [99].

Recognizing the need to boost livelihood protection in India's villages, the Government of India launched the Aajeevika—National Rural Livelihoods Mission (NRLM). The Mission's goal is to "create productive and successful rural poor institutional platforms that enable them to increase household income through sustainable livelihood improvements and better access to financial services." The government encourages the use of solar

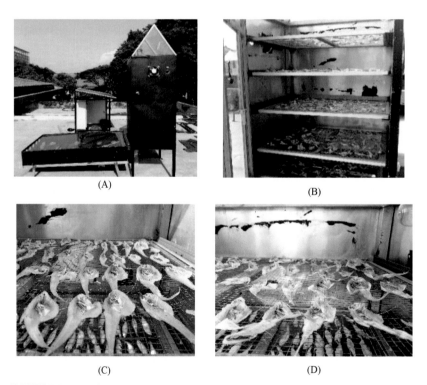

(A)   (B)

(C)   (D)

FIGURE 9.21 (A) Solar dryer setup, (B) tray inside the drying chamber, (C) fish before drying, (D) Fish after drying [99].

dryers for agricultural and aquatic goods. Odisha, Assam, and Madhya Pradesh are the three states targeted by the proposed project. When compared to open-sun drying, solar dryers have several benefits. Solar drying is 30%−40% simpler and can be done on both cloudy and rainy days if there is an electrical backup. Solar dryers for herbs, berries, and fruits have the benefit of preserving the dried product's aroma and flavor. Solar dryers have low operating and maintenance costs with a lifespan of up to 15 years. Fish drying using solar dryers has been attempted in many locations around the world. However, various types of fish have different drying conditions, such as drying time and rate, and the catch caught daily by fishermen contains a diverse mix of fish species. 115 SHGs for solar fish drying have been created in Odisha as part of the Integrated Coastal Zone Management Project 50, focusing on two coastal belts of Paradeep to Dhamra and Gopalpur to Chilka. Solar dryers have already been ordered and sold to SHGs in the sum of 50 units. As a result, the initiative will have complete financial assistance [100].

The aim of the Technological Advancement for Rural Areas (TARA) scheme, which is part of the Science for Equity Empowerment and Development (SEED) programme, is to provide long-term core support to science-based Voluntary Organizations/Field Institutions to promote and foster them as "S&T Incubators"/"Active Field Laboratories" in rural and other marginalized areas to work and provide technological solutions. Solar drying technology was combined with food processing as well as horticultural plant technology under this scheme. As a result, Solar Food Processing Technology has emerged, with a focus on dehydrating fruits and vegetables into fruit bars and dried fruits and vegetables. Mango, Guava, Chikku, Fig, Mixed Fruit, No Added Sugar Mango, and Khatta-Meetha Mango Bars were created. Tomato Powder, Curry Leaf Powder, Carrot Powder, Amla Powder, and Amla Supari are examples of vegetable products. Ragi Malt and Wheat Grass Powder are examples of health products. The procedure was successfully standardized and commercialized. Micro companies, or SHGs, were given access to the processing technology. These products were commercially available in nearby supermarkets [101].

9.11   Conclusions and prospects

Solar drying is an affordable, cost-effective alternate way to fix the shortcomings of conventional open sun drying in low-income countries. Using solar drying for agricultural and marine products has tremendous technological and energy-saving potentials. Various types of solar dryers have been designed and built in different parts of the world, providing different levels of technical performance. For a better quality of dried agriculture and marine products, the temperature and velocity of airflow within the dryer should be controlled properly. The performance of solar dryers can be improved by using a hybrid system and TES system. Various hybrid solar dryers are now developed which provide a better-quality dried product with reduced drying time. PCMs with improved thermal conductivity are used to enhance the performance of the solar drying system that also makes possible drying process to continue when solar radiation is not available. Since cost is a major factor in the development of solar dryers, it is essential to measure and promote the short-term and long-term cost benefits of solar dryers to owners. In conclusion, the PP of solar dryers lies within 5 years with a long working life of 15−30 years. Solar dryers can also reduce carbon emissions to the environment as their long-term benefits. The technical guidance in designing solar-assisted systems for agriculture and marine goods is compact collector sizes, high-performance advanced energy storage systems, and long-life drying systems. Finally, further deployment of technology in developing countries often needs close cooperation and coordination between relevant parties in research and development (R&D) extension relations.

Acknowledgment

The author Abhishek Anand is highly obliged to the University Grants Commission (UGC) and Ministry of Education, Government of India, New Delhi for providing the Senior Research Fellowship (SRF).

References

[1] Lingayat AB, Chandramohan VP, Raju VRK, Meda V. A review on indirect type solar dryers for agricultural crops − dryer setup, its performance, energy storage and important highlights. Appl Energy 2020;258:114005. Available from: https://doi.org/10.1016/j.apenergy.2019.114005.

[2] Gorjian S, Hosseingholilou B, Jathar LD, Samadi H, Samanta S, Sagade AA, et al. Recent advancements in technical design and thermal performance enhancement of solar greenhouse dryers. Sustainability 2021;13:7025. Available from: https://doi.org/10.3390/su13137025.

[3] Thamkaew G, Sjöholm I, Galindo FG. A review of drying methods for improving the quality of dried herbs. Crit Rev Food Sci Nutr 2021;61:1763−86. Available from: https://doi.org/10.1080/10408398.2020.1765309.

[4] Gorjian S., Singh R., Shukla A., Mazhar A.R. On-farm applications of solar PV systems. In: Gorjian S., Shukla A., eds Photovoltaic solar energy conversion. 1st (ed.), London: Elsevier; 2020, p. 147−190. <https://doi.org/10.1016/B978-0-12−819610-6.00006-5>.

[5] Visavale GL. Solar drying: fundamentals, applications and innovations. ISBN:978-981-07-3336-0; 2012.

[6] Bala BK. Principles of drying techniques, 2016, p. 320.

[7] Treybal RE. Mass—transfer operations book description; 1980.

[8] Callahan CW, Elansari AM, Fenton DL. Psychrometrics. Postharvest technology of perishable horticultural commodities, Elsevier; 2019, p. 271−310. <https://doi.org/10.1016/B978-0-12-813276-0.00008-0>.

[9]   Ekechukwu OV, Norton B. 99/02111 Review of solar-energy drying systems II: an overview of solar drying technology. Fuel Energy Abstr 1999;40:216. Available from: https://doi.org/10.1016/s0140-6701(99)97881-5.

[10] Dincer I. Evaluation and selection of energy storage systems for solar thermal applications. Int J Energy Res 1999;23:1017−28 https://doi.org/10.1002/(SICI)1099-114X (19991010)23:12 < 1017::AID-ER535 > 3.0.CO;2-Q.

[11] Le VV, Le TH, Nguyen TP, Duong XQ. A review of solar dryer with phase change material as sensible heat storage mediums. J Mech Eng Res Dev 2021;44:202−14.

[12] Shadab M, Mastan SI, Motarique S, Ahmed SMS, Suman SS. Design and modification of solar seed dryer. Int J Innov Eng Technol 2020;7:115−19.

[13] Udomkun P, Romuli S, Schock S, Mahayothee B, Sartas M, Wossen T, et al. Review of solar dryers for agricultural products in Asia and Africa: an innovation landscape approach. J Env Manage 2020;268:110730. Available from: https://doi.org/10.1016/j.jenvman.2020.110730.

[14] Chauhan YB, Rathod PP. A comprehensive review of the solar dryer. Int J Ambient Energy 2020;41:348−67. Available from: https://doi.org/10.1080/01430750.2018.1456960.

[15] Sengar N. Experimental studies on developed direct solar dryer for conversion of grapes into raisins with temperature control; 2021.

[16] Ayua E, Mugalavai V, Simon J, Weller S, Obura P, Nyabinda N. Comparison of a mixed modes solar dryer to a direct mode solar dryer for African indigenous vegetable and chili processing. J Food Process Preserv 2017;41:e13216. Available from: https://doi.org/10.1111/jfpp.13216.

[17] Sontakke MS, Salve SP. Solar drying technologies: a review. Int Ref J of Eng Sci 2015;4 (4):29−35.

[18] Obayopo SO, Alonge OI. Development and quality analysis of a direct solar dryer for fish. Food Nutr Sci 2018;09:474−88. Available from: https://doi.org/10.4236/fns.2018.95037.

[19] Kumar S, Bhattacharya SC. Technology Packages: Solar, Biomass and Hybrid Dryers, Renewable Energy Technologies in Asia; Regional Energy Resources Information Center (RERIC), AIT, Thailand, 2005: 44−60.

[20] Hallak H, Hillal J, Hilal F, Rahhal R. The staircase solar dryer: design and characteristics. Renew Energy 1996;7:177−83. Available from: https://doi.org/10.1016/0960-1481(95)00127-1.

[21] Hajar E, Rachid T, Najib BM. Conception of a solar air collector for an indirect solar dryer. Pear drying test. Energy Procedia 2017;141:29−33. Available from: https://doi.org/10.1016/j.egypro.2017.11.114.

[22] Kumar M, Sansaniwal SK, Khatak P. Progress in solar dryers for drying various commodities. Renew Sustain Energy Rev 2016;. Available from: https://doi.org/10.1016/j.rser.2015.10.158.

[23] Hatami S, Payganeh G, Mehrpanahi A. Energy and exergy analysis of an indirect solar dryer based on a dynamic model. J Clean Prod 2020;244:118809. Available from: https://doi.org/10.1016/j.jclepro.2019.118809.

[24] Dissa AO, Desmorieux H, Bathiebo J, Koulidiati J. A comparative study of direct and indirect solar drying of mango. Glob J Pure Appl Sci 2011;17:273−294−294.

[25] Tiwari G, Katiyar VK, Dwivedi V, Katiyar AK, Pandey CK. A Comparative study of commonly used solar dryers in India. Int J Curr Eng Technol 2013;3:994−9.

[26] Lingayat A, Chandramohan VP, Raju VRK, Kumar A. Development of indirect type solar dryer and experiments for estimation of drying parameters of apple and watermelon. Therm Sci Eng Prog 2020;100477. Available from: https://doi.org/10.1016/j.tsep.2020. 100477.

[27] Malakar S, Arora VK, Nema PK. Design and performance evaluation of an evacuated tube solar dryer for drying garlic clove. Renew Energy 2021;168:568−80. Available from: https://doi.org/10.1016/j.renene.2020.12.068.

[28] Varghese J, Rupesh S, Augustine J, Nair A, Prajith. Design and analysis of a solar drier with a parabolic shaped dish type collector for drying peanut. IOP Conf Ser Mater Sci Eng 2021;1132:012046. Available from: https://doi.org/10.1088/1757-899X/1132/1/012046.

[29] Lakshmi DVN, Muthukumar P, Layek A, Kumar P. Performance analyses of mixed mode forced convection solar dryer for drying of stevia leaves. Sol Energy 2019;188:507−18. Available from: https://doi.org/10.1016/j.solener.2019.06.009.

[30] Ekka JP, Bala K, Muthukumar P, Kanaujiya DK. Performance analysis of a forced convection mixed mode horizontal solar cabinet dryer for drying of black ginger (*Kaempferia Parviflora*) using two successive air mass flow rates. Renew Energy 2020;. Available from: https://doi.org/10.1016/j.renene.2020.01.035.

[31] Shaikh TB, Kolekar AB. Review of hybrid solar dryers. Int J Innov Eng Res Technol [Ijiert] 2015;2:1−7.

[32] Suherman S, Susanto EE, Zardani AW, Dewi NHR. Performance study of hybrid solar dryer for cassava starch, 2020:080003. <https://doi.org/10.1063/1.5140943>.

[33] Amer BMA, Hossain MA, Gottschalk K. Design and performance evaluation of a new hybrid solar dryer for banana. Energy Convers Manag 2010;51:813−20. Available from: https://doi.org/10.1016/j.enconman.2009.11.016.

[34] Hamdani, Rizal TA, Muhammad Z. Fabrication and testing of hybrid solar-biomass dryer for drying fish. Case Stud Therm Eng 2018;12:489−96. Available from: https://doi.org/ 10.1016/j.csite.2018.06.008.

[35] Ndukwu MC, Simo-Tagne M, Abam FI, Onwuka OS, Prince S, Bennamoun L. Exergetic sustainability and economic analysis of hybrid solar-biomass dryer integrated with copper tubing as heat exchanger. Heliyon 2020;6:e03401. Available from: https://doi.org/ 10.1016/j.heliyon.2020.e03401.

[36] Slimani MEA, Amirat M, Bahria S, Kurucz I, Aouli M, Sellami R. Study and modeling of energy performance of a hybrid photovoltaic/thermal solar collector: Configuration suitable for an indirect solar dryer. Energy Convers Manag 2016;125:209−21. Available from: https://doi.org/10.1016/j.enconman.2016.03.059.

[37] Poonia S, Singh AK, Jain D. Design development and performance evaluation of photovoltaic/thermal (PV/T) hybrid solar dryer for drying of ber (*Zizyphus mauritiana*) fruit. Cogent Eng 2018;5:1−18. Available from: https://doi.org/10.1080/23311916.2018. 1507084.

[38] Sajith KG, Muraleedharan C. Economic Analysis of a hybrid photovoltaic/thermal solar dryer for drying Amla. Int J Eng Res Technol 2014;3:907−10.

[39] Ananno AA, Masud MH, Dabnichki P, Ahmed A. Design and numerical analysis of a hybrid geothermal PCM flat plate solar collector dryer for developing countries. Sol Energy 2020;196:270−86. Available from: https://doi.org/10.1016/j.solener.2019.11.069.

[40] Anum R, Ghafoor A, Munir A. Study of the drying behavior and performance evaluation of gas fired hybrid solar dryer. J Food Process Eng 2017;40:1−11. Available from: https://doi.org/10.1111/jfpe.12351.

[41] Boughali S, Benmoussa H, Bouchekima B, Mennouche D, Bouguettaia H, Bechki D. Crop drying by indirect active hybrid solar - electrical dryer in the eastern Algerian Septentrional Sahara. Sol Energy 2009;83:2223−32. Available from: https://doi.org/10.1016/j.solener.2009.09.006.

[42] Prakash O, Kumar A. Solar greenhouse drying: a review. Renew Sustain Energy Rev 2014;29:905−10. Available from: https://doi.org/10.1016/j.rser.2013.08.084.

[43] Ahmad A, Prakash O. Performance evaluation of a solar greenhouse dryer at different bed conditions under passive mode. J Sol Energy Eng 2020;142:1−10. Available from: https://doi.org/10.1115/1.4044194.

[44] Prakash O, Kumar A. Environomical analysis and mathematical modelling for tomato flakes drying in a modified greenhouse dryer under active mode. Int J Food Eng 2014;10:669−81. Available from: https://doi.org/10.1515/ijfe-2013-0063.

[45] Chauhan PS, Kumar A, Gupta B. A review on thermal models for greenhouse dryers. Renew Sustain Energy Rev 2017;75:548−58. Available from: https://doi.org/10.1016/j.rser.2016.11.023.

[46] Pankaew P, Aumporn O, Janjai S, Pattarapanitchai S, Sangsan M, Bala BK. Performance of a large-scale greenhouse solar dryer integrated with phase change material thermal storage system for drying of chili. Int J Green Energy 2020;17:632−43. Available from: https://doi.org/10.1080/15435075.2020.1779074.

[47] Azaizia Z, Kooli S, Hamdi I, Elkhal W, Guizani AA. Experimental study of a new mixed mode solar greenhouse drying system with and without thermal energy storage for pepper. Renew Energy 2020;145:1972−84. Available from: https://doi.org/10.1016/j.renene.2019.07.055.

[48] Zachariah R, Maatallah T, Modi A. Environmental and economic analysis of a photovoltaic assisted mixed mode solar dryer with thermal energy storage and exhaust air recirculation Int J Energy Res 2020;er.5868. Available from: https://doi.org/10.1002/er.5868.

[49] Prakash O, Kumar A, Laguri V. Performance of modified greenhouse dryer with thermal energy storage. Energy Rep 2016;2:155−62. Available from: https://doi.org/10.1016/j.egyr.2016.06.003.

[50] Blakeney M. Food loss and food waste. Edward Elgar Publishing; 2019. <https://doi.org/10.4337/9781788975391>.

[51] Waheed Deshmukh A, Wasewar KL, Verma MN. Solar drying of food materials as an alternative for energy crisis and environmental protection. Int J Chem Sci 2011;9:1175−82.

[52] Mercer DG. An introduction to the dehydration and drying of fruits and vegetables; 2014.

[53] University of Georgia. Preserving food: Drying vegitable and fruits; 2016.

[54] Grain crop drying, handling and storage n.d.

[55] Luke LaBorde. Drying Herbs; 2013.

[56] Kant K, Shukla A, Sharma A, Kumar A, Jain A. Thermal energy storage based solar drying systems: a review. Innov Food Sci Emerg Technol 2016;34:86−99. Available from: https://doi.org/10.1016/j.ifset.2016.01.007.

[57] Fudholi A, Zohri M, Rukman NSB, Nazri NS, Mustapha M, Yen CH, et al. Exergy and sustainability index of photovoltaic thermal (PVT) air collector: a theoretical and experimental study. Renew Sustain Energy Rev 2019;100:44−51. Available from: https://doi.org/10.1016/j.rser.2018.10.019.

[58] Hidalgo LF, Candido MN, Nishioka K, Freire JT, Vieira GNA. Natural and forced air convection operation in a direct solar dryer assisted by photovoltaic module for drying of green onion. Sol Energy 2021;220:24−34. Available from: https://doi.org/10.1016/j.solener.2021.02.061.

[59] Goud M, Reddy MVV. A novel indirect solar dryer with inlet fans powered by solar PV panels: drying kinetics of *Capsicum annum* and *Abelmoschus esculentus* with dryer performance V.P. C, S. SSol Energy 2019;194:871−85. Available from: https://doi.org/10.1016/j.solener.2019.11.031.

[60] Fterich M, Chouikhi H, Bentaher H, Maalej A. Experimental parametric study of a mixed-mode forced convection solar dryer equipped with a PV/T air collector. Sol Energy 2018;171:751−60. Available from: https://doi.org/10.1016/j.solener.2018.06.051.

[61] Tiwari S, Agrawal S, Tiwari GN. PVT air collector integrated greenhouse dryers. Renew Sustain Energy Rev 2018;90:142−59. Available from: https://doi.org/10.1016/j.rser.2018.03.043.

[62] Gorjian S, Ebadi H, Calise F, Shukla A, Ingrao C. A review on recent advancements in performance enhancement techniques for low-temperature solar collectors. Energy Convers Manag 2020;222:113246. Available from: https://doi.org/10.1016/j.enconman.2020.113246.

[63] Shakouri M., Ebadi H., Gorjian S. Solar photovoltaic thermal (PVT) module technologies. In: Gorjian S., Shukla A., (eds.) Photovoltaic. Solar energy conversion 1st (ed.), London: Elsevier; 2020, p. 79−116. <https://doi.org/10.1016/B978-0-12-819610-6.00004-1>.

[64] Rabha DK, Muthukumar P. Performance studies on a forced convection solar dryer integrated with a paraffin wax−based latent heat storage system. Sol Energy 2017;149:214−26. Available from: https://doi.org/10.1016/j.solener.2017.04.012.

[65] Essalhi H, Tadili R, Bargach MN. Comparison of thermal performance between two solar air collectors for an indirect solar dryer. J Phys Sci 2018;29:55−65. Available from: https://doi.org/10.21315/jps2018.29.3.5.

[66] Singh P, Vyas S, Yadav A. Experimental comparison of open sun drying and solar drying based on evacuated tube collector. Int J Sustain Energy 2019;38:348−67. Available from: https://doi.org/10.1080/14786451.2018.1505726.

[67] Daghigh R, Shahidian R, Oramipoor H. A multistate investigation of a solar dryer coupled with photovoltaic thermal collector and evacuated tube collector. Sol Energy 2020;199:694−703. Available from: https://doi.org/10.1016/j.solener.2020.02.069.

[68] Ssemwanga M, Makule E, Kayondo S. Performance analysis of an improved solar dryer integrated with multiple metallic solar concentrators for drying fruits. Sol Energy 2020;204:419−28. Available from: https://doi.org/10.1016/j.solener.2020.04.065.

[69] Anand A, Shukla A, Sharma A. Recapitulation on latent heat hybrid buildings. International Journal of Energy Research 2020;44(3):1−38. Available from: https://doi.org/10.1002/er.4920.

[70] Alva G, Liu L, Huang X, Fang G. Thermal energy storage materials and systems for solar energy applications. Renew Sustain Energy Rev 2017;68:693−706. Available from: https://doi.org/10.1016/j.rser.2016.10.021.

[71] Anand A, Shukla A, Kumar A, Buddhi D, Sharma A. Cycle test stability and corrosion evaluation of phase change materials used in thermal energy storage systems. Journal of Energy Storage 2021;39:1−18. Available from: https://doi.org/10.1016/j.est.2021.102664.

[72] Vijayan S, Arjunan TV, Kumar A. Exergo-environmental analysis of an indirect forced convection solar dryer for drying bitter gourd slices. Renew Energy 2020;146:2210−23. Available from: https://doi.org/10.1016/j.renene.2019.08.066.

[73] Gorjian S, Calise F, Kant K, Ahamed MS, Copertaro B, Najafi G, et al. A review on opportunities for implementation of solar energy technologies in agricultural greenhouses. J Clean Prod 2021;124807. Available from: https://doi.org/10.1016/j.jclepro.2020.124807.

[74] Braham W, Khellaf A, Mediani A, El M, Slimani A, Loumani A, et al. Experimental investigation of an active direct and indirect solar dryer with sensible heat storage for camel meat drying in Saharan environment. Sol Energy 2018;174:328−41. Available from: https://doi.org/10.1016/j.solener.2018.09.037.

[75] Mishra L., Sinha A., Gupta R. Recent developments in latent heat energy storage systems using phase change materials (PCMs)—a review, Springer Singapore; 2019, p. 25−37. <https://doi.org/10.1007/978-981-13-1202-1_2>.

[76] Atalay H, Cankurtaran E. Energy, exergy, exergoeconomic and exergo-environmental analyses of a large scale solar dryer with PCM energy storage medium. Energy 2021;216:119221. Available from: https://doi.org/10.1016/j.energy.2020.119221.

[77] Faraj K, Khaled M, Faraj J, Hachem F, Castelain C. Phase change material thermal energy storage systems for cooling applications in buildings: a review. Renew Sustain Energy Rev 2020;119:109579. Available from: https://doi.org/10.1016/j.rser.2019.109579.

[78] Mortezapour H, Rashedi SJ, Akhavan HR, Maghsoudi H. Experimental analysis of a solar dryer equipped with a novel heat recovery system for onion drying. J Agric Sci Technol 2017;19:1227−40.

[79] Kesavan S, Arjunan TV, Vijayan S. Thermodynamic analysis of a triple-pass solar dryer for drying potato slices. J Therm Anal Calorim 2019;136:159−71. Available from: https://doi.org/10.1007/s10973-018-7747-0.

[80] Swami VM, Autee AT. Experimental analysis of solar fish dryer using phase change material T R AJ Energy Storage 2018;20:310−15. Available from: https://doi.org/10.1016/j.est.2018.09.016.

[81] Babar OA, Tarafdar A, Malakar S, Arora VK, Nema PK. Design and performance evaluation of a passive flat plate collector solar dryer for agricultural products. J Food Process Eng 2020;43. Available from: https://doi.org/10.1111/jfpe.13484.

[82] Vigneshkumar N, Venkatasudhahar M, Manoj Kumar P, Ramesh A, Subbiah R, Michael Joseph Stalin P, et al. Investigation on indirect solar dryer for drying sliced potatoes using phase change materials (PCM). Mater Today Proc 2021;. Available from: https://doi.org/10.1016/j.matpr.2021.05.562.

[83] Khaing Hnin K, Zhang M, Mujumdar AS, Zhu Y. Emerging food drying technologies with energy-saving characteristics: a review. Dry Technol 2019;37:1465−80. Available from: https://doi.org/10.1080/07373937.2018.1510417.

[84] Chaudhari RH, Gora A, Modi VM, Chaudhari H. Economic analysis of hybrid solar dryer for ginger drying. Int J Curr Microbiol Appl Sci 2018;7:2725−31. Available from: https://doi.org/10.20546/ijcmas.2018.711.312.

[85] EL-Mesery HS, Abomohra AEF, Kang CU, Cheon JK, Basak B, Jeon BH. Evaluation of infrared radiation combined with hot air convection for energy-efficient drying of biomass. Energies 2019;12. Available from: https://doi.org/10.3390/en12142818.

[86] Rigit ARH, Jakhrani AQ, Kamboh SA, Tiong Kie PL. Development of an indirect solar dryer with biomass backup burner for drying pepper berries. World Appl Sci J 2013;22:1241−51. Available from: https://doi.org/10.5829/idosi.wasj.2013.22.09.2724.

[87] Ndukwu MC, Onyenwigwe D, Abam FI, Eke AB, Dirioha C. Development of a low-cost wind-powered active solar dryer integrated with glycerol as thermal storage. Renew Energy 2020;154:553−68. Available from: https://doi.org/10.1016/j.renene.2020.03.016.

[88] Yahya M, Fudholi A, Sopian K. Energy and exergy analyses of solar-assisted fl uidized bed drying integrated with biomass furnace. Renew Energy 2017;105:22−9. Available from: https://doi.org/10.1016/j.renene.2016.12.049.

[89] Srivastava A, Anand A, Shukla A, Kumar A, Buddhi D, Sharma A. A comprehensive overview on solar grapes drying: Modeling, energy, environmental and economic analysis. Sustainable energy technologies and assessments 2021;47. Available from: https://doi.org/10.1016/j.seta.2021.101513.

[90] ELkhadraoui A, Kooli S, Hamdi I, Farhat A. Experimental investigation and economic evaluation of a new mixed-mode solar greenhouse dryer for drying of red pepper and grape. Renew Energy 2015;77:1−8. Available from: https://doi.org/10.1016/j.renene.2014.11.090.

[91] Kurt M, Aksoy A, Sanin FD. Evaluation of solar sludge drying alternatives by costs and area requirements. Water Res 2015;82:47−57. Available from: https://doi.org/10.1016/j.watres.2015.04.043.

[92] Nabnean S, Janjai S, Thepa S, Sudaprasert K, Songprakorp R, Bala BK. Experimental performance of a new design of solar dryer for drying osmotically dehydrated cherry tomatoes. Renew Energy 2016;94:147−56. Available from: https://doi.org/10.1016/j.renene.2016.03.013.

[93] Dejchanchaiwong R, Arkasuwan A, Kumar A, Tekasakul P. Mathematical modeling and performance investigation of mixed-mode and indirect solar dryers for natural rubber sheet drying. Energy Sustain Dev 2016;34:44−53. Available from: https://doi.org/10.1016/j.esd.2016.07.003.

[94] Tiwari S, Tiwari GN. Exergoeconomic analysis of photovoltaic-thermal (PVT) mixed mode greenhouse solar dryer. Energy 2016;114:155−64. Available from: https://doi.org/10.1016/j.energy.2016.07.132.

[95] Haque T, Tiwari M, Bose M, Kedare SB. Drying kinetics, quality and economic analysis of a domestic solar dryer for agricultural products. Ina Lett 2019;4:147−60. Available from: https://doi.org/10.1007/s41403-018-0052-1.

[96] Sandali M., Boubekri A., Mennouche D. Thermal and economical study of a direct solar dryer with integration of different techniques of heat supply, Springer Singapore; 2020, p. 585−595. <https://doi.org/10.1007/978-981-15-5444-5_73>.

[97] Oppong Akowuah J, Bart-Plange A, Agbeko Dzisi K. Financial and economic analysis of a 1-tonne capacity mobile solar-biomass hybrid dryer for maize drying. Int J Agric Econ 2021;6:98. Available from: https://doi.org/10.11648/j.ijae.20210603.11.

[98] Lalitta B. CASE STUDY-India Solar Dryers for Income Generation; 2016.

[99] Hantoro R, Hepriyadi SU, Izdhiharrudin MF, Amir MH. Solar dryer and photovoltaic for fish commodities (Case study in fishery community at Kenjeran Surabaya). AIP conference proceedings 2018;1977:060013. <https://doi.org/10.1063/1.5043025>.

[100] Vinet L, Zhedanov A. UNDP project report. vol. 44. <https://doi.org/10.1088/1751-8113/44/8/085201>; 2015.

[101] SEED DST TARA — Development of fruit bars and vegetable products by solar dehydration technology (2016–17). Phase II 2016-17:1–5.

Chapter 10

Applications of robotic and solar energy in precision agriculture and smart farming

Amir Ghalazman E.[1], Gautham P. Das[1], Iain Gould[1], Payam Zarafshan[2], Vishnu Rajendran S.[1], James Heselden[1], Amir Badiee[3], Isobel Wright[1] and Simon Pearson[1]

[1]*Lincoln Institute for Agri-Food Technology, University of Lincoln, Lincoln, United Kingdom,*
[2]*Department of Agro-Technology, College of Aburaihan, University of Tehran, Tehran, Iran,*
[3]*School of Engineering, University of Lincoln, Lincoln, United Kingdom*

10.1 Introduction

Pathways limiting global warming require rapid and far-reaching changes in many sectors. There are further concerns over the risk of exceeding "tipping points" beyond which further change, for example, ice sheet melt might become unstoppable [1]. Change in land use forms part of numerous pathways to mitigate and adapt to climate change and these include area changes for the forest, energy crop, pasture, and cropped land, and management changes including sustainable intensification of land, ecosystem restoration, as well as changes toward less resource-intensive diets [2]. Modeled pathways for energy use tend to show a requirement for low-carbon technologies, enhanced energy efficiency, and faster electrification of energy end-use. In 2015, global food-system emissions represented 34% of total greenhouse gas (GHG) emissions [3]. Of this, 71% came from agriculture and land-use change, with the remaining from supply chain activities. The prefarm gate emissions [including fishing, aquaculture, and agriculture, including emissions from the production of inputs such as fertilizers, and excluding land use and lane use change (LULUC)] contributed 39% in 2015, and LULUC to about 32%. The significance of GHG emissions from agriculture varies between countries. Unlike in most industries where carbon dioxide (CO_2) is the principle GHG, for prefarm gate agriculture significant emissions can be from methane (CH_4) and nitrous oxide (N_2O). In Britain, where GHG emissions made up 10% of total emissions in 2017, only 12% of these was

Solar Energy Advancements in Agriculture and Food Production Systems.
DOI: https://doi.org/10.1016/B978-0-323-89866-9.00011-0

derived from CO_2 (in $CO_{2\text{-eq}}$), 56% from CH_4, and 31% from N_2O [4]. The CH_4 comes from manure, but the main source is ruminant enteric fermentation whilst N_2O is produced in soil from the breakdown of synthetic fertilizers and animal manure. From this, it might seem that seeking renewable energy and solar-powered replacements for CO_2 emitting energy sources addresses a highly significant but relatively smaller component of GHG emissions. However, not only solar energy could form part of the replacement for fossil fuel-powered devices, recent developments of smart solar-powered technologies may create opportunities for reductions in other gas emissions.

Simultaneously, growth in population and changes in food demand toward a more protein-based diet creates continuing pressures for food production and land-use change. This impacts ecosystems and functions such as the aquatic environment [5], soil quality, including erosion, and biodiversity [6]. The global Living Planet Index [6] shows an average 68% decrease in population sizes of mammals, birds, amphibians, reptiles, and fish between 1970 and 2016. Drivers linked to food production are responsible for 70% of global terrestrial biodiversity loss. Similarly, the State of Nature report [7] documents a decline in species abundance, range, and other factors in recent years in the United Kingdom. The decline is significant and linked to agriculture, as one of the causative effects and a significant one since agriculture covers 72% of land in the United Kingdom. Biodiversity loss threatens food security through impacts such as loss of pollinator services or soil services. Land-use change to allow for more food production, for example from forest to arable land is not an acceptable approach if we are to limit global warming. The biodiversity crisis can only be resolved by changes in land management. Therefore the drive must be to produce more from "less," with fewer external impacts. Precision agriculture (PA) is a solution to minimize the use of resources, including land, water, herbicides, pesticides, and energy. Robots are designed to deliver precision in repetitive tasks in diverse workspaces. This chapter examines an aspect of delivering necessary precision achievement and the potential for solar-powered devices. Robotic technologies in PA include robotic selective harvesting, robotic crop care, and robotic phenotyping. Hence, this chapter overviews these robotic technologies in PA and how PV-powered units can contribute to sustainability in this sector.

10.2 Precision agriculture

Agriculture is better placed than most industries in the drive for reduced GHG emissions since soils can sequester carbon. To be truly sustainable, however, agriculture must meet the needs of present and future generations, while ensuring profitability, environmental health, and social and economic equity [8]. The Food and Agriculture Organization (FAO) [8] describes 20 interconnected actions to transform agriculture to meet the sustainable development goals (SDGs). Whilst not explicitly included, many of the proposed actions may be assisted by solar-powered devices. Arising from the search for more sustainable farming systems, there is a growing movement to follow more sustainable actions on farms (putting aside the wider agrifood change impacts). Known as regenerative agriculture, the principles include maintaining diverse rotations, using resources as efficiently as possible, reduce cultivations, protecting soils by maximizing green cover and the presence of living roots in the soil. Following regenerative agriculture, principles require progressive farming innovation, and this may be using technology to farm efficiently and with precision, and also monitoring impacts on the wider environment. Solar-powered devices have the potential to help in both these areas. The agricultural industry needs to reduce emissions if we are to meet sustainability goals. Nonetheless, agriculture also needs to supply the food requirements of a growing population. Sustainable intensification—the process in which agricultural yields are improved without negative impacts to the environment [9] is thought to be one such pathway to address these challenges.

PA encompasses a range of techniques that enhance the efficiency of agricultural production and distributed sensing technology (Fig. 10.1), whether that be by increasing yields, or by reducing waste [11], and is one such approach to enable sustainable intensification through enhancing yields without compromising ecosystem services [12]. Since the 1980s, PA has witnessed increasing adoption, as technologies such as sensing technologies and global positioning system (GPS) have evolved [13]. Precision approaches have

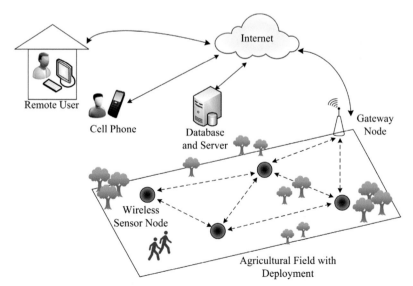

FIGURE 10.1   A typical wireless sensor network for agricultural applications as detailed in [10].

enabled management practices such as variable-rate planting, fertilizer and herbicide applications for more targeted effectiveness, reducing chemical costs and environmental degradation [14], and crop forecasting and yield mapping to inform growers, buyers, and external stakeholders as to expected yields (e.g., [15]). PA relies on an effective system to assess spatial and temporal variability across a farm, a process that fundamentally relies on data acquisition from a range of sensing technologies, from in situ data collection "on the ground" to remote data collection mentionedabove. Here, we highlight the role of technologies employed in PA (e.g., distributed sensing and robotized agriculture) and discuss current, or potential, applications of solar power.

10.2.1 In situ sensing

Many measurements are acquired by sensors "on the ground," providing key information about environmental and plant conditions to inform management. Such sensing may be referred to as proximal sensing [16], with data being either obtained from *fixed sensors* in a field, "on-the-go" with *mobile measurements* [17], or *remote sensing* (e.g., using autonomous ground or aerial vehicles), with each type of sensing method having its own merits for high spatial and temporal resolution data collection. Energy sources for these sensors can be said to be "active"—requiring energy from an external source, or "passive" utilizing solar radiation directly [18].

A widespread example of in situ *fixed position sensing* is that of weather stations. At national scales, networks of metrological stations constantly monitor environmental conditions via multiple sensors, to inform forecasting. In automated weather stations, sensors are either powered by solar photovoltaic (PV), a battery, or a combination of solar PV and battery. Similar networks of stations have been set up to monitor other key environmental properties, such as the COSMOS-UK network monitoring soil moisture, and the UK-SCAPE network monitoring carbon flux [19,20]. Fixed position sensors also play a role in agriculture and informing farm management, albeit sometimes facing challenges in terms of sensor placement and disturbance. For example, sensors monitoring soil properties in annual crops may have to be installed seasonally, or face damage and disruption from cultivations. Nevertheless, in situ fixed sensors see increasing use in modern agriculture to inform precision management. One example of a fixed sensor system in PA is soil moisture monitoring for irrigation [21]. Soil moisture monitoring can provide more accurate data to inform irrigation scheduling, which can lead to optimal plant growth and reductions in water usage for the grower. This can work via a single low power sensor, or array of sensors, deployed in the field linked to a wireless communication system to inform growers [10].

Fixed sensors can provide high-resolution data over time, however, one of the key challenges of sensing in agriculture is the representation of field variability. Agricultural fields exhibit much spatial variability in terms of soil conditions, topography, biological activity, and weed burden, which can be seen over the scale of meters [22,23]. As such, "on-the-go" or "mobile" sensing has become a key approach adopted in PA, enabling growers to obtain data with greater spatial coverage than fixed sensors would allow. Mobile sensing has seen some uptake in agricultural systems, often concerning soil mapping. Assessing variability of soil properties across a field can help inform decisions such as fertilizer application based on soil nutrient status [24] and liming requirements based on soil pH [25]. Sensors that allow online data collection can include those using mechanical sensing (e.g., soil penetration resistance), electrical resistivity, or spectral sensing, although with varied uptake in the commercial sector. In the past, such online data has been obtained by sensors operated by a farm worker on foot or by all-terrain vehicles, mounted to existing farm machinery, however recent studies have shown the potential for robotics systems to integrate with sensors for autonomous field mapping [26].

Remote sensing refers to data collection from a sensor at a greater distance from the field, allowing much larger areas to be covered [16]. The use of UAVs can provide growers with information on the spatial distribution of weeds, with an ability to cover larger areas than any mobile sensors on the ground [27]. Whilst at a further scale, Earth Observation satellites can provide large spatial datasets on crop performance at the local, regional, or national level [28]. However, often remote and in-situ, or proximal, sensings are not employed in isolation, and combined use of tools provide more comprehensive data acquisition for decision making [29].

One of the more recently developed innovations in agriculture is the adoption of wireless sensor networks (WSNs). WSNs consist of a series, or network, of small and relatively cheap sensors which connect to a communication system to provide real-time data acquisition for the end-users. WSN has thus many applications to enhance agricultural production, such as informing irrigation scheduling and nutrient management, PA, and GHG monitoring [29]. A typical setup for a WSN system comprises a series of sensor nodes each with four key capacities: (1) having a power source (e.g., PV) source of energy), (2) ability to sense environmental data, (3) computing capacity, and (4) communications capacity [30]. Each node will comprise a power source and sensor system, for example, a soil moisture sensor, which communicates with one another and also to a nearby gateway node to connect the sensor network with the outer world (Fig. 10.1). Technologies for this communication system vary depending on locality and availability, power consumption, and band frequency, but can include ZigBee, Bluetooth, Wibree, WiFi, GPRS, and WiMax [30].

There are many benefits to introducing WSN systems into agricultural applications. Multiple sensors in one field address the challenges of field variability and soil type in assessing potentially localized properties such as soil moisture and nutrient availability. The capacity that WSN gives for real-time data acquisition can also provide substantial benefits to farmers working in small weather windows or time-pressured operations. Nonetheless, although WSN systems have seen increasing adoption and deployment, several key challenges remain, including the limited battery capacity of the sensor nodes. To address this, effective energy capture and storage systems have been developed, with solar power identified as one of the key sources to utilize. Harvesting solar energy allows energy capture during times of high energy (e.g., daylight), which sensors can run from in daylight hours, utilized stored (battery) energy in low sunlight or at night. Such power systems comprise a solar cell, battery, and control system at each sensor node [30].

Solar-powered WSNs have been deployed for a range of agricultural applications. Gutiérrez et al. [31] developed a WSN network of soil moisture and temperature sensors powered by PV panels. In this instance, the network fed into an automated irrigation system, resulting in significant savings in water consumption. A solar-powered system for monitoring environmental variables in rice paddy systems is presented by Zhang et al. [32] in which solar power was employed to meet the energy demand on cloudy overcast days. Extending to further uses, a solar-powered WSN system was developed by Tuan-Duc Nguyen et al. [33] to measure meteorological parameters in crop fields, with benefits to energy efficiency. Solar-powered WSNs have the potential for deployment in most of the regions around the world with sufficient availability of solar radiation. In temperate climate conditions of higher latitudes, the times of high crop activity and growth, and thus times when high data acquisition is needed, for example soil moisture contents for irrigation scheduling, are in the spring and summer months and correlate with longer daylight hours and more intense solar radiation.

10.2.2 Agriculture and circular economy

Agriculture is one of the core industries which play an important role in supporting the economic development of nations around the world. With increasing global population [34] and enhancing living standards, the global need for healthy food [35] and accordingly cleaner energy sources to support such food production in the agriculture sector increases rapidly. The conventional energy sources are facing grave scrutiny over their environmental impact and GHG emission especially with the recent global movement to meet the Paris agreement with national targets such as UK 2050 Net-Zero. This is driving a continuous search for green energy alternatives to driving future development, particularly in the agrifood sector. In recent years, with the reduction in the cost of the manufacturing of PV modules with improved robustness [36] and efficiency, PV electricity generation has become of interest in the agrifood sector in comparison to other energy sources due to its associated GHG emissions. Throughout the different stages of agricultural development in history, energy was/is the key factor in system transformation. PV integrated agriculture has recently received extensive global attention from the industry [37−42]. In the proposed novel energy and food production concept, the marriage between agriculture and energy can ultimately create double incomes from the food-energy shared land, which can significantly improve land utilization and therefore biodiversity. However, this requires further research and system optimization before full implementation.

Europe is generating 1.3 billion tons of waste annually including 700 million tons of agricultural waste [43]. The future sustainable food production requires a significant waste reduction to meet increasing food demand and reduce the industry's carbon footprint toward global Net-Zero targets. Among various approaches, the circular economy (CE) [43−45] ensures that there is a reduction in the use of natural resources and waste production [44], which effectively reduces the wastes and by using their coproduct values in the systems to develop a model that has no net effect on the environment. A recent example of applying CE to the agriculture sector in AgroCycle [46] which performs an integral analysis of the agrifood value chain, including livestock and crop production, food processing, and retail sector (Fig. 10.2), providing mechanisms to achieve an increase in the recycling and valorization of agricultural waste by maximizing the use of by-products and co-products via the creation of new sustainable value chains. To achieve this, renewable energy

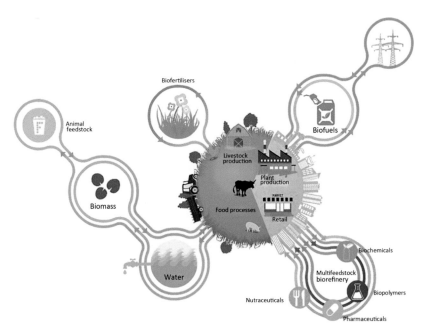

FIGURE 10.2   AgroCycle innovations in the agricultural production chain [46].

sources in particular solar energy, robots, distributed sensors, and accordingly big data analysis will play a crucial role to reduce the carbon footprint of the involved processes by reducing waste, improving the life cycle of the products, and efficiently recycling strategies and by-product use.

Running and maintaining a farm requires the use of heavy machinery and electricity. Modern agriculture is heavily dependent on the energy supply mainly from conventional energy sources, for example, fossil fuels. However, climate change and the finite resources of fossil fuels imply the use of sustainable energy sources in the agriculture sector. Agriculture is among the main users of fossil fuels in Western countries [47−50] and the price of agricultural commodities has a direct relation with the price of conventional fuels [51]. Agriculture contributes to around 14% of the world's GHG emissions [52] owing to conventional farming activities [53], direct use of fossil fuels in the agricultural process, and livestock raising and deforestation [54]. Since the largest proportion of electricity is still created by the burning of fossil fuels, electricity generated by PV panels is an answer to this challenge of sustainable agriculture.

In recent years, solar panels have become cheaper due to technological advances. The farmers can feed the additional electricity generated on-farm back to the grid and receive a payback. The energy return for energy invested (EROEI)[1] of fossil fuels used to be much larger than that of renewable technologies such as PV and wind and that was seen as a fundamental obstacle in the transition to renewable energy in any sector including agriculture. However, the situation is changing fast. The EROEI of PV technology is increasing, owing to technological progress, while limited sources of fossil fuels are left [55]. Today, PV technology has reached a high enough EROEI efficiency to be able to compete with fossil fuels in providing an abundant supply of green energy [56−58]. A comprehensive review of PV systems shows the suitability of PV for a variety of agricultural applications [59] such as water pumping and irrigation, greenhouses, dryers, livestock, and crop protection systems (Fig. 10.3).

In theoretical quantitative terms, renewable technologies can act as primary energy sources to produce large amounts of energy without negatively affecting agriculture (except for biofuels). However, to be able to use the full capacity of PV generated electricity two possible solutions should be considered: (1) changing the energy carrier converting electricity into liquid or gaseous fuels, for example, by producing hydrogen fuel and (2) electrification of the agriculture industry by powering the related applications (e.g., irrigation, tractors, and transport) by electric motors. The latter case has less complexity with more environmental advantages. Electrification of the

---

1. EROEI: the ratio of the amount of energy that will be produced by a plant over its lifetime and the amount of energy that needs to be invested to build the plant, operate it, and finally dismantle it.

PV-powered pumping system   PV-powered solar dryer   PV-powered dairy system

PV-powered greenhouse   PV-powered crop protection system

FIGURE 10.3   Some common agricultural operations powered by photovoltaics [59].

agriculture industry would also enable the exploitation of the high EROEI of modern renewables. The direct application of generated electricity on farms, for example, PV, is mainly related to mechanical power. Mechanical power is essential for almost all farm operations such as using machinery like tractors. A conventional tractor can be easily substituted by an electrical lighter one [60,61] which opens up the possibility of precision via robotic technology to cover a vast range of agricultural tasks. These then would use only sustainable energy and generate no emissions during operation by exploiting solar energy produced on the farm or in a field in the vicinity of the farm. The operation of electric motors would be perfectly feasible in farms particularly with the recent developments in the battery industry and future emergence of solid-state batteries where the lighter weight would be more compatible with the industry's needs.

10.3   Robotic technology for precision agriculture

Most robotic systems in agriculture workspaces require a mobile platform to cover the entire field for selective harvesting, crop care, monitoring, or phenotyping. The first commercial industrial robot (the Unimate) was showcased in 1961, demonstrating a range of its capabilities [62]. While much of Unimate's functionalities were preprogrammed, the advancement of sensors, communication, and computational capabilities have enabled recent robots to perform such dexterous movements autonomously responding to the changes in the environment. Although preprogrammed robotic arms are still used in many manufacturing industries, there has been an increased interest in the field deployment of robots with autonomous sensing and decision-making capabilities. Mobile robots are a major fraction of such autonomous robots, and they have been instrumental in expanding the workspace of even robotic arms, previously confined to individual confined cells, to anywhere in the environment by providing a mobile base for such arms. An autonomous mobile robot may be defined as a robotic platform with enough sensory, computational, and communication capabilities to enable autonomous navigation in an environment. On-board sensors of such a mobile robot enable monitoring the environmental changes whereas the computational resources enable responding to the relevant environmental changes and navigate in the environment. The communication capabilities allow sensory measurements/ information to be processed on a remote computing machine, and to receive control or coordination signals from other agents. Such autonomous mobile robots can be broadly classified as unmanned ground vehicles (UGVs), unmanned aerial vehicles (UAVs), and unmanned underwater vehicles, respectively operating in terrestrial, aerial, and underwater environments. This section discusses UGVs, used specifically in precision agricultural applications, and challenges associated with their field deployments such as localization and navigation planning. Due to the vast spatial coverage required in agricultural fields, when moving from the research phase to the deployment phase, a fleet of autonomous robots would be required. This section gives a glimpse of robotic interventions in performing agriculture tasks and the various challenges associated with the coordination of a fleet of autonomous robots when deployed in agriculture environments.

10.3.1 Unmanned aerial vehicles for agricultural tasks

UAVs have been used in agricultural research mostly for monitoring and spraying. This flying system can be used to detect the effect of water stress on the quality of fruit gardens using aerial photographs. For instance, the amount of carotene in the leaves of the grape garden was estimated using aerial photographs [63,64]. Fig. 10.4A shows images taken by a microspectral camera on a UAV with a resolution of 40 cm which shows grape trees. Fig. 10.4B is an image that separates trees from shadows and light reflections from the soil [64].

Mapping vegetation is an important step in remote sensing applications for precision farming. Conventional aerial platforms such as airplanes and satellites are not suitable for this purpose due to their kinematic-dynamic complexities. Therefore a UAV equipped with a commercial camera in the high-resolution visible spectrum can be used to capture images of several wheat fields during the first growing season. Using these images, researchers can make a good detection of vegetated areas. It should be noted that the use of UAVs to monitor tropical areas and forests is much easier and less

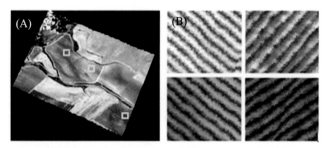

FIGURE 10.4 Images taken by a unmanned aerial vehicle: (A) the grape garden, and (B) shadows and reflections [64].

FIGURE 10.5 Fleets of unmanned aerial vehicles for harvesting apples [74].

expensive than ground surveillance and traditional methods, enabling monitoring and estimate the structure of the forest canopy [65,66].

Moreover, it is possible to inspect the relationship between olive tree canopy expansion and water stress using images taken by thermal cameras installed on UAVs and to improve stress management, irrigation management, and gardens performance by detecting stresses. On the other hand, using aerial imaging by UAVs, the amount of vegetation in an area can be estimated with 95% accuracy [67,68]. UAVs can also be used to inspect and maintain the almond gardens in collaboration with ground robots). On the other hand, field spraying by UAV systems has received a lot of attention in recent years) [69−73].

Apart from spraying tasks, fleets of UAVs are used for harvesting operations as well. For instance, Tevel Tech has demonstrated the use of a fleet of UAVs for picking apples in an orchard (Fig. 10.5) [74]. The picked apples are brought to the storage chamber on a ground-based mobile platform moving along the rows following the UAVs.

10.3.2   Unmanned ground vehicles for agricultural tasks

There is a wide variety of UGVs used in PA research due to the variations in the environments (open fields, poly tunnels, and greenhouses) and the specific applications being focused on. Agricultural environments vary widely

FIGURE 10.6 Different agricultural environments: (A) An open field, (B) a polytunnel with crop on the ground, (C) a polytunnel with table-top crops [75], and (D) a glasshouse with crop on the ground. *(A) by J Thomas, CC BY-SA 2.0; (B) by Colin Smith, CC BY-SA 2.0; (C) and (D) by Goldlocki CC BY-SA3.0.*

depending on the crop and the geographic location. While most of the farming environments are open fields, poly tunnels and glasshouses can prolong the crop yield over unfavourable seasons (Fig. 10.6). Much of these protected environments are used in farming fruits and vegetables.

The UGVs used in PA can be broadly classified into (1) autonomous mobile robots [76], (2) traditional in-field vehicles (such as tractors) embedded with robotic technologies [77], and (3) custom-built platforms for specific use cases (e.g., moving bench and stationary robot combination in greenhouses [78]). Examples of these classes are shown in Fig. 10.7. Being in the early stage of field deployments and are still being widely researched, mobile robotic platforms are more commonly used in protected environments for weather protection. While some purpose-built robotic platforms have been tested in both open fields and weather-protected environments [76,81−84], there are many instances of customizing off-the-shelf robotic platforms for use in indoor environments [85,86].

Conventional in-field vehicles, although they cause challenges such as soil compaction, are still widely used in open fields (orchards, vineyards) and polytunnels due to their easy availability, large capacity, and robustness [87−91]. However, more recently technology interchange has been taking place with many robotic technologies being used on conventional farm vehicles, making them autonomous [77]. Many custom-built platforms are also discussed in the literature, mainly used in glass houses, with the research focus mostly related to manipulation [83,86,92].

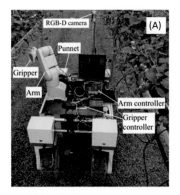
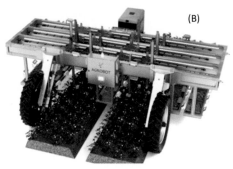

FIGURE 10.7   Unmanned ground vehicle robotic platforms: (A) A robotic platform for general agricultural work [79] and (B) a platform designed for a specific use case [80].

FIGURE 10.8   (A) Cäsar [94] and (B) Seeding robot [95].

The robots used for the land preparations have special devices for mowing, and plowing, which are attached to the mobile platform that navigates itself in the field of operations, for example, "Greenbot" [93] and "Cäsar" [94] (Fig. 10.8). Most of the Greenbot activities [93] can be performed autonomously by a teach and playback mode while for certain other activities, it plans activities without any human interventions. Moreover, Greenbot company has also developed an "X-pert robotic package" using centimeter precise RTK-GPS that can be retrofitted to existing tractors and mowers to enable autonomy in their operation. "Casar" makes use of a preloaded route plan of the target area, and for autonomous movements, it uses standard GPS and radio-transmitted correction signals. Both the robots are equipped with collision-avoiding sensory frameworks to enhance their usability in the field.

The placement of seeds follows land preparation. In precision seeding, the seeds are placed with a particular spacing and depth. The seed emergence and germination rates are very much dependent on the seeding depth [96]. Hence maintaining a controlled depth and spacing is a very crucial requirement to be met by the seeding robots. A modular seeding robot equipped with a seed selector and planter mechanism for planting a single seed into the plant bed is

FIGURE 10.9 Weeding robots: (A) Weeding robot for sugar beet field [98,102], (B) chemical weeding robot using a drop-on-demand methodology [100], and (C) robot capable of performing chemical and mechanical weeding [101,103].

presented in [97]. It uses low-cost ultrasonic sensors to guide the robot along the planting row. The robot shown in Fig. 10.8B is a battery-powered four-wheel-steered precision seeding machine that uses a motor and vacuum fan arrangement for performing the seeding [95]. Robot implementations are widely reported to perform crop caring tasks like weeding, pest control, disease monitoring, spraying, and irrigation. These are the very crucial tasks that need to be performed on regular basis to bring up the crop to the harvesting stage.

Weeding involves the removal of unwanted plants from the crop-grown area. Apart from the general sensory framework of sensors for driving the robots autonomously in a field, selective weeding robots use dedicated vision systems to detect the presence of weeds in between the crops. Upon detection, they either use some mechanical means (mechanical weeding) or by spraying chemicals (chemical weeding), laser, or their combinations to remove the weed. Fig. 10.9A shows an autonomous robot with a dual-camera system for weeding operation in sugar beet fields [98]. The forward-facing black and white camera with a near-infrared (IR) filter helps to identify the rows for guiding the robot motion, while the downward-facing camera captures the images of the plants grown in the row. This image feed helps to classify the crop from the weed and hence leading to the removal of weed by the mechanical tool. The autonomous field robot "Weedy" [99] performs inter and intrarow weed removal using chemicals directed to weeds through a set of spray nozzles. The robot uses a combination of camera, ultrasonic sensors, IR sensors, and encoders for navigation and weed detection. The "drop on demand" method of spraying the herbicide on weeds presented in [100] is the key technology in the chemical weeding robot shown in Fig. 10.9B. This system produces larger herbicide droplets and flushes on the identified weeds. This approach helps in preventing the formation of harmful aerosols which may occur while using conventional sprayers. The robot shown in Fig. 10.9C performs both mechanical and chemical weeding where large weeds are sprayed and the small weeds are stamped mechanically [101]. Weed detection and tracking are performed by a multicamera system with nonoverlapping fields of view.

Apart from these weeding robots, other reported robotic systems perform intrarow weeding without any vision system. Instead, the weeding mechanism is guided to traverse between the crops to remove the weeds, without interfering with the crops. This noninterfering traversing is made possible with different techniques. The autonomous tiller-based intrarow weeding robot reported by Reiser et al. [104] explored the scope of using a feeler and sonar sensor for detecting the trunk of the crop during the weeding action. Nørremark et al. [105] developed an autonomous tractor-based intrarow weeding system using a cycloidal hoe. As a prerequisite of deploying this system, it needs geo-referencing of each plant in its field of operation. And it uses RTK GPS for autonomous guidance during which the cycloidal hoe performs weeding without interfering with the crops.

Along with the timely removal of weeds, crops need regular visual inspections to detect the presence of diseases or any pest/insect attack. Un-identified crop diseases and delayed remedial actions can affect the entire crop productivity and result in decreased yield. Monitoring robots use vision systems to detect changes in crop features caused by any microbial or pest attacks. Some of these robots can also take remedial actions like selective spraying, etc., upon such detections. For instance, e-AGROBOT [106] which is an autonomous agriculture field robot designed for the real-time detection of plant diseases at the microlevel. After detecting any infections, it performs controlled spraying of pesticide to the infected area. Schor et al. [107] proposed a manipulator-based crop disease monitoring system. This system can detect powdery mildew and tomato spotted wilt virus usually found on bell pepper grown in greenhouses. It has an RGB camera and a laser sensor attached to the end effector of the manipulator for detection purposes. A push-cart using a dual camera-based vision technique is presented in [108] to detect the powdery mildew disease among strawberries. A remote-controlled disease monitoring robot is presented in [109] which can do early detection of Xylella fastidiosa in olive groves on the plant to the leaf level. As shown in Fig. 10.10B, this robot is provided with a liftable platform to allow scanning of the height of the olive trees. The camera configuration includes two DSLR cameras, a multispectral camera, a hyperspectral camera, and a thermal camera. Also, the

FIGURE 10.10 Disease monitoring robots: (A) a platform for detecting powdery mildew disease detection in strawberries [108], (B) a field robot for detecting Xylella Fastidiosa infection on olive trees [109], and (C) "RoboHortic"—a field robot to detect pests and diseases [110].

robot is provided with light detection and ranging sensor (Lidar), GPS, and inertial measurement units (IMUs) sensors. All the captured plant data has been geolocated using the GPS and IMUs combination and is presented on a map for visualization. A similar camera configuration except for an additional DSLR camera has been used on a teleoperated field robot for detecting pest and crop diseases (Fig. 10.10C) [110]. All these ground-looking cameras are placed on the mobile platform to capture the image of the crops in a row. To avoid the influence of sunlight, the scene is covered using a tarp and is provided with an additional illumination source. The captured crop data are geo-referenced in a map with the aid of a global navigation satellite system (GNSS) receiver with real-time kinematic (RTK) correction. Beyond detecting plant diseases, an autonomous robot for detecting the presence of pests in the crop has been proposed by Liu et al. [111]. The robot can detect the presence of the Pyralidae pest which can be a potential threat to the economic crops.

Followed by the identification of weeds, pests, or plant diseases, some remedial actions are required like spraying pesticides or herbicides, etc. Precision spraying is a key component to move toward sustainability in agricultural workspaces. Since precision spraying can reduce the usage of chemicals to a considerable extent without negotiating the crop requirements, most of the spraying robots are reported to perform the precision spraying technique.

Also, the reduced usage of chemicals will have a very less impact on the environment compared to uniform spraying methods. In [112], an autonomous weed control robot with a vision-controlled microdosing system has been presented. This webcam-based vision system with the dosing system has been successful in targeting the weeds in subcentimeter level accuracy at indoor test conditions. The modular robot (shown in Fig. 10.11A) uses a six-degree of freedom (DOF) manipulator with a precision spraying end-effector to target the disease-affected area of crop detected by the multispectral image analysis [113]. An operator-driven tractor with a vision-based target spray platform is shown in Fig. 10.11B [114]. A five-DOF polar manipulator with a spray nozzle and ultrasonic sensor mounted on a crawler traveling device has been studied to be capable of uniform and precise spraying of chemicals at the target [115]. Another autonomous crawler mobile robot has multiple nozzles that are

FIGURE 10.11   Spraying robots: (A) A modular agricultural robot for selective spraying for grapevine diseases [113], (B) a target spray robotic platform [114].

arranged on two vertical bars by which chemicals are sprayed on crops while it traverses through the crop rows [116].

Like the precision methodology followed by spraying robots, irrigation robots are also being reported to perform precision irrigation for the optimum utilization of water for meeting the requirements. Field irrigation requirements are usually defined based on the data measured by field sensors installed on crop environments or by real-time probing methods, etc. For instance, an intelligent robot with a sprinkler device performs precise irrigation in the cropping areas depending on the degree of drought [117]. During irrigation, the sprinkler device can expand horizontally and vertically to enable larger area irrigation. Moreover, robotic systems are also used to adjust the water emitters kept in crop environments forming a robot-assisted precision irrigation system [118].

Crop harvesting is considered to be a highly repetitive and labor-intensive task in the agricultural production cycle [119,120]. Nowadays, the demand for autonomous systems for harvesting is very high as it is expected that they can support the timely picking of crop production which is disrupted by lack of skilled laborers. The harvesting robots are usually developed to perform crop harvesting in a selective way or bulk [120]. The selective harvesting robots primarily include a manipulator mounted on a mobile robot with dedicated vision systems for fruit detection and localization and sensors for autonomous navigation. The manipulator will be fitted with a custom end effector to cater to the picking requirements of a crop. Since the robot end-effector must pick crops without damaging the quality, the end-effectors and picking actions are driven by sensory feedback. The end-effectors use different techniques to detach the fruit from the plant, such as a suction cup, rigid or soft fingers to hold the fruit and blades, twisting actions, or hot elements for detachment from the stem [121]. The vision system used in harvesting robots can be binocular vision sensors, laser vision sensors, Kinect cameras, multispectral cameras, or other visual sensors [122]. Because of the complexities involved in selective harvesting processes, robots are still far away from the efficiency level of human pickers. But continuous research is happening to improve the performance of such systems. Abundant robotics has reported the world's first commercial apple harvesting robot [123]. This robot uses a suction-based approach to pick apples and moves to the storage through the flexible tube from the suction head. Some harvesting robots are equipped with multiple manipulators to improve the picking rate. FF Robotics has developed an apple-picking robot that uses twelve telescopic arms fitted with three-jaw grippers to pick the apples [126]. The six arms are arranged on either side of the mobile platform to pick the fruits from both sides of a row. The telescopic arms have a two-dimensional sliding base to increase the reachability of the manipulator to cover the entire tree. After harvesting, the apples are gently placed on the conveyor that moves them to storage. Multiple arms picking systems have

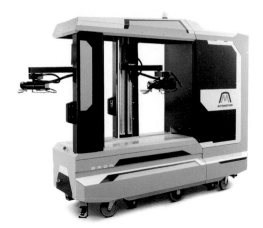

FIGURE 10.12 Selective harvesting robot (GRoW) for tomato harvesting [125].

been also developed for strawberry [80] and Kiwi picking [124]. Agrobot E series is the first precommercial strawberry harvesting three-wheeled robot. This robot has 24 independent arms for simultaneous picking of the strawberries (Fig. 10.7B) [80]. The autonomous kiwi picking robot is a four-wheel-drive UGV equipped with four independently actuated arms and V-shaped grippers to pick the fruit. Moreover, some robots use a bimanual approach, where both manipulators perform picking actions either in a collaborative or independently (Fig. 10.12) [75,125,127−129]. When the robot must perform harvesting on the various heights of plants, the mobile platform will be housed with a liftable platform on which the manipulator is mounted. The sweet pepper harvesting robot namely "SWEEPER," uses a scissor lift mechanism to make the manipulator reach the different heights of the plant [130]. Also, certain harvesting robots have vertical or slanting slides on which the manipulator moves to reach different heights [131,132]. As mentioned before, certain crops, such as rice and wheat, are harvested in bulk. Robots are also used for bulk harvesting the crops, e.g. retrofitted robotic combine harvester [133,134].

10.3.3 Autonomy in farm environments

Irrespective of the precision agricultural tasks assigned to the UGVs, the main research challenge associated with them is autonomous navigation in field conditions. The navigation planning of a UGV continuously tries to find answers to the three basic questions in a loop:

1. Where am I (the robot location)? (self-localization);
2. Where am I going (the target location)? (rask-allocation);
3. How can I reach the target (the route)? (path-planning).

FIGURE 10.13   Thorvald robot in a strawberry production polytunnel with too narrow rows.

Although much of these navigation-related challenges are addressed in indoor environments, agricultural fields involve many new challenges and the simultaneous existence of many of them together. Some of the most impactful challenges, occurring frequently in open-fields and polytunnel environments, include:

1. *Visual inconsistencies:* Constant plant growth and dynamic changes can make localization with perception alone much more difficult [135,136].
2. *Environmental constraints:* Crop rows that constrain the UGVs to use specific areas of the environment for navigation can result in long and sometimes narrow paths (Fig. 10.13) [137,138].
3. *Uneven ground:* Uneven and soft ground conditions can make navigation difficult with some weather conditions making wheel slippage far from negligible [78,138]. Among these, the first challenge is often resolved by using localization methods that rely less on the features from the environment, the second challenge is addressed by environment discretization and efficient path planning algorithms, and the third challenge is resolved by using different locomotion systems that drive and steer UGVs in these environments.

Self-localization: To navigate in an environment, a UGV needs to know where it is in the environment. Many PA operations require high localization capability. PA tasks have been classified into three bands based on the degree of accuracy: (1) low accuracy (meter level), (2) medium accuracy (submeter level), and (3) high accuracy (centimeter-level) [139]. Some environments such as polytunnels and some vineyards need highly accurate localization, while lower localization accuracy can be accommodated in orchards with much spacing available for safe navigation.

Path planning: Modern-day agricultural environments are mostly large and have crops grown in rows (Fig. 10.14). This environment structure constrains the navigation of UGVs to these tracks between the crop rows. Depending on the type of crop and spacing between the crop rows, UGVs

FIGURE 10.14 Crop rows constrain the area that can be navigated by unmanned ground vehicles. Examples of crop rows: (A) Apple orchard, (B) Strawberry field, and (C) Vineyard. *(A) by Oast House Archive, CC BY-SA 2.0; (B) by Rosalind Mitchell, CC BY-SA 2.0; (C) by Joachim Kohler, CC BY-SA 4.0.*

may navigate between or over the crop rows. Many guided autonomous navigation systems have been developed and are extensively used for tractor-like vehicles in environments such as orchards and vineyards [30]. Many times, these applications performed by a single vehicle are defined as a mission, where the UGV will be following a predefined route passing through a finite set of way-points marked by their positions in a given order [140]. However, when working with a mobile robot for PA, the granularity of applications is much smaller. A similar approach for mobile robot navigation uses a topological map representation of discrete way-points (nodes) in the environment and edges connecting them. Any path planning performed on this topological map will be in the discrete space of nodes and edges that have been commonly used in indoor robotic applications, with less computational complexity than traditional path planning in the continuous space. Classical routing algorithms for graph-based planning such as A* and Dijkstra [141] can then be used to plan a path on the topological map representation, combining nodes and edges to a route to the target location as a solution of the path planning problem. However, real-time crop row detection is necessary for finer navigation planning between the way-points to intelligently avoid any static or dynamic obstacles in the environment, as well as performing plant-level agricultural applications. On-board sensors such as LiDAR and RGBD (Red, Green, Blue, Depth) sensors that are used for localization are also commonly used in such crop-row detection and obstacle avoidance approaches [133,142]. Similarly, in open arable fields, guidance lines are the area used repeatedly as tractor tracks There are many reasons for these repeated usages of tracks in open fields:

1. To maximize the area where crops are planted;
2. To optimize field coverage by these vehicles and their attachments (e.g., sprayer boom); and
3. To minimize soil compaction only to these tracks.

Many times, these tracks are selected for optimizing various objectives such as maximum coverage of the field and the minimum number of turns needed for the vehicle while avoiding obstacle areas [143]. However, each use case may differ depending on the objective being optimized and the constraints such as the shape and size of the fields and characteristics of the farm vehicles being used (e.g., capacity of the farm vehicle). Due to the large solution space and computational complexities involved in solving such complex optimization problems, often the use of an evolutionary algorithm [143] or a swarm optimization algorithm [144] is preferred.

Driving and steering system: The locomotion system (driving and steering system) of a UGV is responsible for interacting with the ground surface and navigating the robot in the desired direction and speed. This interaction with the ground terrain is an important aspect affecting the navigation of a UGV. The efficiency of a UGV driving and steering system varies depending on the terrain of the environment. There is a general push toward using smaller UGVs instead of large all-in-one agricultural vehicles in open fields. The major motivations for this are related to sustainability and reducing the impact on nature. For example, a smaller agricultural robot causes less soil compaction which in turn can affect the water and nutrient absorption capability of soil. However, some fields (e.g., vineyards and orchards) may be on sloped terrains. This may need a heavier UGV, and hence a bigger wheel or a tracked system. But such heavier UGVs can negatively affect the soil structure. So many agricultural robot manufacturers support modular driving and steering systems, which can be selected depending on the characteristics of the environment and the applications [76,83]. As most agricultural environments are mostly flat surfaces (sometimes with small slop), wheeled locomotion systems are more commonly used (e.g., Thorvald, see Fig. 10.13). Tracked locomotion systems are also used in UGVs to address the varying softness of the soil and uneven surfaces [145,146]. Although bioinspired legged robotic systems are maturing (e.g., AgAnt [147]), they are not yet commonly used in PA.

Among the UGVs with wheeled locomotion, different steering systems are used. The most commonly used one is a four-wheel drive and steer system or a two-wheel drive and steer with two support wheels [76,148−151]. They allow very precise steering required for many PA applications. Although it is uncommon, some UGVs have more than four wheels [146]. In these, it is common for the outermost wheels to be in control of driving and steering, while the other wheels are nondriven support wheels to distribute the weight of the vehicle. Skid-steer is also commonly used in agricultural mobile robots [146]. In this, instead of steering individual wheels, the wheels are divided into two groups with one side moving forward and the other backward offering a turning motion. While effective in minimizing translation during turning, they can cause more damage to the structure of the topsoil during steering motion, compared with a four-wheel drive and steer system. Most traditional agricultural vehicles and their autonomous counterparts still rely on the Ackermann steer system, most commonly seen in road vehicles, where to turn smoothly, the inner wheel turns on a smaller radius than the outer [152]. Many existing sensors and actuators (e.g., soil sensing and autonomous fertilizer dispenser systems) can be connected to a UGV and pulled like a trailer such as Veris trailers [153], designed to measure soil variability by analyzing pH levels, water content, and some able to take deep soil cores. UGVs with such trailers often have adaptations to their navigation control when used [154].

10.4 Solar-powered robots for agriculture

10.4.1 Photovoltaic technology

PV directly converts sunlight into electricity. When the semiconductors of PV cells are exposed to sunlight create an electric field across which is used to generate electricity [155]. Despite its efficiency, PV installations provide only 0.1% of the world's total electricity generation. Nonetheless, PV installations growth of 40% on average annually has made it a fast-growing industry where it can deliver 1081 GW by 2030. That will make PV-generated electricity the primary global energy source [156]. As shown in Fig. 10.15, installation of renewable power generation sources globally had grown approximately 8% over 5 years before 2019, with solar PV technology significantly leading over the other technologies [157].

One important factor in the continued growth of the PV industry in the fast-changing energy market is the *efficiency of solar cells*. The recent

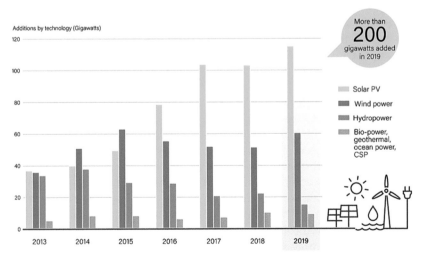

FIGURE 10.15 Annual additions of renewable power capacity by technology, from 2013 to 2019 [157].

studies have resulted in significant improvements in the efficiency of solar cells and the development of new materials, the significant enhancement in the efficiency, and the development of the new materials. It can be noticed that multijunction cells have the highest efficiency among all solar cell materials with 47.1% efficiency achieved in 2020. The emerging materials have still lower efficiency with 24.2% efficiency for perovskite in 2020. The monocrystalline solar cell has 27.6% efficiency, multicrystalline cells with 23.3%, and thin-film technology with 29.1% in 2020.

10.4.2 Life cycle of photovoltaic materials and waste management

For solar PV electricity generation, various manufacturing processes are needed to produce PV modules which are consist of different materials. From a manufacturing viewpoint, various materials such as silicon, copper indium selenide (CIS), cadmium telluride (CdTe), etc. are used to fabricate solar cells. To get the raw materials for solar cell production, mining operation are required which face serious environmental and labor safety issues. In addition, mining machinery heavily relies on carbon-based fuels consumption with obvious environmental impacts [158]. It is reported that dye-sensitized PV cells emit a high volume of sulfur oxide (SO_x), nitrogen oxide (NO_2), and CO_2 compared to other PV materials. Among all, SO_x and NO_2 are responsible for acid rains which harm wildlife, corrode and deteriorate many other materials, while CO_2 is one of the contributors to global warming [159] as shown in Table 10.1. The thin-film technology is expected to be the future of PV technology with a lifetime between 25 and 35 years. However, their end-of-life (EOL) causes serious environmental concerns if not disposed of properly [161]. Increased PV installations, EOL may produce hazardous waste. This will require a strategy for recycling and recovery of the PV waste materials [162]. The worldwide ratio of solar PV waste to new installations is expected to increase considerably over time as shown in Fig. 10.16.

Solar PV development is experiencing *complex challenges* despite the fast development to reduce global reliance on conventional carbon-based energy sources. Some of these challenges include land use conflict, social resistance as the world is moving toward large-scale solar farms (similar challenges were associated with wind turbines) large-scale development, and the negative impact of environmental factors on PV material degradation [163−165]. Growth in large-scale PV farms and competition between land for agriculture versus energy production can create land use disputes. This history and growing concern over land use highlight the challenge of meeting the soaring demands for solar power while protecting rural and agricultural lands [166]. One of the challenges in maintaining the efficiency of the whole PV system is the material degradation as a result of exposure to harsh environmental conditions [167,168] which leads to system degradation and efficiency drop.

TABLE 10.1 GHG emission for fabrication of solar cells using different materials [155,160].

Solar cell material	Amorphous	Monocrystalline silicon	Polycrystalline silicon	Dye-sensitized	CIS	CdTe
GHG emission (g CO_2/kWhe)	15.6–50	44–2820	9.4–72.4	19–47	16.5	15.6

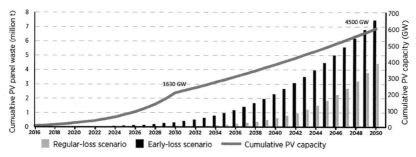

FIGURE 10.16   The historic and forecast data for worldwide solar photovoltaic waste [162].

The main environmental factors affecting a PV system are (1) tempera-ture, (2) dust accumulated on solar cells, (3) humidity, and (4) ultraviolet (UV). It is widely accepted that the efficiency of solar cells decreases with an increase of temperature, and cooling is necessary at high illumination con-ditions such as concentrated sun-light, or cosmic or tropical conditions. [169]. Dust also results in the degenerated performance of PV systems. The power output drops drastically as dust density increase and airborne dust reduce the short circuit current. Where the efficiency drops linearly with the dust deposition density. The components of the gaseous atmosphere, such as water vapor, cause hydrolysis resulting in a reduced lifetime of a PV module [170]. Polymeric materials in a PV module (e.g., encapsulant and back sheet) undergo photothermal degradation since they are exposed to UV. This results in discoloration which develops light yellow to dark brown in time depend-ing on module configuration and operating temperature. The discoloration reduces the light transmission of the encapsulant and reduces the efficiency of the whole system [171].

Off-grid renewable energy systems, batteries, and information and commu-nications technology (ICT) can transform the agricultural practice to meet the rising need for food [172]. The transformation needs a cut production cost, for example, the cost of PV dropped by 88% between 2009 and 2018 in the United States [173] and increased storage *capacity of batteries*. A drastic drop in the PV module price over time has occurred from 2010 to 2020 [174]. The reasons for the drop in PV module price are mainly due to (1) identification of novel materials, (2) increased efficiency, (3) increased lifetime of PV cells, and (4) improvement in government policies [175]. Lithium-ion batteries have a smaller ratio of the size/weight to power storage capacity compared to other technologies. They discharge most of their energy without compromising their life, have low losses while charging and discharging, and have a longer life-span. The cost of lithium-ion battery (LIB) storage has decreased by 87% in the recent 10 years [176]. PV technology and lithium batteries at affordable prices create an attractive alternative to fossil fuel-based generators. This is highly important as most agricultural lands may not be connected to the regional electricity grid. A low-cost energy source lowers production costs in the agriculture sector and enables precision even in remote agricultural workspaces [172]. In the case of LIBs, liquid electrolytes limit battery voltages to approximately 4.3 V, were above this limit, liquid electrolytes start to dissociate. Solid-state batteries (SSEs) represent an opportunity to improve the gravimetric and volumetric energy density of LIBs by enabling the use of Li metal [177]. SSEs may make another revolution in the battery industry. Ag-robots increase food production efficiency and reduce the usage of pesticides/herbicides through PA. Self-charging robots either with on-board PV modules or via a PV charging point, called off-board PV charging are considered to have a high impact on PA in near future.

10.4.3 Solar-powered unmanned ground vehicles and unmanned aerial vehicles in agriculture

PV on robots (onboard-PV robots) are deployed with a fixed solar panel attached to the frame of the robot. These types of robots are limited for outdoor use to exploit solar energy. Onboard PV robots are used for less power-hungry crop-care operations such as weed detection, crop health monitoring, and spot spraying. For instance, ecoRobotix has developed multiple platforms for autonomous farming, including both PV-onboard robots [178,179], and trailer attachments for tractors of other autonomous towing vehicles [180]. The solar-powered weeding robotic platform (Fig. 10.17A) is a four-wheel-drive autonomous robot utilizing a spray bar with 52 nozzles to give centimeter precision spraying [178]. In optimal conditions, it can perform weeding over 10 ha/day, using less than 10% of herbicide compared to

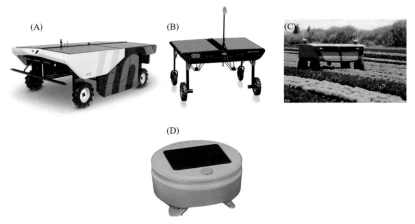

FIGURE 10.17 Examples of solar-powered agricultural unmanned ground vehicle platforms for weeding tasks: (A) AVO [178], (B) sprayer robot [179], (C) RIPPA [181], and (D) Tertill [183]. *(A) Courtesy ecoRoboticx; (B) courtesy ecoRoboticx; (C) courtesy ACFR; (D) courtesy Tertill.*

conventional methods. The ecoRobotix has another autonomous weeding platform (Fig. 10.17B), which can cover 3 ha/day, with two mounted PV modules [179]. The robot uses a parallel manipulator to target weeds for both inter-row and intrarow spot spraying with the capability to use two different products stored in its two tanks. The Australian Centre for Field Robotics has developed a spot spraying robot Robot for Intelligent Perception and Precision Application (RIPPA) (Fig. 10.17C) which can perform 24/7 due to its light design and two large PV modules [181].

The weeding robot developed by the National Academy of Agricultural Science, Suwon, South Korea [182] works utilizing a hybrid approach with the use of a hydrogen fuel cell as a primary supply, with onboard-PV as an eco-friendly power assistant for weed management in rice paddies. As the use of spot-spraying herbicide is unviable in paddy fields due to the flooding, this robot uses three robotic arms which pull physical weeding mechanisms. The Tertill [183] robot (Fig. 10.17D) is the smallest onboard-PV weeding robot. With its careful energy use and long-term autonomy, it tackles weeds through the combined use of a spinning string trimmer for seedling weeds and wearing down paths within the beds.

FIGURE 10.18 Examples of solar-powered agricultural unmanned ground vehicle platforms for monitoring and phenotyping tasks: (A) Phenomobile scouting robot [184], (B) Ladybug [181], (C) Vinescout [185]. *(A) Courtesy ecoRoboticx; (B) courtesy ACFR; (C) courtesy Vinescout.*

EcoRobotix has developed Phenomobile [184] (Fig. 10.18A) for monitoring and phenotyping tasks. Without a parallel manipulator and herbicide tanks, this robot is more lightweight and able to operate for longer durations than its predecessor. ACFR has also developed a monitoring robot, named Ladybug [181] (Fig. 10.18B) which is equipped with multiple sensors such as hyperspectral, thermal infrared, panoramic vision, stereo vision with strobe, and LIDAR, to detect, measure, and assess vegetable crops. The solar panels allow for continuous runtime, while the rechargeable batteries allow up to 9 hours of runtime. The Vinescout [185] (Fig. 10.18C) takes on an interrow monitoring task, in which it performs data gathering on environmental conditions in vineyards. The inclusion of a large battery bank and bright lights on the front enables its operation at night. The Vitirover [186] is designed to monitor vineyards whilst simultaneously mowing the grass between rows.

Ag-robots charged in solar charging stations are also considered to deal with the shortage of energy generated by onboard-PV. Solar-powered charging stations are usually placed across the field to store energy and charge batteries of ag-robots [182]. The AgBotII [151] is an example of a system designed with a solar-PV docking station, developed by the Queensland University of Technology. The AgbotII is an autonomous weeding platform offering both mechanical and herbicide weeding methods. The robot holds two batteries able to store a collective of up to 11 hours of power, with the robot docking in a solar-powered charging station when not in use. The charging station can collect energy throughout the day without any expenditure, transferring this energy directly to the robot when docked, making the system grid-independent.

Small trial fields common in rural areas also benefit from *fixed-position solar-powered Ag-robots*. Autonomous farming in these small fields can still be achieved with low-cost fixed-position Ag-robots. FarmBot is an open-source fully autonomous CNC robot where a universal tool mount is moved across a fixed-size planter in the same manner to a three-dimensional printer [187] (Fig. 10.19A), which is viable for medium-sized plots. The FarmBot systems, also offer a wide range of functionality to cultivate the full lifetime

FIGURE 10.19 Examples of FarmBot fixed-position solar-powered platforms: (A) FarmBot Genesis Max [188], and (B) FarmBot Genesis V1.6 with solar-charging [187]. *Courtesy FarmBot.*

of the plants, from soil drilling, seeding, and fertilization, to watering and weeding (Fig. 10.19B).

The ability to dock to a docking port, to charge autonomously is a key feature of *Ag-Robots with the capability to use solar-charging stations.* Autonomous docking and charging are well-researched areas in indoor mobile robotic applications, and those principles can be also utilized for autonomous recharging of Ag-robots at PV charging stations. The potential use of PV-powered charging stations include autonomous selective harvesting, monitoring, weeding robots (such as the Di-Wheel Concept [189], BoniRob [101], or the Swambot Series [190]) and the Hoeing robots from the Naïo Technologies Series [191−193].

Like solar-powered UGVs, UAVs are also reported to use solar energy as their power source. A large wingspan UAV was designed (Centurion) to operate continuously for weeks or months achieving scientific sampling and imaging missions. Centurion's lithium battery provides enough energy for a 2−5 hours flight after sunset. An onboard PV UAV (Helios) was tested later capable of storing energy for night-time flights [194]. The DLR Institute of Flight Systems developed a new solar-powered UAV (Solitair) whose onboard PV provides the energy needs for continuous operations for above 10 hours [195]. Onboard PV UAV (Solong) could flight 24 hours and 11 minutes using only solar energy coming from its solar panels. The British solar-powered UAV (Zephyr) achieved a maximum duration of 6 hours and reaching an altitude of 7925 m [196]. This solar-powered UAV can carry 100 kg for the missions such as urban mapping, oil spill detection, coastal monitoring, and many others. In 2007, DARPA announced the launch of a new solar HALE project to carry a single 453 kg payload [197−199]. While onboard PV UAVs are mostly used in nonagriculture sectors, they can potentially be useful in this sector in near future. Nonetheless, off-board PV-powered UAVs are the immediate advancement of agriculture UAVs.

10.5 Conclusions

Population growth, healthy diet requirements, and changes in food demand toward a more plant-based protein diet increase existing pressures for food production and land-use change. Soil and biodiversity are at risk of losing their health by the growing demands for agriculture products and current agriculture approaches. These will affect the future ecosystem and food production. One of the solutions to the increasing pressure on agriculture is PA that can contribute to maximum yield and minimum use of resources, including land, water, energy, herbicides, and pesticides. The development of PA requires a multidisciplinary approach involving engineering, AI, and robotics. Robots will play a key role in delivering PA and pave the way toward sustainable healthy food production.

Although the PA is a solution to future agricultural challenges, one of its fundamental elements is the energy supply to the distributed intelligent electrical devices. For instance, we need the devices to collect various data and the intelligent agriculture machinery, which need clean energy to ensure sustainability in energy consumption. Renewable energies (such as solar energy), which are widely accessible, also facilitate sustainable growth of the sector. Among renewable energy sources, solar energy and solar PV have shown a great potential to dominate the future of sustainable energy and agriculture developments. For developing PV in rural and off-grid agriculture farms and lands, the use of solar-powered devices is inevitable. Such transition to PV agriculture requires significant changes to the agricultural practices and adoption of smart technologies like IoT, robotics, and WSN.

This chapter presents a view on the role of solar energy in future PA. We discuss PA, the relevant technologies which show some gaps for a viable PA solution, and the extent to which the PVs are used in the current agriculture practice. Among the essential components of the PA (i.e., sensor networks, UGV, autonomous navigation, UAV, AI for fruit perception, and manipulation for crop care and selective harvesting), selective harvesting needs more development for a future PA approach. Beyond technology development, business model development is also critical to make the technology solution commercially viable—either to provide a business model as a service or as pieces of equipment.

Future food production needs to adapt to changing consumer behavior along with the rapidly deteriorating environmental factors. Along with these adaptations, the new technologies should be using green energy sources (i.e., solar energy) for meeting the power requirements for sustainable developments of these smart technologies. Since there is a rapid inflow of robotic technologies into the agriculture sector, increasing power demand is inevitable especially in remote areas where PV-based systems can play a game-changing role. It is expected for the agriculture sector to witness a technological revolution toward sustainable food production which cannot be achieved without solar PV development and support. PV-powered machines employed in PA can minimize the use of diminishing resources and GHG emissions through renewable energy sources.

Acknowledgment

The authors are thankful to their respective universities for the administrative and financial support to research in the domain of sustainable technologies for improving agriculture productivity. This work was partially supported by Centre for Doctoral Training, United Kingdom (CDT) in Agri-Food Robotics (AgriFoRwArdS) Grant reference: EP/S023917/1; Lincoln Agri-Robotics (LAR) Project which is funded by Research England's Expanding Excellence in England Fund (Research England Project

Reference Number 26.18), United Kingdom; and Research England CERES agritech project "Robofruit."

References

[1] Lenton TM, Rockström J, Gaffney O, Rahmstorf S, Richardson K, Steffen W, et al. Climate tipping points—too risky to bet against; 2019.

[2] Parry I. Summary for policymakers. World transit. 3, Routledge; 2014, p. 23−32. <https://doi.org/10.4324/9781315071961-11>.

[3] Crippa M, Solazzo E, Guizzardi D, Monforti-Ferrario F, Tubiello FN, Leip A. Food systems are responsible for a third of global anthropogenic GHG emissions. Nat Food 2021;1−12.

[4] Achieving NET ZERO Farming's 2040 goal. <https://www.nfuonline.com/nfu-online/business/regulation/achieving-net-zero-farmings-2040-goal/> [accessed 16.06.21].

[5] Pihlainen S, Zandersen M, Hyytiäinen K, Andersen HE, Bartosova A, Gustafsson B, et al. Impacts of changing society and climate on nutrient loading to the Baltic Sea. Sci Total Environ 2020;731:138935.

[6] Almond REA, Grooten M, Peterson T. Living planet report 2020-bending the curve of biodiversity loss. World Wildlife Fund; 2020.

[7] Hayhow DB, Eaton MA, Stanbury AJ, Burns F, Kirby WB, Bailey N, et al., Noble PG and SN. State of Nature 2019.

[8] Food, of the United Nations AO. Transforming food and agriculture to achieve the SDGs: 20 interconnected actions to guide decision-makers; 2018.

[9] Pretty J, Bharucha ZP. Sustainable intensification in agricultural systems. Ann Bot 2014;114:1571−96.

[10] Ojha T, Misra S, Raghuwanshi NS. Wireless sensor networks for agriculture: The state-of-the-art in practice and future challenges. Comput Electron Agric 2015;118:66−84.

[11] Pierce FJ, Nowak P. Aspects of precision agriculture. Adv Agron 1999;67:1−85.

[12] Milder JC, Garbach K, DeClerck FAJ, Driscoll L, Montenegro M. An assessment of the multi-functionality of agroecological intensification. Gates Open Res 2019;3.

[13] Mulla D, Khosla R. Historical evolution and recent advances in precision farming. Soil-Specific Farming Precis Agric 2016;1−35.

[14] Bongiovanni R, Lowenberg-DeBoer J. Precision agriculture and sustainability. Precis Agric 2004;5:359−87.

[15] Al-Gaadi KA, Hassaballa AA, Tola E, Kayad AG, Madugundu R, Alblewi B, et al. Prediction of potato crop yield using precision agriculture techniques. PLoS One 2016;11: e0162219.

[16] Teillet PM, Gauthier RP, Chichagov A, Fedosejevs G. Towards integrated earth sensing: Advanced technologies for in situ sensing in the context of earth observation. Can J Remote Sens 2002;28:713−18.

[17] Adamchuk VI, Hummel JW, Morgan MT, Upadhyaya SK. On-the-go soil sensors for precision agriculture. Comput Electron Agric 2004;44:71−91.

[18] Viscarra Rossel R, McBratney A, Minasny B. Proximal soil sensing; 2010.

[19] Cooper HM, Bennett E, Blake J, Blyth E, Boorman D, Cooper E, et al. COSMOS-UK: National soil moisture and hydrometeorology data for empowering UK environmental science. Earth Syst Sci Data Discuss 2020;1−36.

[20] Morrison R, Callaghan N, Cooper H, Coyle M, Cumming A, Evans C, et al. UK-SCAPE flux tower network: monitoring terrestrial greenhouse gas, water and energy balance; 2019.

[21] Sui R, et al. Irrigation scheduling using soil moisture sensors. J Agric Sci 2017;10.

[22] Usowicz B, Lipiec J. Spatial variability of soil properties and cereal yield in a cultivated field on sandy soil. Soil Tillage Res 2017;174:241−50.

[23] Mzuku M, Khosla R, Reich R, Inman D, Smith F, MacDonald L. Spatial variability of measured soil properties across site-specific management zones. Soil Sci Soc Am J 2005;69:1572−9.

[24] Hedley C. The role of precision agriculture for improved nutrient management on farms. J Sci Food Agric 2015;95:12−19.

[25] Bönecke E, Meyer S, Vogel S, Schröter I, Gebbers R, Kling C, et al. Guidelines for precise lime management based on high-resolution soil pH, texture and SOM maps generated from proximal soil sensing data. Precis Agric 2021;22:493−523.

[26] Fentanes JP, Gould I, Duckett T, Pearson S, Cielniak G. 3-d soil compaction mapping through kriging-based exploration with a mobile robot. IEEE Robot Autom Lett 2018;3:3066−72.

[27] Lamb DW, Brown RB. PA precision agriculture: remote-sensing and mapping of weeds in crops. J Agric Eng Res 2001;78:117−25.

[28] Sishodia RP, Ray RL, Singh SK. Applications of remote sensing in precision agriculture: A review. Remote Sens 2020;12:3136.

[29] Donohue RJ, Lawes RA, Mata G, Gobbett D, Ouzman J. Towards a national, remote-sensing-based model for predicting field-scale crop yield. F Crop Res 2018;227:79−90.

[30] Gorjian S, Minaei S, MalehMirchegini L, Trommsdorff M, Shamshiri R.R. Applications of solar PV systems in agricultural automation and robotics. In: Gorjian S., Shukla A., eds. Photovoltaic solar energy conversion. First, London: Elsevier; p. 191−235, 2020. <https://doi.org/10.1016/B978-0-12-819610-6.00007-7>.

[31] Gutiérrez J, Villa-Medina JF, Nieto-Garibay A, Porta-Gándara MÁ. Automated irrigation system using a wireless sensor network and GPRS module. IEEE Trans Instrum Meas 2013;63:166−76.

[32] Zhang X, Du J, Fan C, Liu D, Fang J, Wang L. A wireless sensor monitoring node based on automatic tracking solar-powered panel for paddy field environment. IEEE Internet Things J 2017;4:1304−11. Available from: https://doi.org/10.1109/JIOT.2017.2706418.

[33] Nguyen T-D, Thanh TT, Nguyen L-L, Huynh H-T. On the design of energy efficient environment monitoring station and data collection network based on ubiquitous wireless sensor networks. In: Proceedings of the IEEE RIVF international conference on computing communication technology—research innovation and vision future, IEEE; 2015, p. 163−8. <https://doi.org/10.1109/RIVF.2015.7049893>.

[34] World population projected to reach 9.7 billion by 2050. < https://www.un.org/en/development/desa/news/population/2015-report.html> 2015. [accessed 16.07.21].

[35] Vinet L, Zhedanov A. A "missing" family of classical orthogonal polynomials. J Phys A Math Theor 2011;44:85201.

[36] Badiee A, Ashcroft IA, Wildman RD. The thermo-mechanical degradation of ethylene vinyl acetate used as a solar panel adhesive and encapsulant. Int J Adhes Adhes 2016;68:212−18.

[37] Chen J, Liu Y, Wang L. Research on coupling coordination development for photovoltaic agriculture system in China. Sustain 2019;11. Available from: https://doi.org/10.3390/su11041065.

[38] Moretti S, Marucci A. A photovoltaic greenhouse with variable shading for the optimization of agricultural and energy production. Energies 2019;12. Available from: https://doi.org/10.3390/en12132589.

[39] Pardo MÁ, Manzano J, Valdes-Abellan J, Cobacho R. Standalone direct pumping photo-voltaic system or energy storage in batteries for supplying irrigation networks. Cost analysis. Sci Total Environ 2019;673:821−30.

[40] Patel B, Gami B, Baria V, Patel A, Patel P. Co-generation of solar electricity and agriculture produce by photovoltaic and photosynthesis dual model by Abellon, India. J Sol Energy Eng 2019;141.

[41] Ravi S, Macknick J, Lobell D, Field C, Ganesan K, Jain R, et al. Colocation opportunities for large solar infrastructures and agriculture in drylands. Appl Energy 2016;165:383−92.

[42] Fraunhofer ISE. Harvesting the sun for power and produce agrophotovoltaics increases the land use efficiency by over 60 percent; 2017.

[43] Muscio A, Sisto R. Are agri-food systems really switching to a circular economy model? Implications for European research and innovation policy. Sustainability 2020;12:5554.

[44] Murray A, Skene K, Haynes K. The circular economy: an interdisciplinary exploration of the concept and application in a global context. J Bus Ethics 2017;140:369−80.

[45] Xia X, Ruan J. Analyzing barriers for developing a sustainable circular economy in agriculture in china using grey-DEMATEL approach. Sustainability 2020;12:6358.

[46] Toop TA, Ward S, Oldfield T, Hull M, Kirby ME, Theodorou MK. AgroCycle − developing a circular economy in agriculture. Energy Procedia 2017;123:76−80. Available from: https://doi.org/10.1016/j.egypro.2017.07.269.

[47] Bardi U, El Asmar T, Lavacchi A. Turning electricity into food: the role of renewable energy in the future of agriculture. J Clean Prod 2013;53:224−31.

[48] Pereira AG, Functowicz S. Science for policy: new challenges, new opportunities. OUP Cat; 2009.

[49] Giampietro M. Energy use in agriculture. Encycl Life Sci 2003.

[50] Woods J, Williams A, Hughes JK, Black M, Murphy R. Energy and the food system. Philos Trans R Soc B Biol Sci 2010;365:2991−3006.

[51] Hendrickson J. Energy use in the US food system: a summary of existing research and analysis. Cent Integr Agric Syst Univ Madison Madison; 2004.

[52] Pimentel D, Food GM. Land, population and the U.S. economy; 1994.

[53] Alghalith M. The interaction between food prices and oil prices. Elsevier; 2010.

[54] Baker JS, Murray BC, McCarl BA, Rose SK, Schneck J. Greenhouse gas emissions and nitrogen use in US agriculture: historic trends, future projections, and biofuel policy impacts. Nicholas Instituted Environ Policy Solut Report, NI; p. 11, 2011.

[55] Murphy DJ, Hall CAS. Adjusting the economy to the new energy realities of the second half of the age of oil. Ecol Model 2011;223:67−71.

[56] Kubiszewski I, Cleveland CJ, Endres PK. Meta-analysis of net energy return for wind power systems. Renew Energy 2010;35:218−25.

[57] Muneer T, Asif M, Kubie J. Generation and transmission prospects for solar electricity: UK and global markets. Energy Convers Manag 2003;44:35−52.

[58] Fthenakis V, Kim HC, Held M, Raugei M, Krones J. Update of PV energy payback times and life-cycle greenhouse gas emissions. In: Proceedings of the twenty-fourth European photovoltaic solar energy confeence exhibition; pp. 21−25; 2009.

[59] Gorjian S, Singh R, Shukla A, Mazhar AR. On-farm applications of solar PV systems. In: Gorjian S, Shukla ABT-PSEC, editors. Photovoltaic solar energy conversion. Elsevier; 2020, p. 147−90. Available from: https://doi.org/10.1016/B978-0-12-819610-6.00006-5.

[60] Faircloth WH, Rowland DL, Lamb MC, Davis JP. Evaluation of peanut cultivars for suitability in biodiesel production system. Proc Am Peanut Res Educ Soc, 39, 2007.

[61] Mousazadeh H, Keyhani A, Mobli H, Bardi U, El Asmar T. Sustainability in agricultural mechanization: assessment of a combined photovoltaic and electric multipurpose system for farmers. Sustainability 2009;1:1042−68.

[62] Correll N. Introduction to autonomous robots, vol. 53. Magellan Scientific; 2016.

[63] Suárez L, Zarco-Tejada PJ, González-Dugo V, Berni JAJ, Sagardoy R, Morales F, et al. Detecting water stress effects on fruit quality in orchards with time-series PRI airborne imagery. Remote Sens Environ 2010;114:286−98.

[64] Zarco-Tejada PJ, Guillén-Climent ML, Hernández-Clemente R, Catalina A, González PM. Estimating leaf carotenoid content in vineyards using high resolution hyperspectral imagery acquired from an unmanned aerial vehicle (UAV). Agric Meteorol, 2013;171:281−294.

[65] Torres-Sánchez J, Pena JM, de Castro AI, López-Granados F. Multi-temporal mapping of the vegetation fraction in early-season wheat fields using images from UAV. Comput Electron Agric 2014;103:104−13.

[66] Zahawi RA, Dandois JP, Holl KD, Nadwodny D, Reid JL, Ellis EC. Using lightweight unmanned aerial vehicles to monitor tropical forest recovery. Biol Conserv 2015;186:287−95.

[67] Berni JAJ, Zarco-Tejada PJ, Sepulcre-Cantó G, Fereres E, Villalobos F. Mapping canopy conductance and CWSI in olive orchards using high resolution thermal remote sensing imagery. Remote Sens Environ 2009;113:2380−8.

[68] Pan Y, Zhang J, Shen K. Crop area estimation from UAV transect and MSR image data using spatial sampling method: a simulation experiment. Proc Environ Sci 2011;7:110−15.

[69] Xue X, Lan Y, Sun Z, Chang C, Hoffmann WC. Develop an unmanned aerial vehicle based automatic aerial spraying system. Comput Electron Agric 2016;128:58−66.

[70] Huang Y, Hoffmann WC, Lan Y, Wu W, Fritz BK. Development of a spray system for an unmanned aerial vehicle platform. Appl Eng Agric 2009;25:803−9.

[71] Giles D, Billing R. Deployment and performance of a UAV for crop spraying. Chem Eng Trans 2015;44:307−12.

[72] Ru Y, Zhou H, Fan Q, Wu X. Design and investigation of ultra-low volume centrifugal spraying system on aerial plant protection. Louisville, Kentucky; August 7−10, p. 1, 2011.

[73] Wang Z, Lan Y, Clint HW, Wang Y, Zheng Y. Low altitude and multiple helicopter formation in precision aerial agriculture. Kansas City, Missouri; July 21-July 24, p. 1, 2013.

[74] Tevel-tech. < https://www.tevel-tech.com/ >; 2021 [accessed 16.06.21].

[75] Le TD, Ponnambalam VR, Gjevestad JGO, From PJ. A low-cost and efficient autonomous row-following robot for food production in polytunnels. J F Robot 2020;37:309−21.

[76] Grimstad L, From PJ. Thorvald II—A modular and re-configurable agricultural robot. IFAC-PapersOnLine 2017;50:4588−93. Available from: https://doi.org/10.1016/j.ifacol.2017.08.1005.

[77] Bergerman M, Maeta SM, Zhang J, Freitas GM, Hamner B, Singh S, et al. Robot farmers: autonomous orchard vehicles help tree fruit production. IEEE Robot Autom Mag 2015;22:54−63. Available from: https://doi.org/10.1109/MRA.2014.2369292.

[78] Hayashi S, Shigematsu K, Yamamoto S, Kobayashi K, Kohno Y, Kamata J, et al. Evaluation of a strawberry-harvesting robot in a field test. Biosyst Eng 2010;105:160−71.

[79] Xiong Y, Peng C, Grimstad L, From PJ, Isler V. Development and field evaluation of a strawberry harvesting robot with a cable-driven gripper. Comput Electron Agric 2019;157:392–402.

[80] Agrobot: Agricultural Robots. <https://www.agrobot.com/>; 2021 [accessed 10.08.21].

[81] Auat Cheein FA, Carelli R. Agricultural robotics: unmanned robotic service units in agricultural tasks. IEEE Ind Electron Mag 2013;7:48−58. Available from: https://doi.org/10.1109/MIE.2013.2252957.

[82] Ball D, Ross P, English A, Milani P, Richards D, Bate A, et al. Farm workers of the future: vision-based robotics for broad-acre agriculture. IEEE Robot Autom Mag 2017;24:97−107. Available from: https://doi.org/10.1109/MRA.2016.2616541.

[83] Grimstad L, Pham CD, Phan HT, From PJ. On the design of a low-cost, light-weight, and highly versatile agricultural robot. In: Proceedings of the IEEE international workshop on advanced robotics and its social impacts, ARSO; 2016:1−6. <https://doi.org/10.1109/ARSO.2015.7428210>.

[84] Hall D, Dayoub F, Perez T, McCool C. A transplantable system for weed classification by agricultural robotics. In: IEEE international conference on intelligent robotic system 2017; 2017, p. 5174−9. https://doi.org/10.1109/IROS.2017.8206406.

[85] De Preter A, Anthonis J, De Baerdemaeker J. Development of a robot for harvesting strawberries. IFAC-PapersOnLine 2018;51:14−19. Available from: https://doi.org/10.1016/j.ifacol.2018.08.054.

[86] Qingchun F, Xiu W, Wengang Z, Quan Q, Kai J. A new strawberry harvesting robot for elevated-trough culture. Int J Agric Biol Eng 2012;5:7384−91.

[87] Jensen MF, Nørremark M, Busato P, Sørensen CG, Bochtis D. Coverage planning for capacitated field operations, part i: task decomposition. Biosyst Eng 2015;139:136−48. Available from: https://doi.org/10.1016/j.biosystemseng.2015.07.003.

[88] Jensen MF, Bochtis D, Sørensen CG. Coverage planning for capacitated field operations, Part II: Optimisation. Biosyst Eng 2015;139:149−64. Available from: https://doi.org/10.1016/j.biosystemseng.2015.07.002.

[89] Miranda-Fuentes A, Rodríguez-Lizana A, Gil E, Agüera-Vega J, Gil-Ribes JA. Influence of liquid-volume and airflow rates on spray application quality and homogeneity in super-intensive olive tree canopies. Sci Total Environ 2015;537:250−9. Available from: https://doi.org/10.1016/j.scitotenv.2015.08.012.

[90] Moshou D, Bravo C, Oberti R, West JS, Ramon H, Vougioukas S, et al. Intelligent multi-sensor system for the detection and treatment of fungal diseases in arable crops. Biosyst Eng 2011;108:311−21. Available from: https://doi.org/10.1016/j.biosystemseng.2011.01.003.

[91] Tabor T, Pezzementi Z, Vallespi C, Wellington C. People in the weeds: pedestrian detection goes off-road. In: SSRR 2015—2015 IEEE international symposium safety, security rescue robot; 2016, p. 1−7. <https://doi.org/10.1109/SSRR.2015.7442951>.

[92] Vakilian KA, Massah J. A farmer-assistant robot for nitrogen fertilizing management of greenhouse crops. Comput Electron Agric 2017;139:153−63. Available from: https://doi.org/10.1016/j.compag.2017.05.012.

[93] Home - Precision Makers. <https://precisionmakers.com/en>; 2021 [accessed 10.08.21].

[94] Autonomous system for agricultural purposes such as plant protection, tillage, etc/ Raussendorf GmbH situated in Obergurig close to Bautzen. <https://www.raussendorf.de/en/fruit-robot.html>; 2021 [accessed 10.08.21].

[95] Haibo L, Shuliang D, Zunmin L, Chuijie Y. Study and experiment on a wheat precision seeding robot. J Robot 2015;2015. Available from: https://doi.org/10.1155/2015/696301.

[96] Kirkegaard Nielsen S, Munkholm LJ, Lamandé M, Nørremark M, Edwards GTC, Green O. Seed drill depth control system for precision seeding. Comput Electron Agric 2018;144:174−80. Available from: https://doi.org/10.1016/j.compag.2017.12.008.

[97] Hassan MU, Ullah M, Iqbal J. Towards autonomy in agriculture: design and prototyping of a robotic vehicle with seed selector. In: Proceedings of the second international conference robotic artificial intelligent ICRAI 2016; 2016, p. 37−44. <https://doi.org/10.1109/ICRAI.2016.7791225>.

[98]   Åstrand B, Baerveldt A-J. An agricultural mobile robot with vision-based perception for mechanical weed control. Auton Robot 2002;13:21−35.

[99]   Klose R, Thiel M, Ruckelshausen A, Marquering J. Weedy−a sensor fusion based autonomous field robot for selective weed control. In: Proceedings of the sixty-sixth international conference agriculture and engineering. Stuttgart-Hohenheim, VDI-Verlag; 2008, p. 167−72.

[100]   Utstumo T, Urdal F, Brevik A, Dørum J, Netland J, Overskeid Ø, et al. Robotic in-row weed control in vegetables. Comput Electron Agric 2018;154:36−45.

[101]   Wu X, Aravecchia S, Lottes P, Stachniss C, Pradalier C. Robotic weed control using automated weed and crop classification. J F Robot 2020;37:322−40.

[102]   Åstrand B, Baerveldt AJ. A vision based row-following system for agricultural field machinery. Mechatronics 2005;15(2):251−69.

[103]   Fawakherji M, Potena C, Pretto A, Bloisi DD, Nardi D. Multi-Spectral image synthesis for crop/weed segmentation in precision farming. Robot Auton Syst 2021;146:103861.

[104]   Reiser D, Sehsah E-S, Bumann O, Morhard J, Griepentrog HW. Development of an autonomous electric robot implement for intra-row weeding in vineyards. Agriculture 2019;9:18.

[105]   Nørremark M, Griepentrog HW, Nielsen J, Søgaard HT. The development and assessment of the accuracy of an autonomous GPS-based system for intra-row mechanical weed control in row crops. Biosyst Eng 2008;101:396−410.

[106]   Pilli SK, Nallathambi B, George SJ, Diwanji V. eAGROBOT—a robot for early crop disease detection using image processing. In: 2015 2nd International Conference on Electronics and Communication Systems (ICECS); 2015, p. 1684−9. Available from: <https://doi.org/10.1109/ECS.2015.7124873>.

[107]   Schor N, Bechar A, Ignat T, Dombrovsky A, Elad Y, Berman S. Robotic disease detection in greenhouses: combined detection of powdery mildew and tomato spotted wilt virus. IEEE Robot Autom Lett 2016;1:354−60.

[108]   Mahmud MS, Zaman QU, Esau TJ, Price GW, Prithiviraj B. Development of an artificial cloud lighting condition system using machine vision for strawberry powdery mildew disease detection. Comput Electron Agric 2019;158:219−25.

[109]   Rey B, Aleixos N, Cubero S, Blasco J. XF-ROVIM. A field robot to detect olive trees infected by Xylella fastidiosa using proximal sensing. Remote Sens 2019;11:221.

[110]   Cubero S, Marco-Noales E, Aleixos N, Barbé S, Blasco J. RobHortic: a field robot to detect pests and diseases in horticultural crops by proximal sensing. Agriculture 2020;10:276.

[111]   Liu B, Hu Z, Zhao Y, Bai Y, Wang Y. Recognition of pyralidae insects using intelligent monitoring autonomous robot vehicle in natural farm scene. ArXiv 190310827; 2019.

[112]   Søgaard HT, Lund I. Application accuracy of a machine vision-controlled robotic microdosing system. Biosyst Eng 2007;96:315−22.

[113]   Oberti R, Marchi M, Tirelli P, Calcante A, Iriti M, Tona E, et al. Selective spraying of grapevines for disease control using a modular agricultural robot. Biosyst Eng 2016;146:203–15.

[114]   Zhou M, Jiang H, Bing Z, Su H, Knoll A. Design and evaluation of the target spray platform. Int J Adv Robot Syst 2021;18. 1729881421996146.

[115]   Ogawa Y, Kondo N, Monta M, Shibusawa S. Spraying robot for grape production. F Serv Robot 2003;539−48.

[116]   Sánchez-Hermosilla J, Rodriguez F, Guzman JL, Berenguel M, Gonzalez R. A mechatronic description of an autonomous mobile robot for agricultural tasks in greenhouses. INTECH Open Access Publisher; 2010.

[117]   Chen M, Sun Y, Cai X, Liu B, Ren T. Design and implementation of a novel precision irrigation robot based on an intelligent path planning algorithm. ArXiv 200300676; 2020.

[118] Gravalos I, Avgousti A, Gialamas T, Alfieris N, Paschalidis G. A robotic irrigation system for urban gardening and agriculture. J Agric Eng 2019;50:198−207.

[119] Oliveira LFP, Moreira AP, Silva MF. Advances in agriculture robotics: a state-of-the-art review and challenges ahead. Robotics 2021;10:1−31. Available from: https://doi.org/10.3390/robotics10020052.

[120] Fountas S, Mylonas N, Malounas I, Rodias E, Hellmann Santos C, Pekkeriet E. Agricultural robotics for field operations. Sensors 2020;20:2672.

[121] Davidson JR, Bhusal S, Mo C, Karkee M, Zhang Q. Robotic manipulation for specialty crop harvesting: a review of manipulator and end-effector technologies; 2020.

[122] Tang Y, Chen M, Wang C, Luo L, Li J, Lian G, et al. Recognition and localization methods for vision-based fruit picking robots: A review. Front Plant Sci 2020;11:1−17. Available from: https://doi.org/10.3389/fpls.2020.00510.

[123] Abundant Robotics. <https://www.abundantrobotics.com/> [accessed 10.04.21].

[124] Scarfe AJ, Flemmer RC, Bakker HH, Flemmer CL. Development of an autonomous kiwifruit picking robot. In: Proceedings of the fourth international conference on automation of robotic agents; 2009, p. 380−4.

[125] GRoW. <https://metomotion.com/> [accessed 16.06.21].

[126] FFRobotics - the future of fresh fruit harvest. <https://www.ffrobotics.com/>; 2021 [accessed 10.08.21].

[127] Sepulveda D, Fernandez R, Navas E, Armada M, Gonzalez-De-Santos P. Robotic aubergine harvesting using dual-arm manipulation. IEEE Access 2020;8:121889−904. Available from: https://doi.org/10.1109/ACCESS.2020.3006919.

[128] Ling X, Zhao Y, Gong L, Liu C, Wang T. Dual-arm cooperation and implementing for robotic harvesting tomato using binocular vision. Rob Auton Syst 2019;114:134−43.

[129] Davidson JR, Hohimer CJ, Mo C, Karkee M. Dual robot coordination for apple harvesting. In: Proceedings of the ASABE annual international meeting; 2017, p. 1.

[130] Arad B, Balendonck J, Barth R, Ben-Shahar O, Edan Y, Hellström T, et al. Development of a sweet pepper harvesting robot. J F Robot 2020;37:1027−39.

[131] Armada MA, Muscato G, Prestifilippo M, Abbate N, Rizzuto I. A prototype of an orange picking robot: past history, the new robot and experimental results. Ind Robot An Int J 2005.

[132] Lehnert C, English A, McCool C, Tow AW, Perez T. Autonomous sweet pepper harvesting for protected cropping systems. IEEE Robot Autom Lett 2017;2:872−9.

[133] Zhang Z, Noguchi N, Ishii K, Yang L, Zhang C. Development of a robot combine harvester for wheat and paddy harvesting. IFAC 2013;1. <https://doi.org/10.3182/20130327-3-jp-3017.00013>.

[134] Iida M, Suguri M, Uchida R, Ishibashi M, Kurita H, Won-Jae C, et al. Advanced harvesting system by using a combine robot. IFAC 2013;1. <https://doi.org/10.3182/20130327-3-jp-3017.00012>.

[135] Aguiar AS, Dos Santos FN, Cunha JB, Sobreira H, Sousa AJ. Localization and mapping for robots in agriculture and forestry: a survey. Robotics 2020;9:1−23. Available from: https://doi.org/10.3390/robotics9040097.

[136] Dong J, Burnham JG, Boots B, Rains G, Dellaert F. 4D crop monitoring: Spatio-temporal reconstruction for agriculture. In: Proceedings of the IEEE international conference on robotic automation. IEEE; 2017, p. 3878−85. <https://doi.org/10.1109/ICRA.2017.7989447>.

[137] Khan MW, Das GP, Hanheide M, Cielniak G. Incorporating spatial constraints into a bayesian tracking framework for improved localisation in agricultural environments. In: IEEE nternational conference on intelligent robotic systems, 2020, p. 2440−5. <https://doi.org/10.1109/IROS45743.2020.9341013>.

[138] Hague T, Marchant JA, Tillett ND. Ground based sensing systems for autonomous agricultural vehicles. Comput Electron Agric 2000;25:11−28.

[139] Pérez Ruiz M, Upadhyaya S. GNSS in precision agricultural operations. Intech; 2012.

[140] Bochtis D, Griepentrog HW, Vougioukas S, Busato P, Berruto R, Zhou K. Route planning for orchard operations. Comput Electron Agric 2015;113:51−60. Available from: https://doi.org/10.1016/j.compag.2014.12.024.

[141] Introduction to the A* Algorithm. <https://www.redblobgames.com/pathfinding/a-star/introduction.html>; n.d. [accessed 10.08.21].

[142] Ponnambalam VR, Fentanes JP, Das GP, Cielniak G, Gjevestad JGO, From PJ. Agricost-maps − integration of environmental constraints into navigation systems for agricultural robots. In: Proceedings of the sixth international conference on control, automation and robotics; 2020.

[143] Hameed IA, Bochtis D, Sørensen CA. An optimized field coverage planning approach for navigation of agricultural robots in fields involving obstacle areas. Int J Adv Robot Syst 2013;10. Available from: https://doi.org/10.5772/56248.

[144] Zhou K, Bochtis D. Route planning for capacitated agricultural machines based on ant colony algorithms. HAICTA 2015;163−73.

[145] Bogatcbev A, Koutcberenko V, Malenkov M, Matrossov S. Developments of track locomotion systems for planetary mobile robots. IFAC Proc 2004;37:153−8. <https://doi.org/10.1016/s1474-6670(17)32140-7>.

[146] Green O, Schmidt T, Pietrzkowski RP, Jensen K, Larsen M, Jørgensen RN. Commercial autonomous agricultural platform: Kongskilde Robotti. In: Proc. Second Int. Conf. Robot. Assoc. High-Technologies Equip. Agric. For. New trends Mob. Robot. Percept. actuation Agric. For. 2014, p. 351−6.

[147] Grift T. Robotics in crop production. Encycl Agric Food, Biol Eng 2007. Available from: https://doi.org/10.1081/E-EAFE-120043046.

[148] Agrobot-Robotic harvesters. <https://www.agrobot.com/e-series> [accessed 16.06.21].

[149] EcoRobotics. Phenomobile scouting robot <https://www.ecorobotix.com/wp-content/uploads/2019/09/ECOX%7B%5C_%7DFlyerPres19-EN-3.pdf > [accessed 16.06.21].

[150] Underwood JP, Burnett C. Agriculture and the environment at ACFR - latest developments - our robots; 2020.

[151] Bawden O, Kulk J, Russell R, McCool C, English A, Dayoub F, et al. Robot for weed species plant-specific management. J F Robot 2017;34:1179−99. Available from: https://doi.org/10.1002/rob.21727.

[152] Gonzalez-de-Santos P, Ribeiro A, Fernandez-Quintanilla C, Lopez-Granados F, Brandstoetter M, Tomic S, et al. Fleets of robots for environmentally-safe pest control in agriculture. Precis Agric 2017;18:574−614.

[153] Erickson B. Site specific management center newsletter; 2006. < https://www.agriculture.purdue.edu/ssmc/Frames/SSMCnewsletter7%7B%5C_%7D2006.pdf > [accessed 16.06.21].

[154] Smith S. Here come the robots: precision and regenerative farming; 2018. < Available: https://thisissamsmith.com/blog/robots-precision-and-regenerative-farming/> [accessed 16.06.21].

[155] Tyagi VV, Rahim NAA, Rahim NA, Selvaraj JAL. Progress in solar PV technology: research and achievement. Renew Sustain Energy Rev 2013;20:443−61. Available from: https://doi.org/10.1016/j.rser.2012.09.028.

[156] Xu Y, Li J, Tan Q, Peters AL, Yang C. Global status of recycling waste solar panels: a review. Waste Manag 2018;75:450−8.

[157] Global overview. REN21. <https://www.ren21.net/gsr-2020/chapters/chapter_01/chapter_01/>; 2021 [accessed 10.07.21].

[158] Aguado-Monsonet MA. The environmental impact of photovoltaic technology. EUR-OP; 1998.

[159] Şengül H, Theis TL. An environmental impact assessment of quantum dot photovoltaics (QDPV) from raw material acquisition through use. J Clean Prod 2011;19:21−31.

[160] Sherwani AF, Usmani JA, Varun. Life cycle assessment of solar PV based electricity generation systems: A review. Renew Sustain Energy Rev 2010;14:540−4. Available from: https://doi.org/10.1016/j.rser.2009.08.003.

[161] Berger W, Simon F-G, Weimann K, Alsema EA. A novel approach for the recycling of thin film photovoltaic modules. Resour Conserv Recycl 2010;54:711−18.

[162] Chowdhury MS, Rahman KS, Chowdhury T, Nuthammachot N, Techato K, Akhtaruzzaman M, et al. An overview of solar photovoltaic panels' end-of-life material recycling. Energy Strateg Rev 2020;27:100431.

[163] Adeh EH, Good SP, Calaf M, Higgins CW. Solar PV Power Potential is Greatest Over Croplands. Sci Rep 2019. Available from: https://doi.org/10.1038/s41598-019-47803-3.

[164] Swain MMC. Managing stakeholder conflicts over energy infrastructure: case studies from New England's energy transition. Massachusetts Institute of Technology; 2019.

[165] Pascaris AS, Schelly C, Burnham L, Pearce JM. Integrating solar energy with agriculture: Industry perspectives on the market, community, and socio-political dimensions of agrivoltaics. Energy Res Soc Sci 2021;75:102023.

[166] Trainor AM, McDonald RI, Fargione J. Energy sprawl is the largest driver of land use change in United States. PLoS One 2016;11:e0162269.

[167] Badiee A. An examination of the response of ethylene-vinyl acetate film to changes in environmental conditions. University of Nottingham; 2016.

[168] Badiee A, Wildman R, Ashcroft I. Effect of UV aging on degradation of Ethylene-vinyl Acetate (EVA) as encapsulant in photovoltaic (PV) modules. Reliab Photovolt Cells, Modul Components, Syst VII 2014;9179. p. 91790O.

[169] Verma S, Mohapatra S, Chowdhury S, Dwivedi G. Cooling techniques of the PV module: a review. Mater Today Proc 2021;38:253−8.

[170] Jiang H, Lu L, Sun K. Experimental investigation of the impact of airborne dust deposition on the performance of solar photovoltaic (PV) modules. Atmos Environ 2011;45:4299−304.

[171] Czanderna AW, Jorgensen GJ. Service lifetime prediction for encapsulated photovoltaic cells/minimodules. AIP Conf Proc 1997;394:295−312.

[172] Bizikova L, Murphy S, Brewin S, Sanchez L, Bridle R. CST. The sustainable agriculture transition: technology options for low- and middle income countries; 2020.

[173] Lazard's levelized cost of energy analysis; 2018. < https://www.lazard.com/media/450784/lazards-levelized-cost-of-energy-version-120-vfinal.pdf > [accessed 16.06.21].

[174] Solar technology got cheaper and better in the 2010s. <https://www.greentechmedia.com/articles/read/solar-pv-has-become-cheaper-and-better-in-the-2010s-now-what>; 2020.

[175] Gorjian S, Ebadi H, Trommsdorff M, Sharon H, Demant M, Schindele S. The advent of modern solar-powered electric agricultural machinery: A solution for sustainable farm operations. J Clean Prod 2021;292:126030. Available from: https://doi.org/10.1016/j.jclepro.2021.126030.

[176] BloombergNEF new energy outlook; 2019. < https://www.gihub.org/resources/publications/bnef-new-energy-outlook-2019/> [accessed 16.06.21].

[177] Tan DHS, Banerjee A, Chen Z, Meng YS. From nanoscale interface characterization to sustainable energy storage using all-solid-state batteries. Nat Nanotechnol 2020;15:170−80. Available from: https://doi.org/10.1038/s41565-020-0657-x.

[178]   EcoRobotix, AVO, weeding robotic platform <https://www.ecorobotix.com/en/avo-autonomous-robot-weeder> [accessed 16.06.21].

[179]   EcoRobotix, sprayer robot, the autonomous robot weeder from ecoRobotix; <https://www.ecorobotix.com/en/autonomous-robot-weeder/> [accessed 16.06.21].

[180]   EcoRobotix, ARA mounted sprayer, ARA the mounted robot weeder by ecoRobotix; <https://www.ecorobotix.com/en/ara_mounted_sprayer/> [accessed 16.06.21].

[181]   Australian Centre for Field Robotics, RIPPA and Ladybug, out robots – agriculture, ACFR - ACFR confluence; <https://confluence.acfr.usyd.edu.au/display/AGPub/Our + Robots> [accessed 16.06.21].

[182]   Kim G-H, Kim S-C, Hong Y-K, Han K-S, Lee S-G. A robot platform for unmanned weeding in a paddy field using sensor fusion. In Proceedings of the IEEE international conference automation science engineering; 2012, p. 904−7.

[183]   Franklin Robotics, Tertill the weeding robot; <https://www.tertill.com/> [accessed 16.06.21].

[184]   EcoRobotix, phenomobile, phenomobile scouting robot; <https://www.ecorobotix.com/wp-content/uploads/2019/09/ECOX_FlyerPres19-EN-3.pdf > [accessed 16.06.21].

[185]   VineScout, news & gallery; <http://vinescout.eu/web/> [accessed 16.06.21].

[186]   Keresztes B, Germain C, Da Costa J-P, Grenier G, David-Beaulieu X, Fouchardière A. Vineyard vigilant & innovative ecological rover (VVINNER): an autonomous robot for automated scoring of vineyards; 2014.

[187]   FarmBot, Genesis V1.6, open-source CNC farming; <https://farm.bot/> [accessed 16.06.21].

[188]   FarmBot, Genesis Max, "It's time for FarmBot express and Genesis Max." <https://www.youtube.com/watch?v = 6XWiTzFPWWc&ab_channel = FarmBot > [accessed 16.06.21].

[189]   University of Sydney, The DI-Wheel Concept; <https://www.digitalfarmhand.org/news/2016/08/13/diwheelconcept> [accessed 16.06.21].

[190]   SwarmFarm, Swarmbots 1, 2 and 3, Media Gallery; <https://www.swarmfarm.com/media/> [accessed 16.06.21].

[191]   Naïo Technologies, DINO, DINO vegetable weeding robot for large-scale vegetable crops; <https://www.naio-technologies.com/en/dino/> [accessed 16.06.21].

[192]   Naïo Technologies, autonomous OZ weeding robot; <https://www.naio-technologies.com/en/oz/> [accessed 16.06.21].

[193]   Naïo Technologies, TED, the vineyard weeding robot; < https://www.naio-technologies.com/en/ted/> [accessed 16.06.21].

[194]   Top 8 solar powered drone (UAV) developing companies. <https://sinovoltaics.com/technology/top8-leading-companies-developing-solar-powered-drone-uav-technology/>; 2021 [accessed 10.08.21].

[195]   Weider A, Levy H, Regev I, Ankri L, Goldenberg T, Ehrlich Y, et al. SunSailor: solar powered UAV. In: Proceedings of the forty-seventh ISR annual conference aerospace science; 2007.

[196]   Zephyr solar-powered HALE UAV—airforce technology. <https://www.airforce-technology.com/projects/zephyr/>; 2021 [accessed 10.08.21].

[197]   Leutenegger S, Jabas M, Siegwart RY. Solar airplane conceptual design and performance estimation. J Intell \& Robot Syst 2011;61:545−61.

[198]   Morton S, Scharber L, Papanikolopoulos N. Solar powered unmanned aerial vehicle for continuous flight: conceptual overview and optimization. In: 2013 IEEE international conference on robotic automation; 2013, p. 766−71.

[199]   Morton S, D'Sa R, Papanikolopoulos N. Solar powered UAV: design and experiments. In: 2015 IEEE/RSJ international conference on intelligent robotic systems; 2015, p. 2460−6.

Chapter 11

Economics and environmental impacts of solar energy technologies

Aneesh A. Chand[1], Prashant P. Lal[1], Kushal A. Prasad[1] and
Nallapaneni Manoj Kumar[2]
*[1]School of Engineering and Physics, The University of the South Pacific, Suva, Fiji, [2]School of
Energy and Environment, City University of Hong Kong, Kowloon, Hong Kong, S.A.R. China*

11.1 Introduction

A considerable amount of effort has been put into sustainable development, emphasizing environmental issues, specifically those associated with energy usage and production. This has finally given the scope for renewable energy (RE) development. The modern energy sector across different nations has given much priority to RE projects. Though the countries have given much importance, only a few RE technologies became popular, especially solar and wind. Among these two, solar energy is widely opted due to the ease in installation, operation and maintenance, and abundant solar energy resources. Typically, two types of solar energy conversion devices exist, the first type is solar photovoltaics (SPVs), and the second type is solar thermal collectors (STCs). We group these two types here onwards under one name-calling it solar energy technologies (SETs). The SETs are used for power generation, heating and cooling applications, street lighting, water pumping, etc. These applications were spreading across different sectors, including agriculture.

For a long time, solar energy has been employed in the agricultural sector to drive different operations, providing clean energy for irrigation, crop cultivation, and preservation. The enhancement of this sector plays a critical role in sustainable development, with significant adequate energy savings [1,2]. Highly effective strategies in the direction of the stimulating goals of greenhouse gas (GHG) emissions reduction, and in this context, the SETs have been defined to reduce energy consumption due to water pumping, heating, and cooling in agriculture and food production sectors [3,4]. To reach these goals, RE technologies have emerged as one of the energy innovations

Solar Energy Advancements in Agriculture and Food Production Systems.
DOI: https://doi.org/10.1016/B978-0-323-89866-9.00006-7
391

globally, with the ever-increasing demand for energy and rising concern about the nonrenewability of energy sources as well as environmental pollution. Today the common forms of RE technologies to meet the energy demands in the agricultural sector are solar thermal and SPV, wind, and biomass. Among these energy-generating methods, the SPV system is dominant as one of the fastest-growing technology globally for multiple reasons. A few of them include cost reductions in solar energy materials, technological improvement, improved conversion efficiency, solar-related hybrid solutions, negligible GHG emission in the use phase, and last but not least is the abundant availability of solar resources [5–7]. Within the advancement of RE, the SETs in agro and other sectors advocate political, economic, social, technological, legal, and environmental (PESTLE) justice in a sustainable way [8].

In this context, it is imperative to consider the economic and environmental impacts of SETs. Depending upon the SETs and applications, their performance, economics, and associated environmental impacts will vary. Hence, this chapter aims at providing different types of SETs that could be potentially used in the agriculture and food sectors. In addition, this chapter provides the methods and tools to evaluate the economics and environmental impacts for different SETs.

11.2 Overview and global progress in solar energy technologies

Along with the significant increase in SPV installation, the global trend increased by 22% in 2019 [9]. Given the incredible growth illustrated by Fig. 11.1A and B, the potential solar PV capacity is installed by different counties. In the accelerated case, global SPV installation is given immense importance. Due to registered additions in the United States, utility-scale add-ons will increase by roughly 3% in 2020.

By 2021, with nearly 117 GW added, another global record for SPV additions is expected—almost a 10% jump in 2020. In the accelerated case, China is intended to create nearly 33% additional PV capacity in 2020 than in 2019, as investors race to finish projects before subsidies are phased out. As DISCOMs' financial health problems persist and COVID-19 steps stifle construction operation, India's additions are falling for the second year in a row [9]. Conversely, the global trend is for less distributed PV additions in main markets such as the United States, the European Union, India, and Japan. Developers often commission projects in both nations in the latter quarter of the year, because of policy schedule; the majority of additional, accelerated cases is provided by China and the United States.

SPV systems transform light directly to electricity, whereas the STCs systems use the sun's heat (thermal energy) to move the turbines. The solar heating and cooling (SHC) systems are the other types where they capture

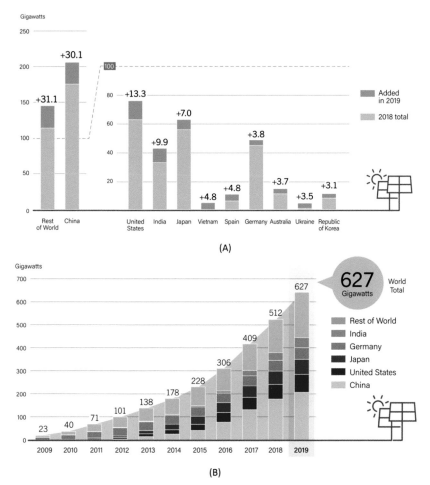

FIGURE 11.1   The global trend of solar energy technologies. (A) Top 10 countries for solar capacity, and (B) solar photovoltaic capacity by country and region, 2009–19 [10].

thermal energy to provide hot water and air heating or cooling. Fig. 11.2 gives a global trend of SPV installation, SHC, and thermal energy storage (TES) systems. As mention earlier, the involvement of solar energy has significantly improved the current energy sector and also taken a milestone in the agricultural sector.

SETs have provided energy services ranging from electricity needs and heating and cooling enhancement to thermal energy storage. In addition, other technologies such as hybrid, concentrating, and air collectors are available to meet specific heat needs. Fig. 11.3 depicts typical applications of SETs in the agriculture sector.

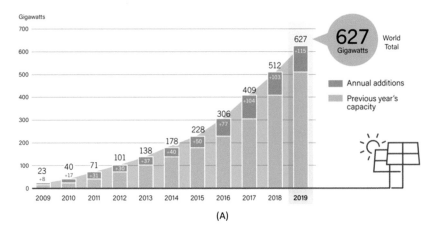

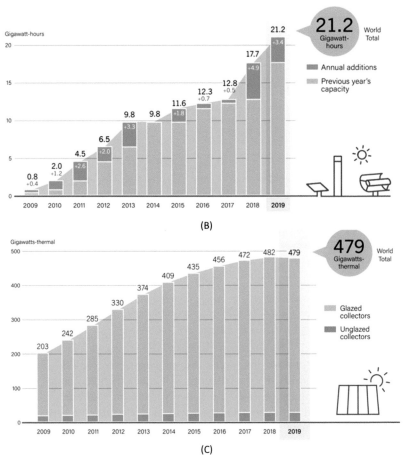

FIGURE 11.2 The global trend of solar energy technologies. (A) Solar photovoltaics global capacity, (B) concentrated solar collectors-based thermal energy storage capacity, and (C) solar water heating collectors, 2009−19 [10].

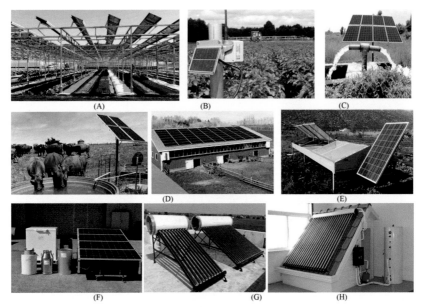

FIGURE 11.3 Common solar energy technologies for (A) photovoltaic powered greenhouse and energy generation, (B) photovoltaic powered weather meter (senor based), (C) photovoltaic powered pumping system, (D) photovoltaic power dairy farms, (E) photovoltaic powered drying system, (F) solar heating and cooling system for power milk chilling, (G) solar water heating system, and (H) solar space heating system.

11.2.1   Solar photovoltaic systems

The SPV plays a crucial role in providing the electricity required to operate the food processing and farm machinery in the agricultural sector. Depending upon the load and where it is used, the SPV system design will vary. For example, in agriculture farms, agrivoltaics could be the best option to serve the water pumping energy needs. Whereas in food processing industries, even a typical rooftop solar and building-integrated photovoltaics could serve [11]. A typical SPV system consists of a monocrystalline PV panel, DC-to-DC charge controller, DC-to-AC inverter, power meter, breaker, and battery. The charge controller (DC/DC) controls the DC electricity produced by the PV panel to produce a controlled DC output. The battery is then charged with the regulated DC. The DC/AC inverter converts the DC output from the battery to AC [7]. The electric current is measured and recorded using a power meter. A simple diagram is shown in Fig. 11.4.

11.2.2   Solar thermal power generation systems

Two basic components exist in solar thermal power generating systems: reflectors (can be called mirrors) that capture and direct sunlight onto a

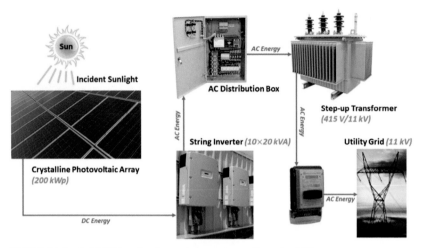

FIGURE 11.4 A 200 kWp solar photovoltaic system configuration [11].

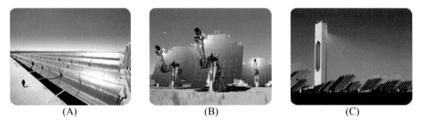

(A)   (B)   (C)

FIGURE 11.5 (A) Parabolic trough, (B) parabolic dish, and (C) solar tower.

receiver [12]. In such systems, solar collectors shown in Fig. 11.5 gather and focus sunlight on generating the high-temperature heat required to generate electricity.

As stated in Fig. 11.5, there are three main types of solar thermal power systems, namely parabolic trough (a most commonly seen solar thermal power generation system), solar parabolic dish, and solar tower. In most solar thermal power systems, the collectors as shown in Fig. 11.5 are used. All these collectors are integrated with a heat-transfer fluid medium where the fluid is heated and circulated in the receiver before being utilized to generate steam [12]. In a turbine, steam is transformed into mechanical energy, which powers a generator, which generates electricity. Most of these thermal systems can be used for power generation on a large scale, and their role in agriculture farms is very minimal. However, such systems in lower capacities can be used in the agriculture sector. In the food processing industries, the role of thermal systems is highly considered depending upon the size of the food processing unit [12,13].

11.2.3   Solar heating and cooling systems

The SHC systems play an essential role in the agriculture and food processing sectors. PV-based reversible heat pumps and solar thermally driven sorption heat pumps are the primary solar-powered heating and cooling technologies. Several studies with various configurations of SHC systems have been conducted to improve performance, increasing economic viability, and optimizing their integration within the house [14,15]. Combining different STCs, PV modules, TES, electrical, and thermally driven chillers/heat pumps will result in various SHC device layouts, as shown in Fig. 11.6. The two leading technologies associated with SHC are as follows:

1. PV systems are utilized to power a reversible heat pump that serves as a source of heat, hot water, and cooling.
2. STCs are used for heating and hot water generation and cooling when used in conjunction with a thermally driven chiller system.

11.2.4   Thermal energy storage systems

Energy storage seems to have become a crucial component of renewable energy sources (RESs). TES is a method of storing thermal energy by heating or cooling a storage medium, which can then be used for heating, cooling, or electric power at a future stage. TES is becoming increasingly essential in the field of electricity storage. When used with STC-based plants, solar heat may be stored and generated when sunlight is scarce. The use of TES in an energy system has many benefits like overall operational efficiency, reliable operation, and resilience support and among others.

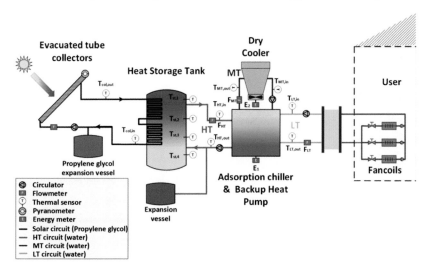

FIGURE 11.6   Generic layout of a solar absorption heating and cooling system [1].

Water is the most widely used material in sensible heat storage systems to operate a TES. Solar thermal energy systems convert solar radiation into heat [16,17]. Energy generation using STCs is regarded as an ecological way of providing domestic hot water. In this sense, it raises the temperature of a heat transfer fluid, which can be air, water, or a specially designed fluid; a system diagram is given in Fig. 11.7. The elements include solar collectors, flow meters, pumps, hot fluid transfer (HTF) tank, ball and valve inspection systems, heat storage devices, heat gage, boiler, and reservoir [1,16]. Here, the two parameters that include temperature and flow rate of the water are measured by the temperature gauge and flow meter, respectively, while the water is circulated in the system by the pump. The heat is conserved between the boiler, the HTF tank and the solar collector and the boiler is utilized to heat water.

These systems increase the temperature of the air, water, or a specially developed liquid HTF. The hot fluid can be used either for hot water or for space heating/cooling requirements or to transfer thermal energy to the ultimate applications using a heat exchanger [13]. The generated heat can also be kept in an appropriate storage tank when the sun cannot be used for hours. Solar thermal technology is also used to produce heat pools and hot water for industrial and commercial facilities.

In addition to TES systems, there are a few other energy storage systems (ESSs) [18]. Table 11.1 summarizes the different types of ESSs used in solar energy systems.

11.3   Political, economic, social, technological, legal, and environmental analysis of solar energy technologies

The PESTLE is the six pillars that are commonly used for analyzing the impact of energy planning. In this chapter, only Economic and Environmental

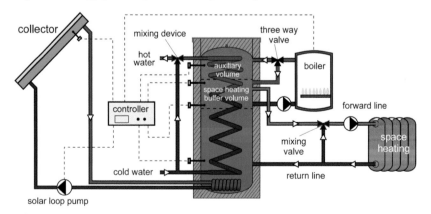

FIGURE 11.7   Generic layout of a solar thermal energy storage heating system [13].

TABLE 11.1 Common energy storage systems for solar energy technologies [18].

Storage type ESS	Storage theory	Storage device
Chemical	Electrochemical energy storage	Batteries such as zinc bromin lead-acid, nickel-metal hydride, lithium-ions and flow-cells
	Chemical energy storage	Metal-air batteries, fuel cells, and molten carbonate fuel cells
	Thermochemical energy storage	Dissociation of solar ammonia—recombination, dissociation of solar methane—recombination of solar metal and sun hydrogen
Mechanical	Kinetic energy storage	Flywheels
	Potential energy storage	Compressed air energy storage and pumped hydroelectric storage
Electrical	Electrostatic storage of energy	Supercapacitors and capacitors
	Magnetic storage of energy	Superconducting magnetic energy storage system
Thermal	Low-temperature energy storage	Aquiferous cold energy storage, cryogenic energy storage
	High-temperature energy storage	Latent heat systems include phase transition materials, graphite, hot rocks, concrete, and sensible heat systems like steam or hot water accumulators

issues will be discussed; however, it is also essential to address other pillars in PESTLE. A graphical representation of PESTLE is given in Fig. 11.8.

- *Political issue*: Corruption, international relationships, the government authorizes, regulatory authority, energy department, and bureaucracy are common political issues.
- *Economic issue:* Gross domestic product (GDP), inflation, economic stability, employment, and local & foreign investments. Economic factors are the principal reason for the deployment of RE; implementation, installation, and maintenance, the labor force (wages), and income division are some of the common costs associated with RE installation.
- *Social issue:* Lifestyles, business and social-cultural structure, local energy demand, demographics, and few other issues which deal with human interaction and behavior.

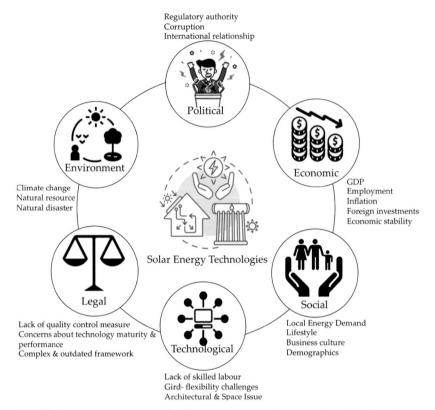

FIGURE 11.8 Systemic diagram of political, economic, social, technological, legal, and environmental representation for solar energy technologies.

- *Technological issue:* The advancement seen in the solar cells and the associated power system (e.g., different cell types, power converters, and among others) is another big challenge and also very important when integrating RE (such as solar, wind, and ocean energy) into the existing electric power grid. On the other side, we have seen the developments that led to the digitalization of solar power systems in recent days. To achieve the benefits of technological advancements seen in digital systems like the internet of things (IoT), artificial intelligence (AI), and advanced control systems can be integrated [19]. All these advancements give a more reliable, stable, and efficient energy generation method. Other aspects of technical issues associated with solar energy are poor network design, poor monitoring, and control system, poor site selection for different RESs, and poor tariff system. SETs increasingly rely on digital technologies, so the digitalization of energy generation is key research that promotes an intelligent, predictable and sustainable, and connected system.

- *Legal issue:* Different counties have policies for accepting different RERs due to their climate and geographical location. Some of the legal issues significantly impacted the solar integration due to land issues, cyclone conditions, and lack of protective devices for the existing grid.
- *Environmental issue:* Most of the RE is dependent on climate and geographical location. Hence, natural resources, climate changes, and space availability for installation need to be considered environmental factors.

The advancement in SETs makes it easier for developing countries to adapt to these solutions and take advantage while not increasing the carbon footprint. Apart from all the benefits stated above, the SETs also have common drawbacks summarized in Table 11.2.

TABLE 11.2 Negative impacts and mitigation methods of solar energy technologies.

SETs	Negative impacts	Mitigation methods
Solar thermal heating and cooling system	Aesthetics and visual impact	• Designing the system with the current standard (i.e., following the standard recommended by national disaster authorize) and regulations. • Design should be environmentally friendly- use of materials which have less impact on the environment. • Building rooftop solutions is one of the promising solutions in terms of space and visual impact.
	Discharge and unplanned releases of chemicals	• Proper maintenance and good disposal practices. • Recycling chemicals.
	Space	• Building PV systems will solve land issues.
Solar PV power generation	Massive Land use: *This is for large projects*	• Applying solution is floating PV systems (installed on dams, quarry lakes, drinking water reservoirs, canals, and ponds). • Rooftops can also be utilized.
	Land Cultivation *Agricultural lands are being used for solar PV installations*	• Large commercial building rooftop.

(Continued)

TABLE 11.2 (Continued)

SETs	Negative impacts	Mitigation methods
		• Avoid fresh lands for installation, that is, ecologically and archeologically sensitive locations. • Agro-solar systems can be used, that is, PV systems, for irrigation purposes.
	Aesthetics and Visual impact	• Designing the system with current standards (i.e., following the standard recommended by national disaster authorize) and regulations. • Building a rooftop is one of the promising solutions in terms of space and visual impact. In this, mirrors and artificial trees can be used for modern architecture.
	Effect on ecosystems *This is applicable for large scale project*	• Avoiding threatened habitats and natural beauty zones, as well as archeological sites.
	Toxic & flammable materials *This is noticed in the construction of the modules*	• Adoption of current safety laws and good practices to avoid the release of potentially dangerous material.
	Old and damaged PV panels	• Skilled personnel are required to operate such a system—use proper protective clothing and handling tools, for example, sunglasses, clothing during construction, and gloves.
Solar thermal energy	Construction Architecture	• Advanced designs are proposed with construction architecture. • Site renovation and maintenance. • The suitable arrangement so that responsive habitats are

(Continued)

TABLE 11.2 (Continued)

SETs	Negative impacts	Mitigation methods
		avoided along with densely populated areas and natural beauty.
	Visual impact-aesthetics	• The suitable arrangement so that responsive habitats are avoided along with densely populated areas and natural beauty.
	Space-Land use	• Suitable arrangement.
	Significant impact on the ecosystem. *Especially for birds such as flora and fauna.*	• The suitable arrangement so that responsive habitats are avoided.
	Impact on water resources water. *Thermal energy causes water pollution: this can be discharge and unplanned releases of chemicals*	• Need suitable and proper construction. • Better methods can be used, such as air as a heat-transfer medium. • Skilled personnel are required to operate such kind of a system.
	Safety issues and hazards	• Skilled personnel are required to operate such a system— use proper protective clothing and handling tools, for example, sunglasses, clothing during construction, and gloves.

11.4   Potential economic and environmental impacts of solar energy technologies

Generating energy and transmitting it to end-users has both economic and environmental impacts. As it is evident, conventional energy generation harms the climate, water, land, environment, atmosphere, wildlife, and landscape and increases harmful radiation levels. SETs are more sustainable, and they can solve a lot of environmental and economic issues encountered with fossil-based power generation and nuclear fuels. The below subsections provide some common SETs solutions and all the SETs' economic (see Table 11.3) and environmental impacts (see Tables 11.4 and 11.5).

TABLE 11.3 Economic role of solar energy technologies.

Economic benefits	Solar PV system	Solar heating and cooling	Solar TES
Energy trade	Grid-connected: generate capital without burning fossil fuel. A stand-alone system reduces massive transmission lines.	Reduces the dependency on grid energy. No fossil fuel is required.	
GDP and Industrial development	Lower the energy bills. Reduction in the importation of fossil fuels.	Lower the utility bill. Establishment of more solar ventures.	
	Although the number of jobs is tied to the size of the solar power plant, rural occupations offer money for other households and boost the utilization of local infrastructure and GDP.		
Employment opportunities and human welfare	Solar offers several types of work, including production, installation, engineering, sales, marketing and other employment. Due to the high demand for solar energy, more factories are built which require employees.		
Investment	Foreign sector Investment in SETs. Gross capital formation and transfer with the country.		

TABLE 11.4 Summary of environmental impact/effect/suggestion from solar thermal heating systems.

Environmental impact and description	Effects	Suggestions to address this issue
Land usage: it is used for low or medium heat systems. For large-scale systems, the design sizes determine the land use with specific characteristics of the preferred system. In the case of communal use, a low-temperature system is designed with significant land use. In this, it can be considered for heat storage,	Soil pollution during construction and maintenance. Damage to the habitat as well as its ecological area.	It is essential to deploy a strategic plan to combat the land issue to assess the area of interest. Use proper installation materials which have less threat to the natural environment.

(Continued)

TABLE 11.4 (Continued)

Environmental impact and description	Effects	Suggestions to address this issue
while the existing structures can be utilized for the collection surfaces. The land use conditions of concentrating collectors supplying process heat are more troublesome for high-temperature systems.		
Visual impact: It poses a poor visual impact for building installation. Poor design can affect the building's internal wiring system. With some benefits associated with solar thermal heating systems, architectural designs sometimes need to be changed. This system shows the aesthetic effect in the sense of personal preference; flat panels are typically built to blend in with the existing roofline and generate minimal blaze. Modern ST systems enable the production of collectors that are aesthetically pleasing and easy to incorporate into buildings. The positioning of these systems in the building could potentially increase fire risks and water penetration into the roof; however, since only four holes per panel are required, such hazards can easily be evaded.	For existing house design, it damages the rooftop. During the architectural design is changed, it causes additional capital, material, and labor.	Proper system foundation needs to be made before installing the system. Preplanning and extension for an upgrade in buildings will surely reduce this impact. Skilled and appealing design can attract more rooftop installation.
Routine and accidental discharges of pollutants: During maintenance, it can cause antifreeze or rust inhibitors will almost	Accidental water pollution due to leakage in the heat transfer liquids. More complex compounds, such as	Proper maintenance is required in terms of refilling coolant.

(Continued)

TABLE 11.4 (Continued)

Environmental impact and description	Effects	Suggestions to address this issue
certainly be present in all indirect systems. Glycol, nitrates, nitrites, chromates, sulfates, and sulfites may all be present in the heat transfer liquids. Solar converters, on the other hand, can reach relatively high temperatures if their coolant is lost. As a result, there is a chance of fire, given the high temperatures. Also, an added complication of out-gassing is on the rise from panel components (insulant, plastic components, epoxies)	aromatic alcohols, CFCs, and oils, would be used in higher temperature practices. The application of SETs on a large scale requires the proper disposal of these substances.	

TABLE 11.5 Summary of environmental impact/effect/suggestion for solar photovoltaic systems.

Environmental impact and description	Effects	Suggestions to address this issue
Land usage: There are several key factors that need to be understood for the land issue associated with the solar PV system, that is, the topography of the landscape, sensitive ecosystems, biodiversity, and the coverage area of the PV system. For large-scale PV systems, massive land area is used for installation and proper feasibility study is required. Even construction events, such as transportation and earth movement, are the primary modification to the landscape at the initial construction phase. Cultivation of agricultural land is a serious threat.	Soil pollution during construction and maintenance. Damage to the habitat as well as its ecological area. A lot of agricultural land is cultivated hence, several social disagreements and displeasure.	To combat the land issue, there is a need to accept other forms of a PV installation, such as floating PV system, the rooftop system is largely being used. Use proper installation materials which have less threat to the natural environment.

(Continued)

TABLE 11.5 (Continued)

Environmental impact and description	Effects	Suggestions to address this issue
Air population: During the construction of Solar PV system, air pollution is produced which cause a lot of health issue. Emission of CO_2 from heavy machines during construction and land cultivation. The manufacturing of PV systems also creates a lot of air pollution.	Air pollution causes health issues for both humans and other habitats.	Cultivate land during rainy sessions. Try to minimize the use of land for solar PV installation.
Visual impact: The arrangement of PV systems depends on the plan of action as well as its surrounding. The visual impact the system will have in conjunction with modern architecture will be highly significant compared to olden architecture. Through an appropriate PV integration scheme, power coverage to rural areas can easily be achieved and the management issues such as regular maintenance, monthly payments, obtaining finance, and providing advice on energy-efficient appliances can also be addressed.	Large-scale solar PV installation needs to have a good visual amenity and building aesthetics or spoil the natural look.	Best architectural designs need to be used to minimize the potential impact on visual amenity and building aesthetics. Advances in the development of multifunctional PV facades. Proper layout and correct selection of material are important. Use of different colors to assemble the PV modules in largescale systems.
Regular and unplanned discharge of pollutants: Normal PV systems have limited pollutants once it is installed, but maintenance causes pollution. During maintenance, cables, PV modules, and foundation structures are replaced, creating a threat of pollution. Small traces of toxic substances are exposed to the atmosphere and the environment.	Slight traces of toxic substances give rise to occupational health issues and pose a small public threat.	Under ideal conditions, PV systems discharge zero gaseous or liquid contaminants as well as zero radioactive matter. A contingency plan and emergency response are required in the event of an unintentional fire or heat exposure. Emissions to soil and water are possible here.

11.4.1 Economic impacts of solar photovoltaic, solar heating and cooling, and thermal energy storage system

In literature, many studies were showing the economics of SETs. Several studies have reported the life cycle metrics of SPV systems. For example, the results demonstrate that the energy return time (ERT) and carbon (CO_2) emissions of both amorphous silicon (a-Si) and multicrystalline silicon (mc-Si) are 1.6, 5.7 years, and 34.3 and 72.4 g/kWh, respectively [20]. In real-life working settings, Ref. [21] reported the outcomes of single crystalline silicon (sc-Si) and mc-Si SPV systems. Sc-Si and mc-Si have both ERT and CO_2 emissions of 1.5 and 15.5 years and 9.4 g-C/kWh and 91 g-C/kWh, respectively [22]. Compared to fossil fuels, CO_2 emissions from PV modules are one-fourth [23]. In Ref. [24], it has been argued that eco-friendly items contribute to the mitigation of CO_2 emissions during the manufacturing processes of SPV module components. The Gobi Desert contains a SPV system of 100 MW crystalline silicone (c-Si). Additionally, 12 g-C/cWh and 1.7 years of CO_2 emission and ERT were observed [25].

In the six climatic situations, Akinyele et al. [26] compared the 1.5 kW of PV systems results. In the Nigerian northeast area, the global temperature warming potential and ERT for this system were respectively 1907 kg of CO_2 and 0.83 years. Coming to solar water heating (SWH), these systems are installed considering economic and environmental benefits. To encourage individuals to use RE and the SWH system, the price of the system must be affordable, particularly for low-income families. As a result, it is vital to assess the system's economic viability [27]. The majority of studies concluded that SWH technology is cost-effective for a variety of applications in terms of installation costs and energy costs over the system's lifetime, as evidenced by performance analysis. This is why SWH systems are widely used in home and industrial applications. With a favorable payback period (PP) of 2−4 years according to the type and size of the system, the SWH systems are cost-efficient and generate substantial studies to further increase SWH thermal efficiency. When solar cooling systems (SCS) are considered, they also appear to be economically viable [27]. In Ref. [28], two SCS systems were compared. The first one is SPV based cooling system, while the second is powered by solar-collected heat. The results show that the SPV-based cooling system outperforms the thermal solar cooling system. The initial investment costs had a significant impact on both the payback and the cost of saved primary energy. A theoretical−practical investigation was carried out in Ref. [29], through a SPV microgrid by feeding a reversible air-water pump. The results indicated that on days with minimum-maximum external temperatures between 1°C and 16°C, the system could maintain a pleasant inside temperature. Global efficiency was also obtained as approximately 18.2%. In an office building in Southern Italy, a study was investigated taking solar electric heat pump as a case, which resulted in reduced demand for energy, as compared with the most frequent Italian form of structure [30]. In comparison with a traditional reference system based on a boiler supplied by natural gas and electric-powered chiller, it was discovered that the heat pump system based on solar energy saved primary energy and corresponding carbon emissions by roughly 81% [30].

11.4.2   Environmental impacts of solar photovoltaic, solar heating and cooling, and thermal energy storage system

11.4.2.1   Environmental impact of solar heating and cooling systems

The production of SHC systems induces a large number of materials, and negligible amounts of these materials are utilized during their operation. The coolant change is the only possible environmental pollutant at that particular time frame. However, this can be eliminated through good working practice. Accidental coolant system leaks can result in fires and gas releases from vaporized coolants, harming public health and safety. The widespread adoption of solar thermal systems, on the other hand, would greatly reduce the combustion of fossil fuels and hence diminish the environmental effects connected with fossil fuels.

11.4.2.2   Environmental impacts of solar photovoltaic power generation

Commonly solar PV systems are considered to be clean and environmentally friendly because they produce zero noise with the added advantage of zero chemical pollutants when compared to the other conventional energy generation systems [31]. The process of power generation using solar panels is completely silent, emitting no noise, making them ideal for residential, urban, and agricultural applications.

11.4.2.3   Environmental impacts of solar thermal power generation technologies

A collection of technologies that collect heat energy from the sun and various forms of electrical heating and generation are used in the term solar thermal. Solar thermal systems are either passive or active for heating. Passive systems are more simple and do not have moving elements; they merely depend on design features to improve the capacity to catch and utilize the rays of the sun such as a greenhouse or a solar oven. Active systems feature mechanical components for circulating hat transport fluids, like fans or pumps. These systems are suitable for household or commercial heating applications.

The concentrating solar power (CSP) systems for the production of electricity use mirrors to focus solar energy on the central collector. This generates enough high temperatures to generate steam to power the generator. CSPs are mainly found in desert places where there is plenty of space and sunlight. One benefit of solar thermal power production is that it is an energy source that is clean and renewable. It employs a free kind of fuel, that is, the Sun. Most solar thermal systems also have low maintenance due to the simple technology they use and the absence of moving parts. Sunshine isn't an extremely concentrated source of energy, thus a wide region of energy, efficiency, and land use can be considered a suitable quantity. Sunshine is likewise intermittent and depends on location and time to be available. There are more benefits and disadvantages to the CSP [32]. For locations using centrally-located power distribution networks, the technology capability for massive generation produces a benefit. That has in the past been an advantage on SPV systems but improvements in photovoltaic technology challenge the idea. One of the primary drawbacks of CSP, as steam turbines are necessary for generating energy, access to water and evaporation are the sites that are usually in isolated areas of the desert. Furthermore, electricity transmission across large distances is expensive and might lead to loss of distribution. Furthermore, practical problems such as cost and awareness of funds in advance can also hinder the application of all sorts of solar thermal systems [32].

It is noticed that some solar thermal energy systems have potentially hazardous fluids for the transfer of heat. As of the limited implementation of solar thermal systems to date, there is no experience of the environmental consequences of such systems. The elemental environmental advantage of solar thermal systems is the elimination of pollution related to traditional electricity production. This system produces no pollution when in use; however, little emission is expected during certain phases of its life cycle (mainly during material processing and manufacture) [33].

11.4.2.3.1 Materials processing and manufacture

Some of the observable gas emissions of solar thermal systems during material processing and manufacture are CO_2, sulfur dioxide (SO_2), and nitrogen oxides (NO_x). The effects of such gasses vary by location and are less than that of traditional fossil fuel technologies.

11.4.2.3.2 Construction

During the construction stage, these projects have the standard environmental impacts which are normally related to any engineering venture. These impacts include disturbance to the ecosystem, virtual intrusion, noise, occupational accidents, change in landscape, temporal blindness, etc.

11.4.2.3.3 Land use

Amongst other SETs, solar thermal electric systems are the most efficient systems on land (annual production ranges around 4−5 GWh/ha). To the present day, the ideal selection site for the physical location of the solar thermal system has been arid desert areas, particularly due to plant communities and fragile soil.

11.4.2.3.4 Visual impact

The tower of the central receiver systems is the main visual impact in conjunction with the collector systems. Areas with low population density have the idea of the atmospheric requirement for such systems, considering no disturbance to the natural beauty is brought about.

11.4.2.3.5 Water resources

Cooling water is needed for parabolic trough and central tower systems that use a traditional steam plant to produce electricity. The downside in arid regions is that the water resources will be relatively low. Furthermore, water resources can be contaminated due to the unintentional release of plant chemicals and thermal discharges, but these falls can be eliminated through good practice.

11.4.2.3.6 Health and safety

Heat transfer fluids, namely oil and water, may be released accidentally from the parabolic trough and central receiver systems, posing a serious health risk. Some central tower systems use either liquid sodium or molten salts, acting as a heat transfer channel. In case of an accident, the outcome will be substantial. However, if the volumetric system approach is implemented, these risks can be eliminated as the heat transfer channel will be air. Central tower systems can absorb light to levels that may be harmful to the eyes. Under ideal conditions, it has zero threat to operators; however, if the tracking systems malfunctions, straying beams can occur, posing a workplace safety risk.

11.5 Tools for modeling the techno-economic and environmental impacts of SETs

11.5.1 Techno-economic modeling tools

Several tools are developed and have been used by many researchers worldwide. Some of the common software used for the analysis, as given in Table 11.6. Apart from these softwares, various online or web version tools are used to analyze system techno-economic and environmental performance. Solar thermal energy system mainly uses different types of software tools as give given in Table 11.7, for analysis purpose while designing and optimizing the final design in a simulation.

The following tools were designed for online purposes:

• *PV Watts*: It is a National Renewable Energy Laboratory (NREL) online calculator [34].

TABLE 11.6 Main highlights of various software tools for solar photovoltaic systems.

Software names	Developed by	Latest version	Operating system	Type of analysis possible	Pricing model and source link
MATLAB® and Simulink	MathWorks	10.2 (part of R2020b)/ September 17, 2020	Linux, macOS, Microsoft Windows	Technically analysis; Statistical analysis; Machine learning; Linear and nonlinear models; Simulate and predict output.	Priced: https://www. mathworks.com/products/ matlab.html
HOMER Pro	NREL, USA (1993)	HOMER Pro Version 3.14.4	Windows Visual and C++	Sensitivity analysis; Technically analysis; Economic analysis; Emission analysis.	Free: http://users. homerenergy.com/pages/ homer_pro Full capabilities required a license subscription
SAM	NRE in collaboration with Sandia National Laboratories in 2005, and at first used internally by the US Department of Energy's Solar Energy Technologies Program.	Version 9.5	macOS; Windows; Linux	Technical analysis; Project coordination; Financials portfolio analysis; Economic analysis.	Free: https://sam.nrel.gov/ about-sam.html Full capabilities required a license subscription
RETSscreen	The Government of Canada.	Version 6.0	Windows	Technical analysis; Energy efficiency analysis, Project feasibility analysis.	Free: https://retscreen-expert. software.informer.com/6.0/ Full capabilities required a license subscription
PVsyst	Institute of Environmental Sciences (ISE), University of Geneva, Switzerland	Version 7.2	Window	PV sizing and simulation; Data analysis as per the built data sets and models; Economic analysis.	Free: https://www.pvsyst.com/ download-pvsyst/ Full capabilities required a license subscription

Software	Developer	Version	Platform/Language	Analysis	Availability
HYBRID2	University of Massachusetts, USA and NREL (Hybrid1 in 1994, Hybrid 2 in 1996)	Version 1.3	Windows XP Visual BASIC	Technical analysis; Economical analysis.	Free: https://www.umass.edu/windenergy/research/topics/tools/software/hybrid2 Full capabilities required a license subscription
RETScreen	Developed by Ministry of Natural Resources, Canada in 1998	RETScreen 4 and RETScreen Plus	Windows 2000, XP, Vista Excel, Visual Basic, C	Financial analysis; Environmental analysis.	Free: https://nrcaniets.blob.core.windows.net/iets/RETScreenExpertInstaller.exe Full capabilities required a license subscription
iHOGA	Electrical Engineering Department of the University of Zaragoza in Spain.	Version 3	C++ Windows XP	Multi or mono objective; Optimization using genetic algorithm.	Free: http://www.unizar.es/rdufo/hoga-eng.htm The PRO version is priced & EDU version is free
INSEL	German University of Oldenburg (1986–1991)	Version 8.1	Windows Fortran and C/C++	Planning, monitoring of electrical and thermal energy systems.	Priced: https://www.insel.eu/en/home_en.html
TRNSYS	University of Wisconsin and University of Colorado (1975)	Version 18 Available for 40 years.	Windows Fortran code	Simulate transient system behavior.	Priced http://www.trnsys.com/
iGRHYSO	University of Zaragoza, Spain	—	Windows C++	Technical analysis; Economic analysis.	Free: https://grhyso.es.tl/

(Continued)

TABLE 11.6 (Continued)

Software names	Developed by	Latest version	Operating system	Type of analysis possible	Pricing model and source link
HYSYS	Wind technology group (CIEMAT), Spain	Version 12.1	Windows	Sizing; Long-term analysis of off grid hybrid systems.	Free: https://www.aspentech.com/en/products/engineering/aspen-hysys Full capabilities required a license subscription
HySim	SNL (late 1980s)	Version 4.7	Windows Vista/X	Financial analysis; Hydrological cycle.	Free: https://hysim.software.informer.com/4.7/ Full capabilities required a license subscription
SolSim	Fachhochschule Konstanz (Germany)	Version 1.5	Windows	Technical analysis; Economic analysis.	Free: https://solsim.software.informer.com/ Full capabilities required a license subscription
ARES	Cardiff school of engineering, University of Wales, UK	—	Windows Linux macOS	Technical analysis; Economic analysis.	Free: https://www.graebert.com/cad-software/download/ares-commander/ Full capabilities required a license subscription

Homer, Hybrid optimization model for electric renewable; *SAM*, system advisor model; *NREL*, National Renewable Energy Laboratory; *iHOGA*, improved optimization of hybrid systems based on genetic algorithm; *TRNSYS*, transient systems simulation; *iCRHYSO*, grid-connected renewable hybrid systems optimization; *SOMES*, society of mechanical engineering students.

TABLE 11.7 Main highlights of various software tools for solar thermal energy systems.

Software names	Developed by	Latest Version	Operating system	Type of analysis possible	Pricing model and source link
SHW	University of Innsbruck, Austria	—	Window	Technical analysis Financial analysis	Free: https://www.uibk.ac.at/bauphysik/forschung/shw.html.en Full capabilities required a license subscription
Viessman	Viessmann Group	Is an online calculator	Any OS with a browser	Technical analysis; Environmental analysis	Free: http://viessmann.solar-software.de/index.php?lang = en Full capabilities required a license subscription
SOLO	Tecsol	Is an online calculator	Any OS with a browser	Energy output analysis; Technical analysis	Free: http://solo.tecsol.fr/ Full capabilities required a license subscription
RETScreen	Developed by Ministry of Natural Resources, Canada in 1998	RETScreen 4 and RETScreen Plus	Windows 2000, XP, Vista Excel, Visual Basic, C	Financial analysis; Environmental analysis	Free: https://nrcaniets.blob.core.windows.net/iets/RETScreenExpertInstaller.exe Full capabilities required a license subscription
Oventrop	Oventrop	Is an online calculator	Any OS with a browser	Environmental analysis; Technical analysis	Free: https://www.oventrop.com/en-GB/downloadsoftware/onlinecalculations/solarcalculation Full capabilities required a license subscription
ScanTheSun	Scientific Mobile	Version 4.2	Android 5.0 or above	Technical analysis	Google Play Store
Transol	Aiguasol and the French research center	Version 3.1	Microsoft Windows macOS	Technical analysis; Cost beneficial analysis	Priced: https://aiguasol.coop/design-of-solar-thermal-systems-with-transol/?nowprocket = 1

(Continued)

TABLE 11.7 (Continued)

Software names	Developed by	Latest Version	Operating system	Type of analysis possible	Pricing model and source link
F-chart	S.A. Klein and W. A. Beckman	Version 1.5.96e	Microsoft Windows	Cost beneficial analysis; Technical analysis	Priced: https://fchartsoftware.com/fchart/index.php/demo.php
GetSolar Professional	Etu Software GmbH	—	Microsoft Windows Android 8.0 or higher IOS 11 or higher	Emission analysis; Technical analysis	Priced: https://www.hotgenroth.de/M/SOFTWARE/Simulation-Solar-PV/GetSolar-Professional/Seite.html,73280,80427
T*Sol	Valentin Software GmbH	Version 5.5.11.32	Microsoft Windows Microsoft.NET Framework 4 or higher	Technical analysis; Cost beneficial analysis	Priced: https://valentin-software.com/produkte/tsol/
Polysun	Vela Solaris	Version 12 or 18.04	Microsoft Windows macOS Linux	Technical analysis	Priced: https://www.velasolaris.com/downloads/?lang = en
TRNSYS	The University of Wisconsin and University of Colorado (1975)	Version 18 Available for 40 years	Windows Fortran code	Simulate transient system behavior	Priced http://www.trnsys.com/

SHW, Simulation software for thermal solar systems; *OS*, operating system.

- *PVGIS*: It is also one of the web-based photovoltaic systems sizing modeling tools. This was produced by the European Commission. In PVGIS, users have numerous options, such as monthly, hourly and daily examination of the performance of grid power PV systems, off-grid solar PV systems, and solar radiation data [35].
- *PV-Online*: Like other tools, it is also a comprehensive web-based tool for analyzing the operation of PV grid plants that are roof-integrated [36].
- *PV*SOL*: It is an online modeling and analysis tool for the PV system. This version is far better than the main PV-online tools [37].
- *Mayfield Design Tool*: The size of module strings is one of the most significant factors of building a solar array for the inverter used [38].
- *Solar Design Tool*: This Web App is an online PV design that facilitates the creation and configuration of appropriate solar panel power systems and layouts for anybody. HelioScope is a tool for the design of online PV systems [39].
- *i-Pals WEB*: This PV simulation software is customizable, including the design and architecture of the system, comprehensive array, module and string settings, and information [40].

11.5.2   Environmental modeling tools

Though few of the above-mentioned tools in Tables 11.6 and 11.7 provide a basic analysis of CO_2 emissions, the obtained result may not be sufficient to assess the environmental impacts. For this reason, we provided the tools that allow the life cycle emissions accounting the all the inputs and outputs from the system [41]. For that reason, a well-ordered life-cycle assessment (LCA) method is used. The LCA is an operative method for evaluating the environmental impact categories shown in Table 11.8. For assessing the environmental impacts, indicators are a must. In general, the LCA practitioners have a well-defined indicator, that is, $tCO_2e/year$ concerning the unit for GHG emissions [42].

The LCA is in line with the ISO 14040:2006 and 14044:2006 standards. In the LCA analysis, the following four fundamental phases are as follows:

- Objective and scope definition to emphasize the LCA target and to identify the boundaries according to ISO 14040 [43].
- The life-cycle inventory is assembled following ISO 14041 [44] where the fluxes of energy, material, and emissions are assembled [44].
- Environmental impact assessment considers the life cycles for which impacts are assessed for seventeen ISO 14042 [44] effect indicators.
- Interpretation of the impact results when impacts acquired are annotated and analyzed according to the ISO 14043 [43] for the LCA.

In Table 11.9, the summary of the tools is provided, and these tools can be used for any type of SETs.

TABLE 11.8 Indicators and their description used in life-cycle assessment [41].

Indicator	Description
Acidification	Soil acidification caused by gasses like oxides of nitrogen and sulfur oxides.
	Acidification of water via production of nitrogen oxides and oxides of sulfur.
Aquatic ecotoxicity	Toxic compounds released into the bodies of freshwater.
	Aquatic bodies effected by the released poisonous chemicals.
Depletion of resources	Nonfossil natural resources depletion.
	Natural fossil fuel resource depletion.
Eutrophication	Enrichment by nutritional elements of the aquatic ecology due to the nitrogen or phosphorus emissions of substances.
Global warming	Global warming from greenhouse gas releases into the air.
Human toxicity	Toxic compounds that ultimately affect humans are released into the environment.
Ozone depletion	Air emissions causing stratospheric ozone layer destruction.
Photochemical ozone creation	Gas emissions that affect the production of sunlight-catalyzed photochemical ozone (smog), in the lower atmosphere.
Terrestrial ecotoxicity	Emissions into the land of harmful substances.
Pollution	Airborne hazardous components.
	Emissions of hazardous substances into water.

The ecological consequences of the c-Si SPVs grid-connected energy generation were explored by Hou et al. [45]. The authors incorporated balance-of-system (BOS) and transportation and assembly of fossil fuels. The results were in ERT and GHG terms. ERT was between 1.6 and 2.3 years, while GHGs were $60.1-87.3$ g CO_2eq/kWh; roughly 84% or more of energy consumption and overall emissions of PV were derived. Authors in [45] also pointed out that efficiency improvements can be achieved with heterojunction of the cell of intrinsic thin-layer (HIT) (from $16\%-18\%$ to 22% or above), which might lead to a GHG emission of 47.5 g CO_2eq/kWh. Luo et al. [46] produced three mc-Si PV setups of the comparative LCA of SPV power generation in Singapore. The FU was a 60-cell PV module of Silicon and the system limit ranged from silica mining to installation of SPV systems. ERT and GHG emissions were the metrics used to assess

TABLE 11.9 Main highlights of various software tools for modeling life cycle assessments.

Software names	Developed by	System requirements	Analysis type	Pricing model and source availability
OpenLCA	GreenDelta and OpenLCA	Microsoft Windows	Life-cycle assessment	Free http://www.openlca.org/
SimaPro	PRé Sustainability	Microsoft Windows	Life-cycle assessment	Priced www.pre-sustainability.com/all-about-simapro
GaBi	Sphera's	Microsoft Windows	Life-cycle assessment	Priced www.gabi-software.com
Umberto	iPoint-systems: iPoint	Microsoft Windows	Material flow analysis & Life-cycle assessment	Priced https://www.ifu.com/umberto/lca-software/

environmental consequences. The three-separate roof-integrated SPV systems consisted of, first, a surface-backed aluminum (Al-BSF) solar cell with a standard module structure, second, a conventionally designed emitter and back-cell (PERC). Third, the PERC solar cell having an unframeable, two-glass module construction. EPBTs of 1.11, 1.08, and 1.01, whereas GHG emissions of 30.2, 29.2, or 20.9 g CO_2eq/kWh were used separately in the scenarios evaluated.

Coming to the thermal systems, the CSP systems are best adapted, without producing other environmental hazards or pollution, to reduce GHG and other pollutants [47], according to a life cycle evaluation. For example, 250−400 kg of CO_2 emissions annually can be avoided per square meter collecting area. The current energy mix in Germany is used as a basis for this LCA of CO_2 emissions. CSP value is relevant for the solar parabola of 80 MW in its operational mode only. The energy return period of solar-concentrated systems is only five months. This compares with their life span of about 25−30 years quite positively. Most resources can be recycled and reused for other plants.

Ref. [48] have examined SHC systems employing a flat-plate collector (FPC) and an absorption chiller difference. The inventory was made up of fabricator data, the ecoinvent database, and the appropriate literature and a standard heat pump reference system, as was the case in the present study.

In the analysis of the SimaPro program, however, the other approach for an effective assessment can be employed, while thermal loading for the considered building was used as a functional unit. The functional unit was employed as 1 kWh of heating and the analysis for the evacuated tube collectors (ETCs) was performed; however, geothermal heat and SPVs were used in this case as the backup scenarios. The assessed solutions included the solar-driven absorption heat pump with the lowest environmental performance, whilst a SPV system found a minimum footprint. In Ref. [49], the cradle was used to seriously analyze the ETC-powered absorption chiller. A traditional heat pump was also utilized as a reference, as in most circumstances. However, the analyzes were carried out in ELISA tool software, and the impact assessment methodologies employed by the IPCC-2013 and the CMD were used. The case studies also assessed, as opposed to the present study, the results of a SPV-powered heat pump and no solar thermal systems.

11.6 Conclusion

As SETs have the potential to play a significant role in the global energy portfolio, they widely adopted the claim that SETs provide a double dividend or "win−win solution" in terms of environmental custodianship and economic stability. In this regard, it has become imperative to look into the economic and environmental impacts of the SETs. Hence, this chapter has looked into various types of SETs used in the food and agricultural sectors. First, we have described various SETs that are playing a key role in decarbonizing the energy-dependent sectors. Second, by applying the PESTLE approach, the SETs are analyzed and compared. Third, we have provided the potential environmental and economic impacts for all the types of SETs disused in this chapter. In the fourth step, we have discussed various software tools, that help in evaluating the techno-economic and environmental performances of SETs. Overall, we believe that this chapter will help to provide an understanding of the sustainability of the SETs. In the summary, we conclude that the SETs are a huge part of the future in decarbonizing the food and agriculture sector by providing energy, heating/cooling, and thermal energy solutions in a much environmentally friendly way without sacrificing huge costs taking the benefits into the account. Finally, SETs are not only environmentally sound, but they contribute to extensive social and economic gains.

References

[1] Roumpedakis TC, Vasta S, Sapienza A, Kallis G, Karellas S, Wittstadt U, et al. Performance results of a solar adsorption cooling and heating unit. Energies 2020;13 (7):1630.
[2] Chandel SS, Sharma A, Marwaha BM. Review of energy efficiency initiatives and regulations for residential buildings in India. Renew Sustain Energy Rev 2016;54:1443−58.

[3]   Chenari B, Carrilho JD, da Silva MG. Towards sustainable, energy-efficient and healthy ventilation strategies in buildings: a review. Renew Sustain Energy Rev 2016;59:1426–47.

[4]   Ruparathna R, Hewage K, Sadiq R. Improving the energy efficiency of the existing building stock: a critical review of commercial and institutional buildings. Renew Sustain Energy Rev 2016;53:1032–45.

[5]   Kannan N, Vakeesan D. Solar energy for future world: a review. Renew Sustain Energy Rev 2016;62:1092–105.

[6]   Kumar NM, Chopra SS, Chand AA, Elavarasan RM, Shafiullah GM. Hybrid renewable energy microgrid for a residential community: a techno-economic and environmental perspective in the context of the SDG7. Sustainability 2020;12(10):3944.

[7]   Chand AA, Prasad KA, Mamun KA, Sharma KR, Chand KK. Adoption of grid-tie solar system at residential scale. Clean Technol 2019;1(1):224–31.

[8]   Islam FR, Mamun KA. Possibilities and challenges of implementing renewable energy in the light of PESTLE & SWOT analyses for island countries. Smart energy grid design for Island countries. Cham: Springer; 2017. p. 1–19.

[9]   Solar Now. Solar PV report – IEA – renewables 2020 analysis and forecast to 2025. https://now.solar/2021/02/24/renewables-2020-solar-pv/ [accessed 28.03.21].

[10] Ren21. Renewables 2020 global status report. https://www.ren21.net/wp-content/uploads/2019/05/gsr_2020_full_report_en.pdf [accessed 28.03.21].

[11] Kumar NM, Yadav SK, Chopra SS, Bajpai U, Gupta RP, Padmanaban S, et al. Operational performance of on-grid solar photovoltaic system integrated into pre-fabricated portable cabin buildings in warm and temperate climates. Energy Sustain Dev 2020;57:109–18.

[12] Reddy VS, Kaushik SC, Ranjan KR, Tyagi SK. State-of-the-art of solar thermal power plants—a review. Renew Sustain Energy Rev 2013;27:258–73.

[13] Kumar L, Hasanuzzaman M, Rahim NA. Global advancement of solar thermal energy technologies for industrial process heat and its future prospects: a review. Energy Convers Manag 2019;195:885–908.

[14] Diwania S, Agrawal S, Siddiqui AS, Singh S. Photovoltaic–thermal (PV/T) technology: a comprehensive review on applications and its advancement. Int J Energy Environ Eng 2020;11(1):33–54.

[15] Buonomano A, Calise F, Palombo A. Solar heating and cooling systems by absorption and adsorption chillers driven by stationary and concentrating photovoltaic/thermal solar collectors: modelling and simulation. Renew Sustain Energy Rev 2018;82:1874–908.

[16] Sarbu I, Sebarchievici C. Review of solar refrigeration and cooling systems. Energy Build 2013;67:286–97.

[17] Mekhilef S, Saidur R, Safari A. A review on solar energy use in industries. Renew Sustain Energy Rev 2011;15(4):1777–90.

[18] IRENA. Solar heating and cooling for residential applications. https://www.irena.org/-/media/Files/IRENA/Agency/Publication/2015/IRENA_ETSAP_Tech_Brief_R12_Solar_Thermal_Residential_2015.pdf [accessed 03.04.21].

[19] Kumar NM, Chand AA, Malvoni M, Prasad KA, Mamun KA, Islam FR, et al. Distributed energy resources and the application of AI, IoT, and blockchain in smart grids. Energies 2020;13(21):5739.

[20] Pacca S, Sivaraman D, Keoleian GA. Parameters affecting the life cycle performance of PV technologies and systems. Energy Policy 2007;35:3316–26.

[21] Ito M, Kato K, Komoto K, Kichimi T, Kurokawa K. A comparative study on cost and life cycle analysis for 100 MW very large scale PV (VLS-PV) systems in deserts using m-Si, a-Si, CdTe and CIS modules. Prog Photovolt 2008;15:17–30.

[22] Kato K, Murata A, Sakuta K. An evaluation on the life cycle of photovoltaic energy system considering production energy of off-grade silicon. Sol Energy Mater Sol Cell 1997;47:95−100.

[23] Kannan R, Leong KC, Osman R, Ho HK, Tso CP. Life cycle assessment study of solar PV systems: an example of a 2.7 kW distributed solar PV system in Singapore. Sol Energy 2006;80:555−63.

[24] Battisti R, Corrado A. Evaluation of technical improvements of photovoltaic systems through life cycle assessment methodology. Energy 2005;30:952−67.

[25] Ito M, Kato K, Sugihara H, Kichimi T, Song J, Kurokawa K. A preliminary study on potential for very large-scale photovoltaic power generation (VLS-PV) system in the Gobi desert from economic and environmental viewpoints. Sol Energy Mater Sol Cell 2003;75:507−17.

[26] Akinyele D, Rayudu R, Nair N. Life cycle impact assessment of photovoltaic power generation from crystalline silicon-based solar modules in Nigeria. Renew Energy 2017; 101:537−49.

[27] Şerban A, Bărbuţă-Mişu N, Ciucescu N, Paraschiv S, Paraschiv S. Economic and environmental analysis of investing in solar water heating systems. Sustainability 2016;8(12): 1286.

[28] Eicker U, Colmenar-Santos A, Teran L, Cotrado M, Borge-Diez D. Economic evaluation of solar thermal and photovoltaic cooling systems through simulation in different climatic conditions: an analysis in three different cities in Europe. Energy Build 2014;70:207−23.

[29] Izquierdo M, de Agustín P, Martín E. A micro photovoltaic-heat pump system for house heating by radiant floor: some experimental results. Energy Procedia 2014;48:865−75.

[30] Roselli C, Sasso M, Tariello F. Dynamic simulation of a solar electric driven heat pump for an office building located in Southern Italy. Int J Heat Technol 2016;34:496−504.

[31] Khan J, Arsalan MH. Solar power technologies for sustainable electricity generation—a review. Renew Sustain Energy Rev 2016;55:414−25.

[32] Faraz T. Benefits of concentrating solar power over solar photovoltaic for power generation in Bangladesh. In: 2nd international conference on the developments in renewable energy technology (ICDRET). IEEE; 2012. p. 1−5.

[33] Ardente F, Beccali G, Cellura M, Brano VL. Life cycle assessment of a solar thermal collector. Renew energy 2005;30(7):1031−54.

[34] PV Watts calculator. https://pvwatts.nrel.gov/pvwatts.php [accessed 26.07.21].

[35] PVGIS. https://ec.europa.eu/jrc/en/pvgis [accessed 26.07.21].

[36] PV-Online. https://www.pvschools.net/schools/pvonline/home [accessed 26.07.21].

[37] PV*SOL. https://pvsol-online.valentin-software.com/#/ [accessed 26.07.21].

[38] Mayfield design tool. https://www.mayfield.energy/design-tool [accessed 26.07.21].

[39] Solar design tool. https://get.solardesigntool.com/ [accessed 26.07.21].

[40] i-Pals WEB. https://i-pals.nippon-control-system.co.jp/index_en.html [accessed 26.07.21].

[41] Kumar NM, Chopra SS, Rajput P. Life cycle assessment and environmental impacts of solar PV systems. Photovoltaic solar energy conversion. Academic Press; 2020. p. 391−411.

[42] Rajput P, Malvoni M, Manoj Kumar N, Sastry OS, Jayakumar A. Operational performance and degradation influenced life cycle environmental−economic metrics of mc-Si, a-Si and HIT photovoltaic arrays in hot semi-arid climates. Sustainability 2020;12(3): 1075.

[43] Pryshlakivsky J, Searcy C. Fifteen years of ISO 14040: a review. J Clean Prod 2013;57:115−23.

[44] Finkbeiner M, Inaba A, Tan R, Christiansen K, Klüppel HJ. The new international standards for life cycle assessment: ISO 14040 and ISO 14044. Int J Life Cycle Assess 2006;11:80−5.

[45] Hou G, Sun H, Jiang Z, Pan Z, Wang Y, Zhang X, et al. Life cycle assessment of grid-connected photovoltaic power generation from crystalline silicon solar modules in China. Appl Energy 2016;164:882−90.

[46] Luo W, Khoo YS, Kumar A, Low JSC, Li Y, Tan YS, et al. A comparative life-cycle assessment of photovoltaic electricity generation in Singapore by multicrystalline silicon technologies. Sol Energy Mater Sol Cell 2018;174:157−62.

[47] Koroneos CJ, Piperodos S, Tatatzikidis CA, Rovas DC. Life cycle assessment of a solar thermal concentrating system. In: Proceedings of the selected papers from the WSEAS conferences, vol. 2325. Cantabria, Spain.

[48] Batlles FJ, Rosiek S, Muñoz I, Fernández-Alba AR. Environmental assessment of the CIESOL solar building after two years operation. Environ Sci Technol 2010;44(9): 3587−93.

[49] Longo S, Beccali M, Cellura M, Guarino F. Energy and environmental life-cycle impacts of solar-assisted systems: the application of the tool "ELISA". Renew Energy 2020; 145:29−40.

Chapter 12

Emerging applications of solar energy in agriculture and aquaculture systems

Shiva Gorjian[1,2], Fatemeh Kamrani[1], Omid Fakhraei[1], Haniyeh Samadi[1] and Paria Emami[2]

[1]*Biosystems Engineering Department, Faculty of Agriculture, Tarbiat Modares University (TMU), Tehran, Iran,* [2]*Renewable Energy Department, Faculty of Interdisciplinary Science and Technology, Tarbiat Modares University (TMU), Tehran, Iran*

12.1 Introduction

The increasing trend of the global population has been doubled since the 1960s, and it is estimated to reach over 9.8 billion people by 2050 [1]. In this regard, nourishing the ever-increasing population and providing "Food Security" have been always global concerns. The statistics released by the Food and Agricultural Organization (FAO) in 2021 indicated that the number of people suffering from malnutrition has reached 690 million people [2]. To mitigate this issue, investing in the agriculture sector is more crucial than ever. Additionally, other global issues of climate change, water scarcity, and energy security have put excessive pressure on the agriculture and food production sectors, requiring more sustainable and eco-friendly agricultural operations [3,4]. One method to achieve this goal is to supply the energy demand of agricultural tasks using renewable energies that among them solar energy is the most abundant source with the highest adaptability with agricultural applications [5,6].

Although several pieces of research have studied the integration of conventional and modern agricultural operations with solar energy technologies such as solar-powered drying [7], solar-powered greenhouse cultivation [8,9], solar-powered irrigation and water pumping [10,11], solar desalination system [12,13], and solar-powered farm machinery [1,5], the investigation of some other emerging applications have been less considered. Therefore, this chapter presents a comprehensive study of some emerging applications of solar energy technology in agriculture, aquaculture, and food production

Solar Energy Advancements in Agriculture and Food Production Systems.
DOI: https://doi.org/10.1016/B978-0-323-89866-9.00008-0
425

systems. In this case, solar-powered crop cultivation techniques, solar-powered microalgae cultivation, solar-powered crop protection systems, solar aquaculture systems, and solar-powered pasteurization technologies are introduced and discussed. It is expected that investigating these technologies can assist in their global deployment as a step toward sustainable development achievement in the agriculture sector.

12.2   Solar crop production technologies

12.2.1   Soil solarization

Soil-borne diseases are considered as one of the main barriers in the production of economic crops such as vegetables as well as high-value ornamental plants. Soil disinfection has been historically performed using chemical treatments and mechanical methods as the most common approaches, causing environmental hazards and damages to beneficial microorganisms and plants [14,15].

Solarization, also called *solar heating*, *plastic mulching*, or *soil trapping* is a simple nonchemical method that uses solar energy to destroy soil-borne pathogenic and weed seeds in agricultural open fields as well as greenhouses before cultivation [16]. This technology was firstly developed in Israel in the mid-1970s as a preplanting soil treatment and hence was used for many crops such as tomato, strawberry, bell pepper, olive, lettuce, peanut, potato, coffee pulp, melon, and cucumber. The solarization is accomplished mainly during hot months of the year by mulching the moist soil surface. In this way, after sufficient irrigation, the soil surface is covered with uncolored polyethylene (PE) mulching material for 4−6 weeks [17]. During this period, the soil temperature is increased as about 85%−95% of solar penetrated radiation through PE sheets, resulting in killing soil-borne plant pathogens or greatly weakening them due to the excessive moisture state and high-temperature soil [15,17]. Fig. 12.1 shows the use of soil solarization treatment in open field farming and protected cultivation inside the greenhouse. To increase the solar heating effectiveness, enough solar radiation, favorable

FIGURE 12.1   Solarization treatment: (A) A metal-plastic greenhouse [18], (B) An open field cropland [19].

seasonal temperatures, mulching during high temperatures, and wet soil conditions are necessitous. Since the temperature of the soil is lower in shallow layers, the mulching process should be performed at the top 30 cm of the soil to control the pathogens as they are only destroyed at soil temperatures between 35°C and 60°C [20].

The most important factors affecting the soil temperature in the solarization process are the airtight state of the glasshouse, temperature differences between interior and exterior, as well as the duration of the process, and the climatic conditions under which the solarization is performed. In addition, the color and thickness of the PE sheet and the texture and moisture content of the soil can highly affect the solarization effectiveness [21,22]. To obtain the most solar insolation and reduce the longwave radiation return in this process, the color of the plastic film cover should be black, opaque, or made of translucent materials. But these are not suitable for solarization since the radiation could not pass through these materials and therefore, the underlying soil would not be warmed, causing the absorbed solar energy to be radiated back into the air with only a slight warming effect on the surface soil [17]. To solve this problem, plastic films mainly made of ethylene-vinyl acetate and low-density PE with the proven solarizing characteristics are employed in the solarization process [15]. Another effective parameter on the performance of the solarization is the soil moisture since soil organisms would be more sensitive to heat when the soil is wet, resulting in faster and deeper heat transfer into the soil. Additionally, the cellular activities of weed seeds and the growth of soil-borne micro-organisms are prospered by the soil moisture, making them more vulnerable to the lethal effects of high soil temperatures [14].

During solarization, the number of beneficial microbes is also increased over time, and therefore, solarized soils become more resistant to pathogens [23]. The solarization process has the best performance in tropical zones such as deserts and Mediterranean regions that possessing high availability of solar radiation with high summer temperatures [17]. Bonanomi et al. [24] investigated the effect of solar solarization (using biodegradable materials, plastic films, and organic amendments), and soil disinfestation (using fumigants) on soil chemical and microbial parameters, crop productivity, the incidence of soil-borne disease, and weed suppression. In this study, biodegradable spray, a mixture of polysaccharides at a concentration of 1.5%, and the addition of cellulose fibers for mechanical reinforcement were applied in the solarization treatment. The data of experiments demonstrated that the biodegradable solarizing materials have the potential in place of plastic films, but the technology of this treatment should be improved to be comparable in terms of performance with other pest management techniques. In another study by Oz et al. [25], to perform solar solarization, bubbles were filled up with water to concentrate solar radiation using the magnifying-glass effect of water film and create hot-points to stimulate heat transfer to deeper layers of the soil (see Fig. 12.2). This experiment was conducted in Isparta (Turkey) from July to

FIGURE 12.2 Cover material preparation for use in bubble water solarization treatment [26].

FIGURE 12.3 Accomplished soil solarization [27]. (A) In a multispan steel-framed closed greenhouse, (B) in a multispan wooden concrete greenhouse.

August 2011−12 for 40 days. The experimental area was medium and medium-fine textured, deep, and salt-free soil with weak profile development. The results showed that soil nitrate content is increased during solarization and may cause environmental problems that should be considered before planting. They concluded that using organic fertilizer simultaneously with solarization results in the formation of higher mineral nitrogen.

Castello et al. [27] evaluated the effects of soil solarization to control natural infections caused by phytopathogenic pseudomonads on tomatoes cultivated in different greenhouses under climate conditions of Sicily (south Italy) from 2010 to 2013. For this purpose, four experiments were conducted to evaluate the performance of both innovative and traditional films solely or combined with other control measures under wooden concrete and steel-made greenhouses against tomato bacterial infections (see Fig. 12.3). The results indicated that compared to bare plots, evaluated greenhouse covering and mulching films increase the soil temperature up to 9.6°C at 15 cm and 7.7°C at 30 cm, respectively. They concluded that solarization treatments are effective in bacterial infections, resulting in a yield increase of 45% in comparison with bare plots. In a study by Abd-Elgawad et al. [28], different aspects of soil solarization were evaluated against strawberry black root rot fungi. To perform evaluations, the cloth bags were artificially infested with single fungal species and buried into the soil in different depths of 1−10, 11−20, and 21−30 cm at three points of each plot. Then both disease incidence and severity in solarized and unsolarized soils were evaluated and obtained results were compared with the case of fungicide Actamyl application. It was found that solarization for 3, 6, and 9 weeks can considerably decrease the disease incidence and severity and increase the yield of the strawberry. They concluded that under the warm weather conditions of Egypt, solarization can be an effective and safe alternative for methyl bromide and other banned chemicals for the strawberry pests and disease management.

To examine the effect of using different mulch materials and the addition of biochar on soil solarization, several experiments were conducted by ÖZ [17] at the Suleyman Demirel University in Isparta during summer and autumn (July−August). The experiments were carried out under greenhouse conditions to determine the most effective month according to the temperature change of the soil and to evaluate the impact of soil solarization on the lettuce plant's quality, yield, and leaf nutrient content characteristics. For this purpose, two PE mulch materials of uncolored PE and uncolored bubbled PE were tested and four various PE mulch, bubble PE mulch, PE + biochar, bubble PE + biochar, and nonsolarization treatments were applied. The results indicated that the biochar added (150 g/m^2 of the soil surface) bubble mulching film solarization is more effective to increase the soil temperature and keep the heat compared to the traditional mulch material, while the bare soil temperature and the air temperature inside the greenhouse were not/less changed. Additionally, compared to the control case, a yield increase of 100% was obtained for lettuce under biochar added bubble solarization mulch. In areas with scarce irrigation water, most fields are left bare during hot summer months and therefore soil solarization can be a successful method to control weeds. In this case, farmers can cultivate different crops since planting in monoculture systems results in detrimental biotic, abiotic agents' accumulation, and soil sickness [29].

12.2.2   Solar composting

Composting is an eco-friendly method for recycling organic wastes and transmuting them into safe and beneficial products that can then be used as biofertilizers. In this process, the incorporation of heat and aeration provides desirable conditions for microbial growth [30,31]. To carry out the composting process, the existence of four main components including organic matter, moisture, oxygen (O_2), and bacteria is crucial [32]. Under ideal conditions, composting proceeds through three major phases of *mesophilic*, *thermophilic*, and *maturation*. In the initial mesophilic phase, the decomposition occurs by mesophilic microorganisms under moderate temperatures. As temperature rises, the second thermophilic phase initiates under higher temperature values between 50°C and 60°C by various thermophilic bacteria. At the final stage, in the maturation phase, as the supply of high-energy compounds abates, the temperature is reduced and the mesophiles predominate the process again. During the composting process, some offensive smells as a result of hydrogen sulfide (H_2S) are released and greenhouse gases (GHGs) such as CO_2, SO_2, and NO_2 are emitted [33]. The energy source that is utilized to perform forced aeration in composting piles and reactors is an important element in the composting process. Therefore, the introduction of solar energy may help to reduce the energy cost, causing a reduction in the total air emissions of the composting facility by replacing diesel generators (DGs) to drive the aeration equipment [30]. In this case, Lin et al. [34] constructed a laboratory-scale solar composting greenhouse with a volume of 1 m^3 (see Fig. 12.4) to evaluate the physicochemical and microbiological changes occurring during the food and green cellulosic waste treatment in fed-batch mode. In this compost house, the southern side of the roof was covered by double-hollow glass to receive solar radiation. The results indicated that the solar composting greenhouse performs well in the reduction of waste and retention of nutrients by decreasing the feedstock volume of 45.0%−58.8% over 12-day composting cycles, removing 41% of dry matter after three composting batches. Additionally, an increase of 29.5% in the nitrogen, 252.9% in the phosphorus, and 96.6% in the potassium content of the feedstock were observed after 42 days of composting. From the preliminary quality assessments of the compost, it was found that the solar composting greenhouse has a high potential of transforming organic waste into organic fertilizer.

In another study by Poblete et al. [31], the impact of using solar heat and aeration on the waste grass exposed to the composting process was

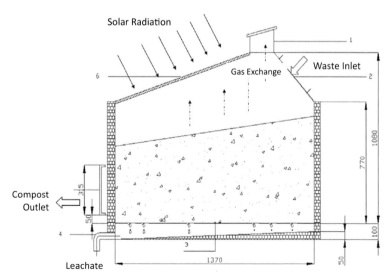

FIGURE 12.4 Schematic of the solar composting greenhouse [34].

evaluated. In this regard, compost piles were subjected to different processes including the application of solar-heated aeration, only-aeration, solar heating with a greenhouse, and control. During 70 days of composting process evaluation, different parameters of temperature, O_2, moisture, organic matter loss, and humification rate were monitored. The comparison between the results obtained from the case with solar-heated air introduced to a compost pile of grass clippings [through polyvinyl chloride (PVC) pipes], and the greenhouse compost system indicated that the highest system temperature of 68.2°C is obtained for the compost pile with the greenhouse on day 15. However, after 70 days of composting, the carbon to nitrogen (C:N) ratios were decreased by 20% and 15% for the greenhouse and the system subjected to the solar-heated air, respectively. Additionally, it was reported that despite the higher temperature of the solar-heated air in comparison with the greenhouse, the required thermophilic temperature levels cannot be reached in the aerated compost pile, representing an excessive aeration cooling effect even with the heated air.

The amount of rural domestic waste with a complex composition and low classification has been increased in recent years due to the rapid population growth and improvement of living standards. In this regard, Chen et al. [35] developed an efficient low-energy aerobic composting reactor to produce organic composting material in which solar energy was used as an auxiliary energy source to supply a part of the required power of the setup (see Fig. 12.5). In this study, considering the effect of ventilation on the efficiency of the rural domestic waste composting, an innovative method of drum hot-air aeration oxygen was developed for use under the cold winter environment. The composting heap's temperature rise is affected by the natural air in cold environments, causing adverse impacts on the composting efficiency. Ensuring the normal operation of aerobic composting, the blowing air was heated and could significantly improve the biochemical reaction rate and the microbial community distribution, solving the low-efficiency problem of conventional composting.

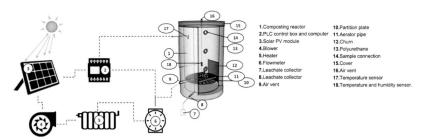

FIGURE 12.5 Schematic diagram of a solar aerobic composting reactor. *Adapted from Chen W, Luo S, Du S, Zhang M, Cheng R, Wu D. Strategy to strengthen rural domestic waste composting at low temperature: choice of ventilation condition. Waste Biomass Valoriz 2020;11:6649–65. https://doi.org/10.1007/s12649-020-00943-4 [35].*

The steady usage of natural fertilizer (*N*-fertilizer) can produce an enhanced yield of up to 20%−30% for the cultivation of all types of crops. In agricultural lands, the conversion of sawdust into *N*-fertilizer is an eco-friendly and low-cost process with a high root absorption rate in agricultural cultivation that can eliminate the widespread use of the existing chemical fertilizers. In this regard, Rajkumar [36] studied a sawdust recycling process using forced air circulation driven by solar power to reduce the composting time. The results indicated that by the composting method and controlling the compost parameters such as pH, temperature, moisture, forced air circulation, and C:N ratio, sawdust can completely be converted into (100%) natural fertilizer. It was also reported that the final compost product is obtained within 32 days in the forced air supply method (-aerobic method) and 47 days in the conventional method with few turns of compost (anaerobic method).

12.2.3 Solar corridor crop system

High inputs of synthetic fertilizers and pesticides cause a reduction in biological and ecological interactions that threaten the ecosystem because of changing the level of nutrient and pest populations. The environmental efficacies associated with mentioned problems have raised concerns about industrial agricultural sustainability. In this regard, some efforts have been made to alternate other management practices to ensure economic and environmental sustainability [37]. Crop production can be improved by effective utilization of sunlight and carbon dioxide (CO_2) since plant productivity highly depends on the photosynthetic process and the efficiency of the sunlight conversion into various products [38]. Growing two or more crops in alternating strips is called intercropping which allows for biological interaction of contiguous crops and includes relay cropping, interseeding, companion planting, and smother cropping. Intercropping can improve the ecosystem processes in comparison with mono-cropping. The solar corridor crop system (SCCS) is an alteration of strip intercropping in which two or more crops are cultivated simultaneously in different strips across the field to enhance the usage of solar radiation, water, and nutrients [39,40]. An example of the solar corridor model, proposed by Deichman [41] is depicted in Fig. 12.6 The consequence of interacting two or more crops in the SCCS is that the growth and yield of the main cash crop are improved and soil productivity maintenance is promoted, resulting in additional forage, grazing, or grain benefits provided by cover or secondary crops established in the corridor [37,43].

The only energy source to produce photosynthates for field crops is solar radiation. Changing the concept of the solar corridor might be beneficial in aerodynamic resistance of the canopy by the arrangement of them which govern the rate of CO_2 exchange to the leaf. A substantial change like a solar corridor in cropping systems requires plant hybrids with optimal response to these changes. The fertility could be increased through crop canopies

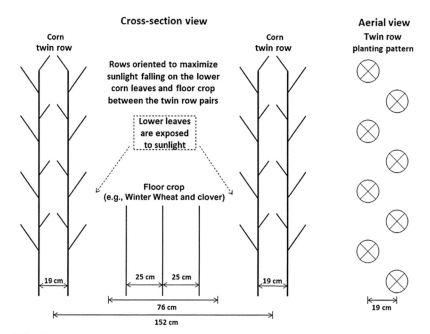

Cross-section view

Corn
twin row

Rows oriented to maximize
sunlight falling on the lower
corn leaves and floor crop
between the twin row pairs

Lower leaves
are exposed
to sunlight

Floor crop
(e.g., Winter Wheat and clover)

Corn
twin row

Aerial view

Twin row
planting pattern

19 cm   25 cm | 25 cm   19 cm

76 cm

152 cm

19 cm

FIGURE 12.6   Solar corridor crop system design [42].

manipulation to capture light and maintain CO_2 levels throughout the canopy to be nonlimiting, resulting in a high efficiency [38]. A solar SCCS includes narrowing crop row widths to capture sunlight and CO_2 or using a wide-row spacing of twin narrow rows for tall-stature crops like corn to form a uniform corridor and a vertical distribution of incident sunlight which is available for whole parts of the leaves. A solar corridor planting arrangement incorporates increased sunlight availability to plants in all rows in the field. Since the photochemical processes that are required for photosynthesis initiate with light energy, solar radiation can be the most productive, low-cost, and abundant form of light for these processes [37].

12.2.4   Phototrophic microalgae cultivation

Algae are various categories of aquatic organisms with the capability of conducting photosynthesis to efficiently convert solar energy, representing a significant group of organisms for biotechnological exploitation. Considering their size, they are classified as (1) macroalgae, and (2) microalgae [44]. It has been estimated that algae include different species from 30,000 to more than 1 million. Although the primary habitats of algae are freshwater and marine environments, some of them also grow in soils, rocks, caves, glaciers, and even buildings. Blue-green algae are used to fix atmospheric nitrogen and are utilized as biofertilizers in rice fields. Algae are also used in pisciculture as food for fish, where some specific species are mostly used as feed for cattle and poultry [45].

12.2.4.1 Microalgae

Microalgae are single-celled or multicellular organisms with microscopic dimensions that require water, CO_2, and light to live and grow. With an estimate of 72,500 species, they are of great ecological importance with the contribution to half of the global organic carbon fixation. They can conduct photosynthesis due to their chlorophyll content with 50% of their dry weight composed of carbon. Microalgae have the potential of transforming 9%–10% of solar energy into biomass with a theoretical yield of about 280 ton/ha/year [46,47]. Microalgae can uptake CO_2 from both the atmosphere and flue gas emissions, converting it to biomass or other organic compounds. For those living in oceans, due to their photosynthetic activity, they can lead to an annual CO_2 fixation of 45–50 Gton in this environment. Currently, due to identified features of microalgae, they can be used as feed material for biofuel production (see Fig. 12.7). Additionally, microalgae synthesize various chemical intermediates and hydrocarbon that can be converted into different fuels such as alcohols, diesel, methane, and hydrogen [44,49]. The conversion of CO_2 into biomass by photosynthesis can be described in a

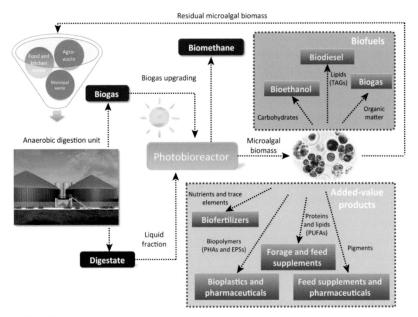

FIGURE 12.7   Microalgal biomass as feedstock for biofuels production and other added-value bioproducts [48].

simplified way as presented in Eq. (12.1). It is estimated that 1 kg of microalgae biomass to fix 1.8 kg of CO_2 [50]:

$$6CO_2 + 6H_2O \rightarrow C_6H_{12}O_6 + 6O_2 \tag{12.1}$$

Large-scale microalgae cultivation is very cost-effective for biomass production and it has a high photosynthetic rate. Producing O_2 by uptaking CO_2 via photosynthesis makes microalgae one of the most productive biological systems that play an essential role in natural ecosystems due to their capability in the fixation of CO_2, contributing to atmospheric equilibrium [48]. The microalgae's potential of interest aspects are as follow [51]:

- Ability to grow in marginal areas;
- Utilization of wastewater as a growth medium that affects wastewater treatment;
- Significant fast growth rates compared to terrestrial plants, doubling their biomass in a day (or less);
- Ability to sequester flue gas emissions under suitable conditions.

12.2.4.2 Requirements for photosynthetic microalgae production

Similar to other microorganisms such as bacteria or fungi, microalgae's productivity and growth rate entail adequate culture conditions. In this regard, the most important parameters are light, nutrients, pH, and temperature that must properly be supplied to the cells [52].

Light availability: Light has a dominant effect on the growth and productivity of microalgae and is considered as the energy input for photosynthetic microalgae that can maximize the yield. Nonetheless, excess reception of light can harm the photosynthetic apparatus, especially when it combines with temperature or O_2 levels over the accepted limits. To supply sufficient light for the cells, choosing proper geometry and orientation of the culture environment is crucial. The photosynthesis rate determines the growth of microalgae and is a direct function of the irradiance received by cells inside the culture [52]. Photosynthetic microorganisms can only utilize photosynthetically active radiation (PAR) in the range from 400 to 700 nm. Thus, using different light sources (lamps, LEDs, sun), only PAR is utilized to perform photosynthesis. Since the light distribution inside the microalgae cultures is not homogeneous, the outer part of the culture is directly exposed to the sun, while the inner part can be quite dark. Therefore, an average irradiance concept is defined as the volumetric integral of the corresponding irradiance in all the points inside the culture. The average irradiance at which the cells are exposed in microalgae cultures ranges from 10 to 1000 $\mu E/m^2$ s [53].

Nutrients supply: The main components of microalgae biomass are carbon (30%−50% dry weight), O_2 (30%−50% dry weight), hydrogen (H_2) (3%−7% dry weight), nitrogen (N_2) (4%−9% dry weight), phosphorus (P) (1%−3% dry weight), and minor values of other components such as sulfur (S), potassium (K), magnesium (Mg), and calcium (Ca) [52]. All these constituents are required to maximize the performance of the cultures. Generally, the nutrient-excess conditions can enhance the productivity of cultures, but both economic and sustainability criteria should also be taken into account since the excess nutrients are commonly released and lost from the system if the recirculation for the cultivation medium is not available [54]. To keep up the microalgae activities, the CO_2 shortage should be avoided. The carbon can alternatively be supplied as bicarbonate or carbonate to the cultivation medium. Also, the higher levels of O_2 above the saturation air (0.2247 mol O_2/m^3 at 20°C) could stop the photosynthesis of some species of microalgae. A severe photo-oxidation can occur at high concentrations of O_2 combined with a high level of received radiation [55].

Culture conditions: Similar to other organisms, microalgae require a proper culture environment for growth, and deviation from the optimal conditions can considerably decrease the yield. Temperature and pH are the most important variables with the optimum pH values ranging from neutral to slightly alkaline (7.0−10)[1] [56]. The most common way to control the pH value inside the microalgae cultures is the CO_2 injection which can simultaneously provide the required carbon for the culture medium. However, the pure CO_2 supply is associated with up to 30% of the overall production cost of microalgae. The carbon loss in photobioreactors (PBRs) has been reported as 67% in open systems to 50% in closed environments. The temperature increase of microalgae culture mediums mainly occurs because of the absorbtion of light. The optimal temperature range for microalgae growth is between 20°C and 35°C, although some species can tolerate the temperature up to 40°C. Therefore, the employment of temperature control, especially in large-scale reactors, is crucial [57].

Mixing: The existence of nutrients gradients in microalgae cultures can be minimized by providing sufficient turbulence inside the medium, avoiding sediment creation, forcing the cells to move between dark and light zones, and enhancing photosynthesis. The turbulence creation is performed using several methods including bubbling, liquid circulation via the pumps, and stirring. However, the mixing is associated with additional costs that should be optimized. Using mechanical and pneumatic devices can harm the cell, affecting the culture performance. Bubbling the air can also be damaging because of the supplied energy especially

---

1. The optimum value for some species is below 3.0.

when bubbles are small. Similarly, the use of mechanical stirring devices can create stress as a function of the type and speed of the stirrer. In liquid circulation using pumps, damage can occur due to the interaction with the wall of the reactor [52].

12.2.4.3 Microalgal cultivation systems

Microalgae can be produced in outdoor conditions either in open pond systems or closed PBRs. Open pond cultivation offers the simplest and cheapest approach for the phototropic production of microalgae. The most common types of open cultivation systems are large shallow ponds, tanks, circular ponds, and raceway ponds [45]. A raceway pond is a rectangular canal in which the algal culture current flows from one end (supply) to the other end (exit). In raceway ponds, the length-to-width ratio is an important parameter that should be optimum since larger values may cause weak current speed which is not desirable for mass transfer and mixing. A paddle wheel is also used to create a steady water flow inside the pond [58]. In open pond cultivation systems, only a few algae species can be grown since there is no control over temperature and light and the microalgae growth is a function of the climate conditions of the location [59]. The main constraints of open pond systems are their poor light utilization by the cells, low productivity (<10 g/m^2/day), high water demand, low cell densities ($^\cdot$0.5 g/L), high probability of contamination, and the requirement for a large space [60].

In contrast, PBRs are vessels with transparent walls that absorb light and produce live microalgae masses through photosynthesis. A PBR is considered as an artificial environment that provides a favorable environment in a fully controlled manner for respective microalgae species, allowing for much higher growth rates and purity levels than anywhere in natural habitats. In a PBR, the phototropic biomass can be derived from nutrient-rich wastewater and CO_2 flue gas [61]. For the most effective utilization of light and surface area, an appropriate configuration of a PBR should be established. As mentioned before, biomass mixing is also required to ensure the homogenous distribution of nutrients and light in the culture medium. To achieve optimal production, a PBR must have the following characteristics [62]:

- *Sufficient levels of light absorption*: Increasing the light absorption enhances productivity.
- *Proper aeration*: To perform photosynthesis, CO_2 in the culture medium should be sufficient. The higher the amount of water-soluble CO_2, the higher the productivity.
- *Proper discharge of generated O_2 from the culture medium*: Accumulation of excess O_2 in the reactor causes light respiration, reducing the efficiency of photosynthesis and production.
- *Adjustable temperature*: The temperature of PBRs should be adjustable to prevent excessive temperature increase or decrease in different seasons.

- *Constant circulating water flows inside the reactor*: The water flow circulation causes the algae to move between the deep dark environment inside the reactor and the light environment on the surface, which in turn increases the photosynthetic capacity of microalgae.
- *Having no adhesion*: Sedimentation and accumulation of algae cause the reactor to malfunction and removing them causes additional problems.
- *Reasonable and practical price*: Reactors should be designed in a way to be implemented under any conditions with reasonable costs.

Compared to open ponds, PBRs require less water, reduce the risk of contamination by birds and dust, and can use artificial light instead of solar radiation. The cost of these systems is high that can be compensated somewhat by their higher productivity and the higher quality of the final product. There are different designs of PBRs utilized mainly for microalgae biomass production including laboratory fermenters, tubular, bubble column, and flat-plate PBRs [62] as shown in Fig. 12.8. Comprehensive studies on different designs of microalgae cultivation systems can be found in Refs. [45,67]. Table 12.1 presents the main benefits and limitations of open and closed microalgae cultivation systems.

12.2.4.4   Economics and prospects

The cost of large-scale biomass production is affected by the cost of PBRs when the closed-type reactors are employed. In this regard, their cost reduction can significantly decrease the final cost of biomass production. The cost

FIGURE 12.8   (A) Prefabricated algae raceways [63], (B) bubble column array photobioreactor [64], (C) tubular photobioreactor [65], and (D) flat-plate photobioreactor [66].

TABLE 12.1 Benefits and limitations of open and closed microalgae production systems [61].

Type of PBR	Benefits	Limitations
Open reactors	• Ten times more economical to build and operate; • Lower energy consumption; • Net energy producer; • Rapid growth of microalgae and resistance to predators; • No accumulation of dissolved oxygen (DO); • Industrial-scale microalgal production in several applications.	• Low biomass productivity ($10-20 \ g/m^2/d$); • Large space is required; • No control on microalgae strain growth; • Poor mixing and weak light and CO_2 utilization; • Cultures are easily contaminated; • Temperature and evaporation cannot be controlled.
Closed reactors	• High biomass production ($20-40 \ g/m2/d$); • Allow the growth of monocultures; • Fewer contamination problems; • Can distribute the sunlight over a large surface area; • Evaporation can be avoided; • Lower space is required; • Flexible technical designs.	• High construction cost; • High energy demand; • Removal of O_2 is crucial; • Higher electricity demand for the mixing and pumping; • Some degree of biofilm growth; • Difficult to scale up.

reduction approaches depend on the type of the PBR, the production technology of biomass, and the type of the algal strain. The major cost factors in PBRs are mixing, microalgae's photosynthetic efficiency, the medium, and the CO_2 costs since the latter is associated with the highest cost in biomass production [45]. In this regard, one method to eliminate/reduce the CO_2 cost is using flue gases from industrial sources. Additionally, the use of wastewater containing mineral nutrients can significantly reduce costs. For biomass production, the calculated cost for three commercial-scale biomass production systems is €4.95 for open ponds, €4.15 for horizontal tubular PBRs, and €5.96 for flat-plate PBRs, all per kg of dry biomass. Using flat-plate or tubular PBRs can reduce the total costs to €0.68 and €0.70/kg of dry biomass, respectively while employing raceways cannot reduce the costs less than €1.28/kg [68]. As a general solution, the best way to reduce the costs of PBRs is to enhance their productivity. Additionally, the large-scale PBRs with more than 150 ton/ha/year biomass production must be operated with low labor costs, using wastewater as a cultivation medium, and utilizing flue gases as the carbon source as much as possible [45]. Microalgae are considered as a sustainable feedstock for important components in food, biofuels, health, and livestock industries. However, producing biomass on a large scale is still an impassable challenge that should be solved considering three main aspects of technology, economy, and ecology. PBRs can be used as the best alternative for the production of high-quality microalgae, but the employed strategies should be efficient and affordable. In this case, there should be a balance between the energy demand and cost when PBRs are designed. Additionally, the PBRs should be adjustable and flexible enough to become suitable for the cultivation of various microalgae species and be employed in different areas with specified climate conditions.

12.3   Solar-powered crop protection systems

12.3.1   Solar-powered bird repellers and pest controllers

In natural ecosystems, birds affect the agri-ecosystem to be survived. Since their attacks are very detrimental to the farmers, they can be considered pests. Some types of birds like pigeons, sparrows, starlings, common myna, jungle myna, crows, and blackbirds, not only give damage to the agricultural farms but also cause problems in human life. To overcome this problem, some mechanical and chemical bird repellents have been introduced. Bird control techniques involve physical or visual deterrents, sonic devices, trained birds of prey (falconry), chemicals, contraceptives, and active barriers. Generally, a bird repellent is a device usually employed by farmers to scare birds, and preventing them from eating planted crops. In this way, audio and visual threats are generated through electronic bird repellers to frighten, irritate, and disorient birds [69]. The economic threat to the planted crops which is caused by birds in farms or at storage facilities requires the deployment of an effective bird repellent. The ultrasonic waves[2] have been suggested in different studies as a novel technology to effectively repel pesky birds from determined places [70]. A model and prototype of an automated bird detection and repeller system using the internet of things (IoT) was proposed by Riya [71] that could be erected on the center of a field track on a raised platform, preventing the birds from entering the field within the radius of approximately 12 m. In this study, the designed model could perform two main functions of motion detection using passive infrared (PIR) and predator's sound generation using an MP3 module and megaphone to scare the birds away. To take the advantage of using solar energy, solar-powered bird repeller electronic devices are designed to keep the farms safe. A solar-powered ultrasonic electronic device can generate, amplify, and broadcast waves in varied frequencies (between 15 and 25 kHz) at a sound pressure level that is high enough. An example of a solar-powered ultrasonic

---

2. Ultrasound is sound waves with frequency higher than the upper audible limit of human hearing.

FIGURE 12.9   (A) A solar-powered ultrasonic bird repeller [72] and (B) solar-powered bird repeller with laser technology [73].

bird repeller (DC 4.5 V, 200 mA) is shown in Fig. 12.9A. The ultrasonic waves create a hostile environment for pest birds with a repellent effect on them. Despite the small radius of the used device, the birds are drowning away from the designated locations [3]. The laser technology can also be adopted as a bird repellent using the instinct of birds. In this method, birds understand the approaching laser beam as a predator and take the flight to seek safety. A novel solar-powered automated laser-based bird deterrent installed in a blueberry farm in Oregon, USA integrating with a photovoltaic (PV) module and battery storage is shown in Fig. 12.9B [73].

Using loud noises for bird scaring usually would make the birds flying far away and stay on a safe perch that might be quite close to the area where they can cause problems. Using distress call bird scarers causes the birds to move further away from the call source as well as the entire inconvenient area. For this purpose, Muminov et al. [74] developed a solar-powered audible bird scarer with an effective operation. To examine due to which sound various species of birds get deterred, different sounds were noticed and studied. The main components of the proposed scarer are a 12 V PV module, an intelligent pulse-width modulation (PWM) solar charge controller, a 12 V battery, an MP3 player, an amplifier, two 20 W speakers, three sonar or PIR sensors, the Arduino UNO microcontroller, and a battery in which the produced electricity by the PV module is stored. Electric fences are other devices that can be utilized as barriers to protect agricultural farms from entering wild animals. In 1888, the first electric fence was developed and installed in Texas. In this system, an overshot wheel was utilized to generate the electricity to charge the top two wires of a four-wire fence. In electric fences, the required electricity can be provided by PV cells and stored in batteries for operations at night or cloudy days as shown in Fig. 12.10. Depending on the type of animal, vegetation load, fence length, and power source, the converter unit produces a different range of voltage pulses [76].

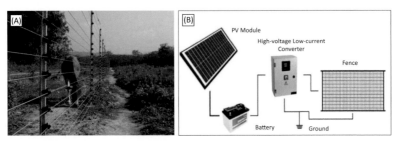

FIGURE 12.10 (A) Solar-powered electric fence installed around a farm [75] and (B) schematic diagram of a photovoltaic-powered electric fence [3].

FIGURE 12.11 (A) Solar-powered automatic pest control system [78] and (B) solar-powered automatic insect trap installed in the cacao plantation [79].

In a study by Nair et al. [77], a solar fence was developed and installed around to protect the farm by preventing the animals to enter. In the proposed system, short electric pulses were transmitted through the fence to create a short shock, causing the animals that intend to enter the farm to be threatened and gone away from the fence. The system was designed in a way to uses solar energy as the power source. They claimed that it is a low-cost system assisting to increase the production of crops.

The crops may also be at risk of pests' attack from planting to maturity stages. The losses associated with pests have been estimated to lie between 5% and 40%. Okoronkwo and Onu [78] developed a solar-powered automatic pest control system as shown in Fig. 12.11A and evaluated the effect of deploying the proposed system on the yield of rice. In the proposed system, three principal signals of motion, sound, and light were at determined intervals to simulate the presence of humans to frighten rodents and bird pests from the farm. In this device, the solar power produced by PV cells was used to drive mechanical, acoustic, and photo devices. The results indicated that the device can improve the rice yield by prohibiting 49% loss previously caused by rodent and bird pest attacks during 6 months of deployment and monitoring.

The steady use of pesticides can result in pest resistance, pest resurgence, and environmental pollution, disrupting the ecosystem due to an increase in toxic residues in plant tissue as well as the soil. To address this issue, some simple technologies such as lamps, yellow binders and attractants have been utilized to repellent the insects. Since the insects are desired to gather around light sources, lights can act as flying pests catchers in agricultural lands. In this regard, Telaumbanua et al. [79] designed a pest trap with an automatic actuator using a microcontroller (Fig. 12.11B). In this automatic insect trap, five units of the infrared sensor (type E18-D50NK) were used to detect insect pests. The TL lamps, yellow lights, and attractants were attached to the trap system. The results indicated that the catching accuracy of the system is 82.74%, insect drop time is 6 min and 33 s, and the actuator response speed turns on the lamps, yellow LEDs, and pumps in ± 10 Ms. They claimed that the proposed system can reduce the utilization of spray pesticides by 20%−50%, saving the purchasing costs of pesticides up to US $5.14/ha of Cacao cultivated land.

The high-frequency and the electromagnetic devices are two versions of repellent devices. High-frequency devices are applied by transmitting sound waves greater than 20,000 Hz. It is hard for pests to withstand when the power of the sound wave is sufficiently large. So, the pests are repelled away from the area without affecting the environment and humans [80]. The insects recognize noise through unique hairs or sensilla positioned upon the antennae (mosquitoes) or genitalia (cockroaches), or through more complex tympanal body parts. Fixed frequency generation by conventional ultrasonic insect repellent system might be a bug because it is not capable of repelling all kinds of insects. The electromagnetic devices emit electromagnetic waves which are unfavorable to pests. Conventional noise systems and chemical pesticides used to repel the pests make sound and air pollution which is harmful to human beings and the environment. Nowadays, smart ultrasonic insect repellent devices are developed and equipped with solar charging units to control the attack of insects on crops during the night. This technology will overcome the farm's manually controlling problems [81].

12.3.2 Solar-powered sprayers

In the agriculture sector, one of the most important operations which is crucial to protect crops from the attack of insects and achieve the highest yield is the spraying of pesticides [79]. In general, hand-operated, engine/fuel operated, and electric motor pump-based are three types of sprayers using in farms considering the type of the cultivated crops [82]. Having some difficulties while spraying is one of the problems associated with the use of hand-operated sprayers. In engine/fuel operated sprayer, petrol is mostly used as the fuel which is not a proper option in terms of the economy. The electric motor pump sprayers store the electricity in their embedded batteries which are used to drive the spray pump [83]. The electric sprayers can be driven using the electricity produced by PV cells and store the generated DC power in embedded batteries to run the pump. A solar-powered pump used in a pesticide sprayer has many benefits over the petrol-powered engine sprayer's pump as the motor is associated with lower operation and maintenance costs with less environmental impacts, resulting in a considerable saving on the consumption of the fuel/petrol. Solar energy is absorbed by PV cells and is converted to DC electricity and stored in batteries. The battery runs the motor pump to spray out fertilizers or pesticides from the tank [82]. The use of solar-powered pesticide sprayers satisfies the farmers because of reducing maintenance costs, using less electrical power and fuel, and reducing running costs and time. In some regions, where there is no easy accessibility to the fuel, the solar-operated sprayers will help the farmers to do pesticide spraying tasks with very little environmental pollution [84]. A typical solar-powered sprayer consists of PV modules, a motor, a control system, and a spray lance with a spinning disc. The PV module absorbs the solar radiation and converts it into useful power which is applied to accomplish the spraying task. A solar charge controller is also utilized to arrange the PV modules' voltage and current values. The controller is located between the PV module and the battery. Maintaining the proper charging voltage to the battery could be met by the controller which helps overcharging/discharging protection [85]. Two samples of solar-powered pesticide sprayers are depicted in Fig. 12.12.

The backpack solar-powered sprayers have received considerable attention from farmers due to their ability to improve the quality of spraying and

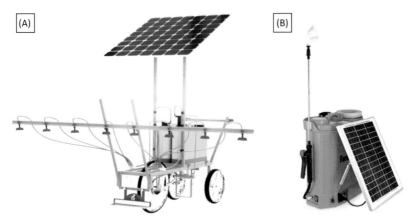

FIGURE 12.12   (A) Solar-powered multinozzle pesticide sprayer [86] and (B) Solar battery-operated knapsack sprayer [87].

FIGURE 12.13   Photo of the developed solar-powered backpack sprayer [88].

reducing the operator's physical efforts. But there are main drawbacks to using solar sprayers in the field which are as follows: (1) the duration that batteries can keep the charge, and (2) the way that batteries are commonly charged. In this case, Shigueaki Sasaki et al. [88] developed a solar-powered backpack sprayer and evaluate its operation in the field. For this purpose, an electric backpack sprayer, MTS brand, model Spritz 18, was employed and an aluminum frame in the form of the "tilt window" was built to allow the PV modules to achieve the maximum solar intensity (see Fig. 12.13). Some experiments were conducted to evaluate the instantaneous power generation of the modules during both static and in the movement by the operator. They reported the average generated instantaneous power as 1.4 and 2.18 W in movement and static state, respectively. They also asserted that by optimizing the longevity of the battery, the sprayer can be utilized in remote or distant locations and charged with solar energy whenever it is required.

12.4   Applications of solar technology in aquaculture systems

According to the report released by the World Fish Centre in 2020, total global capture fisheries production in 2018 reached the highest level ever recorded at 96.4 Mt [89]. The object of global policies related to food production systems is to alleviate the pressure on oceans. The ultimate solution to achieve this goal is aquaculture-based fish farming to satisfy the food requirement of the growing population without destabilizing nature. Fish farming is one of the most beneficial businesses for investment in due to being economical, diverse (with over 500 different fish species), and more healthy compared to poultry and livestock breeding [90]. Recently, fish cultivation has become a valuable activity, especially in developing countries to promote food security and overcome poverty. It has been estimated that hundreds of kilograms of fish cultivation are required per day to feed the fish farming business [91]. The home industry scale bulk feed business is quite promising with a bright perspective for fish farmers in various fish cultivation centers, as long as it guarantees the quality of local raw materials on an ongoing basis [92]. Fish is a kind of seafood that can be considered a valuable source of specified proteins such as essential fats and amino acids as well as minerals which are crucial to gain a healthy balanced diet. However, several factors affect fishery production and the efficiency of the marine systems including the tools and machinery utilized to manage fisheries [91].

12.4.1 Demand fish feeders

One of the most important challenges to the proper design of fish feeding devices is providing artificial feeds by considering the fish's requirements, maintaining a suitable cultural environment. In most automatic fish feeders, adjusting the precise time and amount for feeding as a function of the fish's appetite is tough work and if not perform properly, food wastage, overfeeding, and the pollution of the water pond would occur. Regarding this, the employment of self-demand feeders can be a preferable solution in which the fish themselves decide about the time and amount of food they require [93]. A fish feeder is composed of two main parts: (1) a rotating food hopper and (2) an asynchronous motor and gearbox arrangement to drive the hopper in which the bottom of the hopper fits with the top of the motor housing (see Fig. 12.14). As the hopper is rotated, the pellet bins in its base pass sequentially over the pellet delivery hole. The hopper is retained in its flush-mounted position by a sprung bearing which is mounted inside the top of the motor housing and through which the hopper spindle passes. In a typical demand feeder, feed drops by gravity as a direct function of the angle of repose[3] [91]. Both angles of repose and inclination[4] are the most important parameters that affect the design of feeding devices for aquaculture purposes. For a demand feeder, the mechanism of activation is composed of a feed platform, steel bait rod, feed protecting cover, and a pendulum or feeding tray in case of sinking feed. As mentioned above, when the fish activated the rod, the feed drops by gravitational force into the adjustable acrylic feed platform located at the bottom of the hopper and above the water level. The bait rod is suspended from the conical tube on a V-shaped steel wire that holds the rod. The feeder can be suspended from a pipe stand with the activating mechanism extending into the water through the conical portion of the hopper [94].

The merits of demand feeders are being (1) low-cost with easy fabrication, (2) resistant to corrosion, (3) maintenance-free, (4) easy to install and operate, (5) suitable for ponds/tanks/raceways, (6) stable for adverse climate, and (7) suitable for advanced fingerlings and above size [94].

---

3. Steepest angle of descent (0 and 90 degrees) relative to horizon in which materials can be piled without slumping.
4. Angle between a line and x-axis (0 and 180 degrees) which is measured counterclockwise from the x-axis to right of the line.

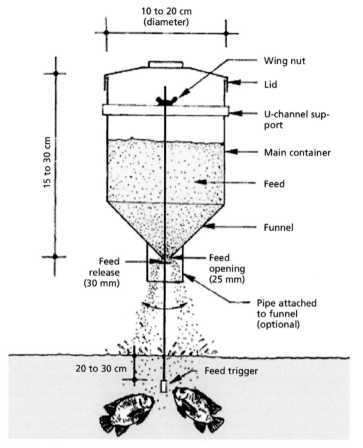

10 to 20 cm
(diameter)

15 to 30 cm

Wing nut

Lid

U-channel sup-
port

Main container

Feed

Funnel

Feed
release
(30 mm)

Feed
opening
(25 mm)

Pipe attached
to funnel
(optional)

20 to 30 cm

Feed trigger

FIGURE 12.14   Diagrammatic representation of a demand feeder [93].

12.4.2   Automated fish feeders

Preservation and conservation of fish are essential since overfishing has destructive effects on ecosystems, causing fish biomass to be declined. In this regard, fish pond ecosystems can be improved by employing a smart self-sustaining system. For this purpose, several systems have been proposed to manage/monitor biological parameters including feeding in fish ponds. Additionally, some other key parameters are pH, water temperature, water salinity, DO, ammonium, and nitrates. In terms of costs, in marine fish farming, feeding accounted for 60% of the total cost, where 8.26% is allocated to food loss [95].

An automated fish feeder is an electronic device that can be fixed or mobile. This system is designed in a way to dispenses proper amounts of pellets at a specified time. Besides, because of their capability to repeat the task accurately every day, they are considered promising devices with high efficiency and productivity in the fish farming field in long term [96]. One of the most important tasks involved in an automated fish feeder is time management to determine when is the fish's mealtime. The mealtime depends on some specific conditions such as type of fish, size of the pond, the density of the fish, and several other parameters which are selected by the programmer to set the time. Maintaining a regular feeding schedule is not an easy task, but to keep fish healthy, taking care based on some principle rules is required, otherwise, a significant loss will occur due to starvation [92,97]. The benefits of automatic fish feeders are, fewer workforces for the owners in handling specified tasks, for example refilling the pellet, cleaning the feeder, and even repairing or maintaining the system. All of these require less time and energy compared to the traditional fish feeding systems. The manual feeders involve difficulty in the management of the whole feeding schedule, especially in large environments [98]. An automated feeder should be capable of working without the need for human supervision even at a certain time interval. There are several different designs and brands of automatic fish feeders on the market that some of them feed the fish by using a timer set by the user and are more suitable for use in modern aquariums [92].

A sustainable aquaculture feed system (SAFS) equipped with a machine vision technology was developed by Lee et al. [99] to estimate the amount of food required by the fish based on their appetite, preventing waste of nutrients. They claimed that this technology can be used to count the number of fish and deduce their size which are critical parameters in determining the required amount of food, ensuring that a proper amount is released. They also noted that the incorporation of a graphical user is essential to provide a data analysis structure to study patterns in the tank [100]. If the physical characteristics of fish are deduced, the farmers would be able to release a proper amount of food in the pond and prevent wastages (see Fig. 12.15). Time management is another crucial parameter that should be considered in the design of fish feeders as the fish stock must be fed at proper times, ensuring that optimal nutritional value is derived from the foods [101].

12.4.3   Solar-powered fish feeders

Solar PV modules can be used to power fish feeders as mobile power. Solar-powered fish feeding technology can satisfy the requirements of ease, comfort, and speed in the providing of fish feed in fisheries ponds. This system can be installed in a limited area, easily moved, or transported to other locations only by truck/pickup [102]. A fish pond management system is generally composed of three main components of (1) measuring node, (2) base node, and (3) server. In these systems, the water level, temperature, food, and amount of chemicals in the water of the pond are monitored and

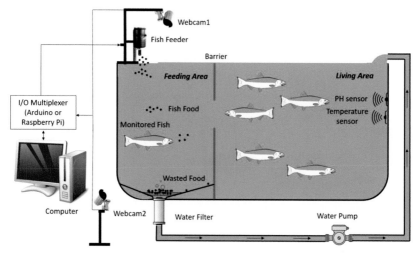

FIGURE 12.15 The smart fish feeding system setup [101].

controlled. The control unit sets the temperature of the environment considering various data received from the wireless sensor node [103].

To supply the fish management node with solar energy, the required
power of the node should be calculated in a way that the battery can supply
the required power for 24 hours once it is charged. The charge controller
also protects the battery and charges it during the day when PV modules produce electricity [104]. Fourie et al. [103] designed an autonomous solar-
powered fish pond management system with the capability of conservation
of fish and enhancing the quality of fish's life in a pond. In this study, a
node that could communicate with a base station was designed to handle the
controller of the pond. The user interface allowed actuators to be fully controlled and supply pH, food, heat, and DO to the fish environment as shown
in Fig. 12.16. They claimed that the system can control the pH, DO and
water temperature of the pond successfully, and dispense the required food
into the water. The solar charging unit was also employed to ensure that
enough power is supplied to the pond during the day.

Certainly, solar-powered productions with high-power battery chargers are
more environmentally friendly, accordingly have less environmental impacts,
less water pollution, and more security. In the absence of the sun, in cloudy or
rainy conditions, the battery can last and survive 5−7 days. Adopting the
method of rotating and throwing bait, the feed crushing rate is reduced and the
costs are saved. Choosing the best components, solar PV modules can be utilized for 25−30 years [105]. Two solar-powered fish feeder machines which
are commercially available on the market are Koisan automatic solar-powered
fish feeder and FIAP solar feeder which are shown in Fig. 12.17.

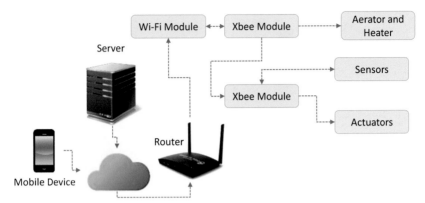

FIGURE 12.16   Flow diagram of the working principle of a fish pond management system. *Adapted from Fourie CM, Bhatt DV, Silva BJ, Kumar A, Hancke GP. A solar-powered fish pond management system for fish farming conservation. In: 2017 IEEE 26th international symposium of industrial electronics, IEEE; 2017. p. 2021–6. https://doi.org/10.1109/ISIE.2017. 8001565 [103].*

FIGURE 12.17   Commercially available solar-powered fish feeders: (A) Koisan automatic solar fish feeder [106] and (B) FIAP solar fish feeder [107].

12.4.4   Solar-powered fish pond filters

Pond filtration is another necessity for fish ponds that comes in several forms and designs. A fish farming system employs pumping systems, sensors, and filtering mechanisms to measure parameters, control variables, and eradicate wastes [108]. Filtration is carried out through two major mechanical and biological processes. In mechanical filtration, solids such as leaves are removed from the pond and protect the pump from clogging up and running inefficiently. The pond skimmer is an example of mechanical filtration. In contrast, in biological filtration, a pressure filter or a biological waterfall is used to remove impurities and toxins from the pond that have been left behind from the decomposition of organic matters such as fish waste, food, plants, and dead bugs. This type of filtration uses beneficial bacteria to break down

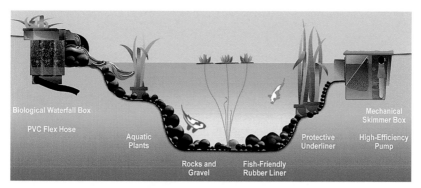

FIGURE 12.18   Mechanical and biological filtration processes in a fish pond [110].

FIGURE 12.19   Commercial solar-powered pond filter kit for small fish ponds [112].

the toxins which are the food for bad organisms and algae that degrade the quality of the water [109]. Fig. 12.18 shows mechanical and biological filtration carried out in a typical fish pond.

A solar-powered pond filter is an economical and reliable entry-level which keeps the fish pond clean with very little involved maintenance and is easy to set up and install. As soon as the sun shines on the solar panel, the pump will start working. The system is mostly operated with high-quality brushless pumps, some risers with different sprinklers (3-step fountain and water bell), and various water games. The PV modules will be connected to the pump and will power the fountain using the day's sunlight [111]. Fig. 12.19 shows a commercial solar-powered pond filter.

12.4.5   Solar-powered fish pond pumps

Generally, there are two types of water pumps as fixed and variable speed. Fixed speed pumps work at a specific speed and consume a constant amount

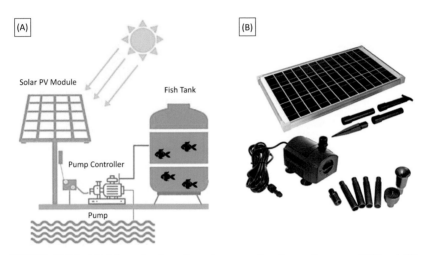

FIGURE 12.20   (A) Schematic view of a solar-powered surface water pump [115] and (B) solar-powered submersible pond water pump kit [116].

of energy, although their effective flow rate may be modified in operation under different conditions. In these pumps, there is not any mechanism to increase or decrease the pumping speed. While variable speed pumps are capable to operate at different pumping levels to supply various flow rates according to the demand [109]. In fish farm environments, two types of pumps can be employed as *submersible pumps* and *surface pumps*. Choosing a proper type depends on the water source and the application. Surface pumps are less expensive than submersible pumps, but they are not well suited for suction and can only draw water from about 6.5 vertical meters. Surface pumps are excellent for pushing water over long distances [113]. The fish pond pump system is used to ensure demanded water level inside the fish pond/tank. The input is taken from fish tank sensors where the output is sent to control the operation level and work duration of the pump [114] (Fig. 12.20).

Solar-powered water pumps are environmentally friendly since they do not emit any pollutants and minimize the dependency on electricity. These pumps allow transferring water from its remote source to the location that is required without the need for power lines. The most effective method of using a solar pump is PV direct in which the pump is directly connected to PV modules without using batteries [117]. The benefits of solar-powered fish pond pumps are their unattended operation with low maintenance and easy installation, while they have also associated with no fuel costs. In return, their main barriers are high initial costs, low output on cloudy days, and the necessity for a good exposure between 9 and 15 a.m. during the day [118].

Koyuncu [119] developed a solar-powered trout fish farming system to cool the water of fish growing floating pools or tanks by pumping cool water

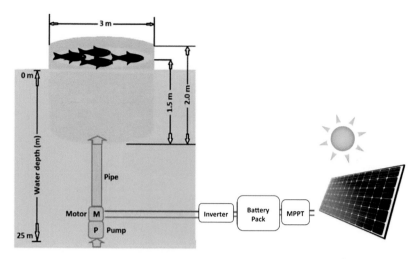

FIGURE 12.21 Schematic of the developed solar-powered water cooling system in Ref. [119].

from the depth of 25 m. The proposed system contained a monocrystalline silicon PV module, MPPT charge controller, battery pack, inverter, and a submersible pump as depicted in Fig. 12.21. In this study, the power demand of the water pump and the fish growing unit during its lifetime, and the required area for PV modules for one fish growing unit (floating pool or water tank) were calculated considering various technical data. The system was evaluated and compared with a naturally cooled (uncooled) system in terms of costs. The results indicated that the affordable functioning of a small-scale or a private fishery enterprise entails a minimum number of 19 fish growing tanks for the solar-powered cooling system, 17 for the on-grid cooling system, and 15 for the uncooled system.

12.4.6 Aquaponics systems

The concern of climate patterns, rising prices of food, and scarcity of water, especially in developing countries, have resulted in significant global challenges [120]. To address these problems, sustainable farming procedures based on water conservation, nutrients recycle, and waste (water) conversion into high-value resources have been emerged [121]. Aquaponics is a sustainable food production method and is formed by the integrating of aquaculture[5] with hydroponics in a soil-less system, in which nitrogen-rich effluent from fish production is utilized for the growth of the plant [122]. In this system, ammonia-rich fish wastes from the aquaculture tanks are pumped to the

---

5. Aquaculture is the captive rearing and production of aquatic organisms including fishes and aquatic plant under controlled conditions.

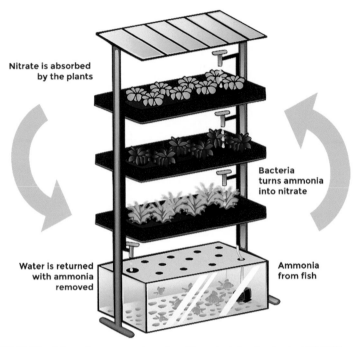

Nitrate is absorbed
by the plants

Bacteria
turns ammonia
into nitrate

Water is returned
with ammonia
removed

Ammonia
from fish

FIGURE 12.22   Schematic representation of working principle of a simple APS [125].

hydroponic beds where they are converted into organic fertilizers by living bacteria in the beds. In return, the plant roots filter and treat the water for the habitation of fishes, which is then recycled back into the aquaculture tanks. The aquaponics system allows plants and fish to coexist in a symbiotic environment, promoting sustainability in agriculture and fisheries [123]. Some of the benefits of aquaponics systems (APSs) are as follows: (1) recovery of nutrients, (2) minimizing the water demand, and (3) increasing profitability by coproduction of two cash crops [124]. The APSs usually contain freshwater, but saltwater systems are also plausible depending on the type of aquatic animals and plants (see Fig. 12.22).

APSs provide an unrivaled opportunity for the year-round production of plants and fish. The production of leafy greens, vegetables, and herbs out of the season is a remarkable source of income for owners of APSs due to the benefit of much higher seasonal prices [126]. The most commonly used APSs according to the type of the grow bed are nutrient film technique (NFT), floating-raft (deep water culture), and media-filled (flood and drain) as shown in Fig. 12.23. The NFT class provides high O_2 levels to plant roots, assisting in producing a high yield of vegetables. In this type, the grow bed cannot provide sufficient space for a high quantity of roots, and therefore it is the only admissible type for small vegetable species. In the NFT, efficient

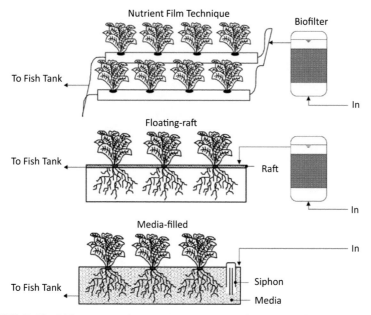

FIGURE 12.23 Different types of aquaponics systems according to types of grow beds [124].

removal of solids is crucial to prevent clogging of the channel of the grow bed [127]. The most adopted class in an APS is the floating-raft type, allowing the plant roots to absorb the water nutrients freely without the probability of clogging the channel of the water. In both APS types of floating-raft and the NFT, the use of a biofilter and a sedimentation tank is crucial to perform nitrification, and to remove solids, respectively [128]. The simplest APS is the media-filled type containing media such as pumice stones or clay beads in the grow bed for nitrification with no requirement for separate biofilters [129]. Additionally, a siphon is utilized for the water filling and draining to supply O_2 by direct contact between the air and plant roots. However, clogging along with insufficient O_2 levels in the grow bed during a long-term operation of APSs is common [124].

12.4.7 Solar-powered aquaponics systems

As presented in Fig. 12.24, solar PV modules can be considered as an ideal source to provide the electricity demand of APSs [132]. In this case, the solar power conversion set-up supplies electricity to the aquaponics, however, ensuring that the system is powered up reliably, PV modules and batteries with higher ratings are required. Using solar energy and the employment of water recycling and waste management can assist in saving the environment, making APSs environmentally sustainable structures [123].

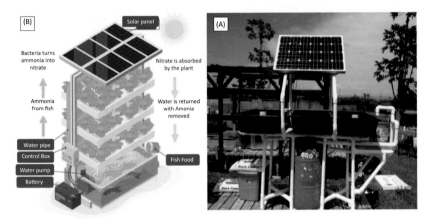

FIGURE 12.24 (A) Schematic view of a solar-powered aquaponics system [130] and (B) a solar-powered aquaponics system setup [131].

Typically, PV modules produce the highest amount of electricity during the middle of the day and somewhat less before and after that. Additionally, as new PV modules can use diffuse solar radiation, they can be employed to produce power in cloudy areas with frequent rain [133].

The power supply for the mechanical equipment in an APS is supplied by installing a PV energy circuit as depicted in Fig. 12.24A. The presented design functions as follows: the sun hits the PV modules, generating a continuous current that passes through the regulator which determines how much electricity can be directed to the accumulation or battery system. Often it might be necessary to install an adaptation system for currents or inverters, which must have the capacity to support the full power of the system. This electronic circuit including the inverter transforms the continuous current into an alternating current and also changes the voltage required for consumptions [132]. Tous et al. [134] designed a single-family APS powered by solar energy. The fish farming in this system was tilapia due to being technically feasible and being tasty with an acceptable flavor for Chad's inhabitants. The cultivated plants were also all types of fruits and vegetables as well as aromatic plants. To operate the system on cloudy days or generally when the available solar radiation is not sufficient, the employment of an accumulation or battery system is required. Therefore, an estimation was performed to calculate periods without sunshine considering the operation of the system 7 days a week and 24 hours a day. The production limitation was 20 kg of fish per 1 m^3 of water for the aquaponic part and nearly 8 L of water per plant for the hydroponic part for the vegetable production. The cultivation of fish and vegetables must be planned to take these limitations into account. Pantazi et al. [135] developed a solar-powered APS monitored by an Arduino microprocessor. A 50-W monocrystalline PV module was utilized to supply the required power for the monitoring system and the water pump. It was claimed that the proposed system is being able to monitor the water temperature, air temperature, and pH of the water, while at the same time control the water debt.

Galido et al. [136] developed a smart solar-powered APS based on Arduino using the IoT. In this study, to compare the growth of plants and fish, three aquaponics setups including Ion-Sensitive Field Effect Transistor (ISFET), pH sensor-monitored setup, glass-electrode pH sensor-monitored setup, and uncontrolled setup were developed. The system was able to detect two essential parameters of pH and temperature using relative sensors integrated into the Arduino YUN microcontroller. In this case, the length and weight of cultivated vegetables (wild chili, cherry tomato, eggplant, and bok choy), as well as those of the fish (Black Nile Tilapia), were measured in two-week intervals. The results indicated that the ISFET-monitored setup can monitor and maintain the optimal quality of the water of the APS much better than the two others.

12.5   Solar-powered pasteurization systems

Most of the industries mainly depend on conventional or nonrenewable energy resources, resulting in considerable carbon emissions and consequently global warming. In this regard, solar energy as a renewable energy source can be harnessed by using solar thermal collectors or PV modules to supply the required power of the dairy and food processing equipment. Specifically, in the milk and food processing industry, the heating and cooling requirements can be supplied by solar thermal collectors [137,138].

In a study by Panchal et al. [139], a solar milk pasteurization system was developed and its performance was evaluated. For this purpose, a solar hot water system, a concentric tube heat exchanger, and a cooling tank were used and three mass flow rates of hot water (2, 3, and 5 L/min) and cold milk (0.5, 0.7, and 1.0 L/min) were considered (see Fig. 12.25). The system was composed of a solar flat-plate collector (FPC) with a surface area of $2 m^2$ consisted of an aluminum absorber plate painted with matt black color, milk passage, insulation material, and cover and case. The milk obtained from the concentric tube-type heat exchanger was tested in the laboratory and satisfactory results were obtained. Additionally, from a series of experiments, it was found that the optimum flow rate values are 5 and 0.5 L/min for the hot water and milk, respectively.

Contaminated water is the main cause of disease and death, especially in developing countries. Water pasteurization is a method used to decontaminate water. In this case, Carielo et al. [140] developed a solar pasteurization system in which a batch of contaminated water was retained inside the solar collector at a specified temperature and for a certain duration. The experimental setup was composed of a solar FPC with an aperture of $2 m^2$, a heat

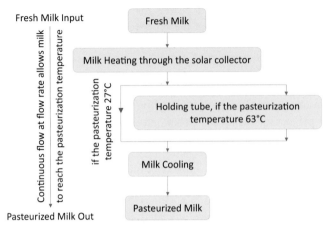

FIGURE 12.25 Passage of the milk using solar milk pasteurization system. *Adapted from Panchal H, Patel R, Sathyamurthy R. Investigation and performance analysis of solar milk pasteurisation system. Int J Ambient Energy 2021;42:522−29. https://doi.org/10.1080/01430750.2018.1557552 [139].*

exchanger, two thermoresistances, two solenoid valves, and electronic control. A series of experiments were performed in various temperature setpoints of 55°C, 60°C, 65°C, 75°C, and 85°C at different internment times of 3600, 2700, 1800, 900, and 15 s, respectively. The results indicated that the heat exchanger can increase productivity by about 110% and decrease the solar radiation cutting level at about 6.6 MJ/m² day. It was also reported that the system can reach the productivity of 80 L of treated water in a clear day. Moreover, the results of the bacterial analysis for the treated water indicated an average reduction of 98.7% in heterotrophic bacteria at different levels of contaminated feed water.

In solar thermal pasteurization systems, intermittent availability of solar radiation limited the steady operation of the system. In this case, Bologna et al. [141] designed a solar thermal water pasteurization system including thermal energy storage (TES) and an auxiliary gas boiler. In this study, a lumped-component model was also developed to determine the size of the components and the operating parameters under a specified delivery flow rate. The experimental setup was designed to pasteurize the water considering two main requirements; (1) retaining the water temperature at a constant value during the process, and (2) making the water heating process fast enough to avoid spurious effects on the disinfection results. As depicted in Fig. 12.26, the pasteurization process in this design occurs in the immersed coil in the thermal storage tank where the heat demand is provided by using solar thermal collectors (solar field), and an auxiliary boiler acts as a heat supplier backup. Additionally, the plate heat exchanger recovers the waste heat from the treated water and preheat the raw water inflow.

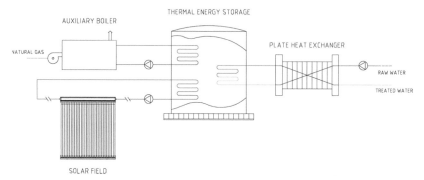

FIGURE 12.26 Schematic representation of the solar pasteurization system using TES and gas boiler [141].

To treat the biologically contaminated water, boiling is not necessary since all microbes that cause human diseases are destroyed in the temperature above 65°C. In this regard, passive solar thermal devices can also be employed. One of the main barriers against the widespread deployment of flow-through solar-powered pasteurizations systems is their high costs. In this regard, Denkenberger and Pearce [142] proposed a system in which the contaminated water reservoir is placed at a higher elevation than the clean water reservoir where the inlet water and the solar collector temperatures are at 30°C and 65°C, respectively. In this study, the mentioned problem was addressed by optimization of the costs of heat exchangers as the most expensive component of the solar pasteurization systems. For this purpose, the cost optimization for a polymer microchannel heat exchanger was performed. In this case, they studied two distinct scenarios; replacement of the coiled copper heat exchanger with the polymer microchannel heat exchanger, and design of a polymer microchannel heat exchanger for a collector with the capability of fitting in an arbitrary build volume. The results indicated that the first scenario can reduce the overall expenditure of the system by a factor of 50 while using the second scenario shows the way that system designers can optimize a heat exchanger for an arbitrary design of a solar pasteurization system.

Sizirici [143] developed and evaluated the performance of a modified biosand filter (MBSF) integrated with a solar pasteurizer to treat the contaminated water effectively. In this design, the MBSF was fabricated using an extra disinfection layer made of zero-valent iron (ZVI) and three underdrain layers to remove *Escherichia coli* (*E. coli*), total coliforms, and turbidity from the feed water. The developed solar pasteurizer was also composed of a solar FPC with a solenoid valve used as an extra disinfection device. The required power for all circuits was also supplied by PV cells. The results from the evaluation of the system indicated its capability in the production of an average volume of the pasteurized water as 4.5 L/day, at 75 °C as the pasteurization temperature. Also, the analysis of the treated water samples revealed total reductions of 96.25% in coliforms, 98% in *E. coli*, and 94.8% in turbidity. Additionally, after pasteurization, a 100% reduction in *E. coli* and 99.99% in the total coliforms were observed in the treated water.

The performance of a closed-coupled solar pasteurization system was evaluated by Dobrowsky et al. [144] to reduce the microbiological load of the harvested rainwater and to evaluate the variation of its chemical components after pasteurization. In this study, the Apollo solar pasteurization system composed of a closed-coupled system was employed. The system consisted of a solar FPC, wherein the water is heated in a direct or indirect mode. The closed-coupled systems are less expensive and easily installed compared to the split[6] ones. The temperature of the pasteurized water tank was ranged from 55°C to 57°C, 64°C to 66°C, 72°C to 74°C, 78°C to 81°C, and 90°C to 91°C. The analysis of the pasteurized water indicated that the cations and anions are within the guidelines for the drinking water, except for iron, aluminum, lead, and nickel. Moreover, heterotrophic bacteria, Escherichia coli, and total coliforms as indicator bacteria were reduced to below the detection limit at pasteurization temperatures of 72°C and above. Fabricating the storage tank of the pasteurization system from other materials rather than stainless steel was suggested due to the capability of the system in the production of large water quantities from roof harvested rainwater. Since pasteurization systems require thermal energy, solar thermal collectors can be a proper option for integration with these facilities to provide their required heat. Especially in remote regions and agricultural farms with no availability of clean water and insufficient accessibility to fossil fuels, the use of solar energy is quite feasible and can be considered as a sustainable solution. Additionally, providing the heat demand of pasteurization systems using solar energy can make these systems self-sustained with no dependency on fossil fuels or grid electricity.

12.6 Conclusions and prospects

This chapter studies the integration of solar energy with agricultural and aquaculture systems to perform various operations including crop production, crop protection, aquaculture, aquaponics, and pasteurization. Some of the mentioned applications benefit from solar energy in its passive form without the requirement for solar collectors or solar PV modules such as soil solarization treatment, SCCSs, and microalgae cultivation. In contrast, other applications require to supply their power demand by integrating with PV systems such as solar-powered sprayers, solar-powered bird repellers, solar-powered fish feeders, and solar-power fish pond pumps. While thermal pasteurization

---

6. Split is a costly system composed of a collector and a storage tank for water treatment in which water directly/indirectly is heated.

systems require to supply their heat demand by employing solar thermal collectors. Due to benefit of being modular, the solar PV modules can be suitably integrated with portable systems working in the farms to protect crops or in aquaculture environments. Some of the introduced technologies in this chapter have become commercial and are available on the market, but the others require more research and improvements in terms of technology and reductions in cost to become commercial. In conclusion, integrating agriculture and food production systems with solar energy as the most abundant and reliable source of renewable energy can make them eco-friendly and self-sustained, paving the way to achieve sustainable development in the agriculture sector.

Acknowledgment

The authors would like to thank Tarbiat Modares University (TMU) (http://www.modares. ac.ir) for the received financial support (grant number IG/39705) for the "Renewable Energies Research Group."

References

[1] Gorjian S, Ebadi H, Trommsdorff M, Sharon H, Demant M, Schindele S. The advent of modern solar-powered electric agricultural machinery: a solution for sustainable farm operations. J Clean Prod 2021;292. Available from: https://doi.org/10.1016/j.jclepro.2021.126030.

[2] World Hunger. Key Facts and Statistics 2021. Action Against Hunger. Available from: https://www.actionagainsthunger.org/world-hunger-facts-statistics; 2021 [accessed 27.07.21].

[3] Gorjian S, Singh R, Shukla A, Mazhar AR. On-farm applications of solar PV systems. In: Gorjian S, Shukla ABT-PSEC, editors. Photovoltaic solar energy conversion. Elsevier; 2020. p. 147−90. Available from: https://doi.org/10.1016/B978-0-12-819610-6.00006-5.

[4] Gorjian S, Ghobadian B, Ebadi H, Ketabchi F, Khanmohammadi S. Applications of solar PV systems in desalination technologies. In: Gorjian S, Shukla A, editors. Photovoltaicc solar energy conversion. 1st ed. London: Elsevier; 2020. p. 237−74. Available from: https://doi.org/10.1016/B978-0-12-819610-6.00008-9.

[5] Gorjian S, Minaei S, MalehMirchegini L, Trommsdorff M, Shamshiri RR. Applications of solar PV systems in agricultural automation and robotics. In: Gorjian S, Shukla A, editors. Photovoltaic solar energy conversion. 1st ed. London: Elsevier; 2020. p. 191−235. Available from: https://doi.org/10.1016/B978-0-12-819610-6.00007-7.

[6] Shakouri M, Ebadi H, Gorjian S. Solar photovoltaic thermal (PVT) module technologies. In: Gorjian S, Shukla A, editors. Photovoltaic solar energy conversion. 1st ed. London: Elsevier; 2020. p. 79−116. Available from: https://doi.org/10.1016/B978-0-12-819610-6.00004-1.

[7] Gorjian S, Hosseingholilou B, Jathar LD, Samadi H, Samanta S, Sagade AA, et al. Recent advancements in technical design and thermal performance enhancement of solar greenhouse dryers. Sustainability 2021;13:7025. Available from: https://doi.org/10.3390/su13137025.

[8] Gorjian S, Calise F, Kant K, Ahamed MS, Copertaro B, Najafi G, et al. A review on opportunities for implementation of solar energy technologies in agricultural greenhouses. J Clean Prod 2021;285:124807. Available from: https://doi.org/10.1016/j.jclepro.2020.124807.

[9] Gorjian S, Ebadi H, Najafi G, Singh Chandel S, Yildizhan H. Recent advances in net-zero energy greenhouses and adapted thermal energy storage systems. Sustain Energy Technol Assess 2021;43:100940. Available from: https://doi.org/10.1016/j.seta.2020.100940.

[10] Mukherjee P, Sengupta TK. Design and fabrication of solar-powered water pumping unit for irrigation system. Lecture Notes in Electrical Engineering, vol. 575. Singapore: Springer; 2020. p. 89−102. Available from: https://doi.org/10.1007/978-981-13-8687-9_9.

[11] Senthil Kumar S, Bibin C, Akash K, Aravindan K, Kishore M, Magesh G. Solar powered water pumping systems for irrigation: acomprehensive review on developments and prospects towards a green energy approach. Materials Today: Proceedings, vol. 33. Elsevier; 2020. p. 303−7. Available from: https://doi.org/10.1016/j.matpr.2020.04.092.

[12] Gorjian S, Ghobadian B. Solar desalination: a sustainable solution to water crisis in Iran. Renew Sustain Energy Rev 2015;48:571−84. Available from: https://doi.org/10.1016/j.rser.2015.04.009.

[13] Gorjian S, Ghobadian B, Tavakkoli Hashjin T, Banakar A. Experimental performance evaluation of a stand-alone point-focus parabolic solar still. Desalination 2014;352:1−17. Available from: https://doi.org/10.1016/j.desal.2014.08.005.

[14] Dai Y, Senge M, Yoshiyama K, Zhang P, Zhang F. Influencing factors, effects and development prospect of soil solarization. Rev Agric Sci 2016;4:21−35. Available from: https://doi.org/10.7831/ras.4.21.

[15] Katan J. Soil solarization: the idea, the research and its development. Phytoparasitica 2015;43:1−4. Available from: https://doi.org/10.1007/s12600-014-0419-0.

[16] Cohen R, Orgil G, Burger Y, Saar U, Elkabetz M, Tadmor Y, et al. Differences in the responses of melon accessions to fusarium root and stem rot and their colonization by Fusarium oxysporum f. sp. radicis-cucumerinum. Plant Pathol 2015;64:655−63. Available from: https://doi.org/10.1111/ppa.12286.

[17] ÖZ H. A new approach to soil solarization: addition of biochar to the effect of soil temperature and quality and yield parameters of lettuce (*Lactuca sativa* L. Duna). Sci Hortic (Amst) 2018;22(8):153−61. Available from: https://doi.org/10.1016/j.scienta.2017.10.021.

[18] Candido V, D'Addabbo T, Basile M, Castronuovo D, Miccolis V. Greenhouse soil solarization: effect on weeds, nematodes and yield of tomato and melon. Agron Sustain Dev 2008;28:221−30. Available from: https://doi.org/10.1051/agro:2007053.

[19] Soil Solarization | Purdue University Vegetable Crops Hotline. Available from: https://vegcropshotline.org/article/soil-solarization/; 2021 [accessed 02.07.21].

[20] Mihajlovic M, Rekanovic E, Hrustic J, Grahovac M, Tanovic B. Methods for management of soilborne plant pathogens. Pestic i Fitomedicina 2017;32:9−24. Available from: https://doi.org/10.2298/pif1701009m.

[21] Al-Kayssi AW, Al-Karaghouli A. A new approach for soil solarization by using paraffin-wax emulsion as a mulching material. Renew Energy 2002;26:637−48. Available from: https://doi.org/10.1016/S0960-1481(01)00120-3.

[22] Castronuovo D, Candido V, Margiotta S, Manera C, Miccolis V, Basile M, et al. Potential of a corn starch-based biodegradable plastic film for soil solarization. Acta Hortic 2005;698:201−6. Available from: https://doi.org/10.17660/ActaHortic.2005.698.27.

[23] Soil solarization − Wikipedia. n.d.

[24] Bonanomi G, Chiurazzi M, Caporaso S, Del Sorbo G, Moschetti G, Felice S. Soil solarization with biodegradable materials and its impact on soil microbial communities. Soil Biol Biochem 2008;40:1989−98. Available from: https://doi.org/10.1016/j.soilbio.2008.02.009.

[25] Oz H, Coskan A, Atilgan A. Determination of Effects of various plastic covers and biofumigation on soil temperature and soil nitrogen form in greenhouse solarization: new

solarization cover material. J Polym Environ 2017;25:370−7. Available from: https://doi.org/10.1007/s10924-016-0819-y.

[26]   Maraveas C. Environmental sustainability of plastic in agriculture. Agriculture 2020;10:310. Available from: https://doi.org/10.3390/agriculture10080310.

[27]   Castello I, D'Emilio A, Raviv M, Vitale A. Soil solarization as a sustainable solution to control tomato Pseudomonads infections in greenhouses. Agron Sustain Dev 2017;37:59. Available from: https://doi.org/10.1007/s13593-017-0467-1.

[28]   Abd-Elgawad MMM, Elshahawy IE, Abd-El-Kareem F. Efficacy of soil solarization on black root rot disease and speculation on its leverage on nematodes and weeds of strawberry in Egypt. Bull Natl Res Cent 2019;43:175. Available from: https://doi.org/10.1186/s42269-019-0236-1.

[29]   Khan MA, Marwat KB, Amin A, Nawaz A, Khan R, Khan H, et al. Soil solarization: an organic weed-management approach in cauliflower. Commun Soil Sci Plant Anal 2012;43:1847−60. Available from: https://doi.org/10.1080/00103624.2012.684822.

[30]   Ayilara MS, Olanrewaju OS, Babalola OO, Odeyemi O. Waste management through composting: challenges and potentials. Sustain 2020;12:1−23. Available from: https://doi.org/10.3390/su12114456.

[31]   Poblete R, Salihoglu G, Salihoglu NK. Incorporation of solar-heated aeration and greenhouse in grass composting. Environ Sci Pollut Res 2021;28:26807−18. Available from: https://doi.org/10.1007/s11356-021-12577-7.

[32]   Composting in the Home Garden - Common Questions. Available from: https://web.extension.illinois.edu/compost/process.cfm; 2021 [accessed 03.07.21].

[33]   Domingo JL, Nadal M. Domestic waste composting facilities: a review of human health risks. Environ Int 2009;35:382−9. Available from: https://doi.org/10.1016/j.envint.2008.07.004.

[34]   Lin H, Ye J, Sun W, Yu Q, Wang Q, Zou P, et al. Solar composting greenhouse for organic waste treatment in fed-batch mode: physicochemical and microbiological dynamics. Waste Manag 2020;113:1−11. Available from: https://doi.org/10.1016/j.wasman.2020.05.025.

[35]   Chen W, Luo S, Du S, Zhang M, Cheng R, Wu D. Strategy to strengthen rural domestic waste composting at low temperature: choice of ventilation condition. Waste Biomass Valoriz 2020;11:6649−65. Available from: https://doi.org/10.1007/s12649-020-00943-4.

[36]   Rajkumar P. A study on recycling of timber waste and fertilizer production using solar powered forced air supply. Shanlax Int J Arts, Sci Humanit 2020;8:116−23. Available from: https://doi.org/10.34293/sijash.v8i2.3446.

[37]   Kremer RJ, LeRoy Deichman C. Introduction: the solar corridor concept. Agron J 2014;106:1817−19. Available from: https://doi.org/10.2134/agronj14.0291.

[38]   Hatfield JL, Dold C. Photosynthesis in the solar corridor system. The solar corridor crop system. Elsevier; 2019. p. 1−33. Available from: https://doi.org/10.1016/B978-0-12-814792-4.00001-2.

[39]   Nelson KA. Corn yield response to the solar corridor in upstate Missouri. Agron J 2014;106:1847−52. Available from: https://doi.org/10.2134/agronj2012.0326.

[40]   Kremer RJ. The solar corridor: a new paradigm for sustainable crop production. Adv Plants Agric Res 2016;4:273−4. Available from: https://doi.org/10.15406/apar.2016.04.00136.

[41]   Deichman CL. A sustainable farming system to maximize photosynthesis. In: Proceedings of the farmings system design symposium. Monterey, CA; 2009.

[42]   LeRoy Deichman C. Variety selection for the solar corridor crop system (SCCS). The solar corridor crop system. Elsevier; 2019. p. 35−56. Available from: https://doi.org/10.1016/B978-0-12-814792-4.00002-4.

[43] Hendrickson MK. The economic and social conditions of the solar corridor cropping system. The solar corridor crop system. Elsevier; 2019. p. 121−43. Available from: https://doi.org/10.1016/B978-0-12-814792-4.00006-1.

[44] Kour D, Rana KL, Yadav N, Yadav AN, Rastegari AA, Singh C, et al. Technologies for biofuel production: current development, challenges, and future prospects. Prospects of Renewable Bioprocessing in Future Energy Systems. Springer; 2019. p. 1−50. https://doi.org/10.1007/978-3-030-14463-0_1.

[45] Gupta PL, Lee S-M, Choi H-J. A mini review: photobioreactors for large scale algal cultivation. World J Microbiol Biotechnol 2015;31:1409−17. Available from: https://doi.org/10.1007/s11274-015-1892-4.

[46] Lu Y, Zhang X, Gu X, Lin H, Melis A. Engineering microalgae: transition from empirical design to programmable cells. Crit Rev Biotechnol 2021;1−24. Available from: https://doi.org/10.1080/07388551.2021.1917507.

[47] Khan MI, Shin JH, Kim JD. The promising future of microalgae: current status, challenges, and optimization of a sustainable and renewable industry for biofuels, feed, and other products. Microb Cell Fact 2018;17:36. Available from: https://doi.org/10.1186/s12934-018-0879-x.

[48] Koutra E, Economou CN, Tsafrakidou P, Kornaros M. Bio-based products from microalgae cultivated in digestates. Trends Biotechnol 2018;36:819−33. Available from: https://doi.org/10.1016/j.tibtech.2018.02.015.

[49] Masojídek J, Ranglová K, Lakatos GE, Silva Benavides AM, Torzillo G. Variables governing photosynthesis and growth in microalgae mass cultures. Processes 2021;9:820. Available from: https://doi.org/10.3390/pr9050820.

[50] Amaral M S, Loures C CA, Naves F L, Samanamud G L, Silva M B, Prata A. MR. Microalgae cultivation in photobioreactors aiming at biodiesel production. Biotechnological applications of biomass. IntechOpen; 2020. Available from: https://doi.org/10.5772/intechopen.93547.

[51] Hallenbeck PC, Grogger M, Mraz M, Veverka D. Solar biofuels production with microalgae. Appl Energy 2016;179:136−45. Available from: https://doi.org/10.1016/j.apenergy.2016.06.024.

[52] Acién FG, Molina E, Reis A, Torzillo G, Zittelli GC, Sepúlveda C, et al. Photobioreactors for the production of microalgae. Microalgae-based biofuels bioproduction. Elsevier; 2017. p. 1−44. Available from: https://doi.org/10.1016/B978-0-08-101023-5.00001-7.

[53] Sánchez Zurano A, Gómez Serrano C, Acién-Fernández FG, Fernández-Sevilla JM, Molina-Grima E. Modeling of photosynthesis and respiration rate for microalgae−bacteria consortia. Biotechnol Bioeng 2021;118:952−62. Available from: https://doi.org/10.1002/bit.27625.

[54] Iasimone F, Panico A, De Felice V, Fantasma F, Iorizzi M, Pirozzi F. Effect of light intensity and nutrients supply on microalgae cultivated in urban wastewater: biomass production, lipids accumulation and settleability characteristics. J Environ Manage 2018;223:1078−85. Available from: https://doi.org/10.1016/j.jenvman.2018.07.024.

[55] Daneshvar E, Sik Ok Y, Tavakoli S, Sarkar B, Shaheen SM, Hong H, et al. Insights into upstream processing of microalgae: a review. Bioresour Technol 2021;329:124870. Available from: https://doi.org/10.1016/j.biortech.2021.124870.

[56] Galès A, Triplet S, Geoffroy T, Roques C, Carré C, Le Floc'h E, et al. Control of the pH for marine microalgae polycultures: a key point for CO2 fixation improvement in intensive cultures. J CO_2 Util 2020;38:187−93. Available from: https://doi.org/10.1016/j.jcou.2020.01.019.

[57] Chaisutyakorn P, Praiboon J, Kaewsuralikhit C. The effect of temperature on growth and lipid and fatty acid composition on marine microalgae used for biodiesel production. J Appl Phycol 2018;30:37−45. Available from: https://doi.org/10.1007/s10811-017-1186-3.

[58] Vale MA, Ferreira A, Pires JCM, Gonçalves AL. CO2 capture using microalgae. Advances in carbon capture. Elsevier; 2020. p. 381−405. Available from: https://doi.org/10.1016/B978-0-12-819657-1.00017-7.

[59] Kumar K, Mishra SK, Shrivastav A, Park MS, Yang J-W. Recent trends in the mass cultivation of algae in raceway ponds. Renew Sustain Energy Rev 2015;51:875−85. Available from: https://doi.org/10.1016/j.rser.2015.06.033.

[60] Masojídek J, Torzillo G. Mass cultivation of freshwater microalgae. Reference module in earth systems and environmental sciences. Elsevier; 2014. Available from: https://doi.org/10.1016/B978-0-12-409548-9.09373-8.

[61] García-Galán MJ, Gutiérrez R, Uggetti E, Matamoros V, García J, Ferrer I. Use of full-scale hybrid horizontal tubular photobioreactors to process agricultural runoff. Biosyst Eng 2018;166:138−49. Available from: https://doi.org/10.1016/j.biosystemseng.2017.11.016.

[62] Lam MK, Loy ACM, Yusup S, Lee KT. Biohydrogen production from algae. Biohydrogen. Elsevier; 2019. p. 219−45. Available from: https://doi.org/10.1016/B978-0-444-64203-5.00009-5.

[63] RW22-101—MicroBio Engineering. Available from: https://microbioengineering.com/rw22-101; 2021 [accessed 16.07.21].

[64] Bubble Column Array - PBR—MicroBio Engineering. Available from: https://microbioengineering.com/bubble-column-array-pbr; 2021 [accessed 16.07.21].

[65] Co2 Bio-Mitigation using Microalgae. Available from: https://www.ctc-n.org/technologies/co2-bio-mitigating-micro-algae; 2021 [accessed 16.07.21].

[66] Photobioreactor-Wikipedia. Available from: https://en.wikipedia.org/wiki/Photobioreactor#/media/File:Photobioreactor_PBR_500_P_IGV_Biotech.jpg; 2021 [accessed 16.07.21].

[67] Suparmaniam U, Lam MK, Uemura Y, Lim JW, Lee KT, Shuit SH. Insights into the microalgae cultivation technology and harvesting process for biofuel production: a review. Renew Sustain Energy Rev 2019;115:109361. Available from: https://doi.org/10.1016/j.rser.2019.109361.

[68] Divjot K, Kusam LR, Neelam Y, Yadav AN, Ali A, Rastegari CS, Puneet N, et al. Prospects of renewable bioprocessing in future energy systems: production by cyanobacteria. In: Asghar A, Nath A, editors. Prospects of Renewable Bioprocessing in Future Energy Systems. Produced by Cyanobacteria. Switzerland AG: Springer Nature; 2019.

[69] Baral SS, Swarnkar R, Kothiya AV, Monpara AM, Chavda SK. Bird repeller − a review. Int J Curr Microbiol Appl Sci 2019;8:1035−9. Available from: https://doi.org/10.20546/ijcmas.2019.802.121.

[70] Arun V, Senthamilan JM, Udayakumar D, Vinithkumar V. Fabrication of mobile ultrasonic bird repeller. Int J Innov Res Adv Eng 2019;6:26−33.

[71] Riya R, Varsha KR, Sonamsi S, Jain D. Automated bird detection and repeller system using iot devices: an insight from indian agriculture perspective. SSRN Electron J 2020;3−5. Available from: https://doi.org/10.2139/ssrn.3563395.

[72] Ultrasonic Bird Repeller Solar Powered - Life Changing Products. Available from: https://lcpshop.net/product/ultrasonic-bird-repeller-solar-powered/; 2021 [accessed 04.07.21].

[73] Lasers solve bird problem on American blueberries. Int Pest Control 2018;60:166−7.

[74] Muminov A, Jeon YC, Na D, Lee C, Jeon HS. Development of a solar powered bird repeller system with effective bird scarer sounds. In: 2017 international confernce on

infformation science and communication technology. IEEE; 2017. p. 1−4. Available from: https://doi.org/10.1109/ICISCT.2017.8188587.

[75] GTS Electric Fence Repairs - Electric Fence Repairs in Centurion. Available from: https://gts-electric-fence-repairss.business.site/; 2021 [accessed 20.07.21].

[76] Kadam DM, Dange AR, Khambalkar VP. Performance of solar power fencing system for agriculture. J Agric Technol 2011;7:1199−209.

[77] Nair G, Chawla M, Bawane N. Automatic farming for minimum water usage and animal protection using solar fencing with GSM. In: 2020 international confernce on innovative trends and infformation technology. IEEE; 2020. p. 1−6. Available from: https://doi.org/ 10.1109/ICITIIT49094.2020.9071530.

[78] Okoronkwo EN, Onu UG. The effect of fabrication and deployment of a solar powered automatic pest control system on the yield level of rice. IOSR J Agric Veterinary Sci 2020;13:52−8. Available from: https://doi.org/10.9790/2380-1310015258.

[79] Telaumbanua M, Haryanto A, Wisnu FK, Lanya B, Wiratama W, Design OF. Insect trap automatic control system for cacao plants. Procedia Environ Sci Eng Manag 2021;8:167−75.

[80] Karunanayake PN, De SW, Jayasundara JACLC, Wanniarachchi YG, Karunarathne API. Intelligent pest repellent system for sri lankan farming industry; n.d. 1−6.

[81] Rashid H, Ahmed IU, Reza SMT, Islam MA. Solar powered smart ultrasonic insects repellent with DTMF and manual control for agriculture. 2017 IEEE international conference on imaging, vissionpattern recognition. IcIVPR; 2017. Available from: https://doi. org/10.1109/ICIVPR.2017.7890869.

[82] Kumawat MM, Wadavane D, Ankit N, Dipak V, Chandrakant G. Solar operated pesticide sprayer for agriculture purpose. Int Res J Eng Technol 2018;5:3396.

[83] Murthy KB, Kanwar R, Yadav I, Das V. Solar pesticide sprayer. Int J Latest Eng Res Appl 2017;2:82−9.

[84] Yallappa D, Palled V, Veerangouda M, Sushilendra. Development and evaluation of solar powered sprayer with multi-purpose applications. In: 2016 IEEE global humanity and technology conference. IEEE; 2016. p. 1−6. Available from: https://doi.org/10.1109/ GHTC.2016.7857252.

[85] Mishra A, Bhagat N, Singh P. Development of Solar operated sprayer for small scale farmers. Int J Curr Microbiol Appl Sci 2019;8:2593−6. Available from: https://doi.org/ 10.20546/ijcmas.2019.802.301.

[86] Aglave K, Bhuse S. Multi nozzle pesticide sprayer for agriculture. Available from: https:// edu.3ds.com/en/projects/multi-nozzle-pesticide-sprayer-agriculture; 2021.

[87] CSIR - CMERI develops two solar powered sprayers for marginal farmers. Available from: http://agrospectrumindia.com/news/105/1222/csir-cmeri-develops-two-solar-power-edsprayers-for-marginal-farmers.html; 2021.

[88] Sasaki RS, Teixeira MM, Filho DO, JúnioCesconetti C, Silva AC, Leite DM. Development of a solar photovoltaic backpack sprayer. Comun Sci 2014;5:395−401.

[89] FAO. The state of world fisheries and aquaculture 2020. FAO; 2020. Available from: https://doi.org/10.4060/ca9229en.

[90] Chang B, Zhang X. Aquaculture monitoring system based on fuzzy-PID algorithm and intelligent sensor networks. In: 2013 cross strait quad-regional radio cience and wireless technology conference. IEEE; 2013. p. 385−8. Available from: https://doi.org/10.1109/ CSQRWC.2013. 6657435.

[91] Wei HC, Salleh S, Mohd Ezree A, Zaman I, Hatta M, Md Zain BA, et al. Improvement of automatic fish feeder machine design. J Phys Conf Ser 2017;914:012041. Available from: https://doi.org/10.1088/1742-6596/914/1/012041.

[92] Nasir Uddin M, Rashid M, Mostafa M, Salam S, Nithe N, Rahman M, et al. Development of an automatic fish feeder. Glob J Res Eng A Mech Mech Eng 2016;16:11.

[93] White PG, Shipton TA, Bueno PB, Hasan MR. Better management practices for feed production and management of Nile tilapia and milkfish in the Philippines; 2018.

[94] Tsutsubuchi M, Hirota T, Niwa Y, Shimasaki T. Application of plastics CAE: vol. II; 2011.

[95] Chen JH, Sung WT, Lin GY. Automated monitoring system for the fish farm aquaculture environment. Proceedings - 2015 IEEE International Conference on Systems, Man, 2016. Cybern SMC; 2015. p. 1161−6. Available from: https://doi.org/10.1109/SMC.2015.208.

[96] Yeoh SJ, Taip FS, Endan J, Talib RA, Siti Mazlina MK. Development of automatic feeding machine for aquaculture industry. Pertanika J Sci Technol 2010;18:105−10.

[97] Prangchumpol D. A model of mobile application for automatic fish feeder aquariums system. Int J Model Optim 2018;8:277−80. Available from: https://doi.org/10.7763/IJMO.2018.V8.665.

[98] Noor MZH, Hussian AK, Saaid MF, Ali MSAM, Zolkapli M. The design and development of automatic fish feeder system using PIC microcontroller. Proceedings - 2012 IEEE Control and System Graduate Research Colloquium, 2012. ICSGRC; 2012. p. 343−7. Available from: https://doi.org/10.1109/ICSGRC.2012.6287189.

[99] Lee JV, Loo JL, Chuah YD, Tang PY, Tan YC, Goh WJ. The use of vision in a sustainable aquaculture feeding system. Res J Appl Sci Eng Technol 2013;6:3658−69. Available from: https://doi.org/10.19026/rjaset.6.3573.

[100] Mustafa FH. A review of smart fish farming systems. J Aquac Eng Fish Res 2016;2:193−200. Available from: https://doi.org/10.3153/jaefr16021.

[101] Alammar M M, Al-Ataby A. An intelligent approach of the fish feeding system. Computer science information technology (CS IT). AIRCC Publication Corporation; 2018. p. 85−97. Available from: https://doi.org/10.5121/csit.2018.81506.

[102] Sukoco A, Nasihien RD, Setiawan MI, Bin Bon AT. Solar powered fish feeding machine, technology for sme in Sidoarjo, East Java, Indonesia. Proc Int Conf Ind Eng Oper Manag 2020;0:2097−101.

[103] Fourie CM, Bhatt DV, Silva BJ, Kumar A, Hancke GP. A solar-powered fish pond management system for fish farmng conservation. In: 2017 IEEE 26th international symposiumon industrial electronics. IEEE; 2017. p. 2021−6. Available from: https://doi.org/10.1109/ISIE.2017.8001565.

[104] Olsen Y. Resources for fish feed in future mariculture. Aquac Environ Interact 2011;1:187−200. Available from: https://doi.org/10.3354/aei00019.

[105] Cotfas DT, Cotfas PA. Multiconcept methods to enhance photovoltaic system efficiency. Int J Photoenergy 2019;2019:1−14. Available from: https://doi.org/10.1155/2019/1905041.

[106] Koisan Automatic Solar Powered Fish Feeder. Available from: https://www.bidorbuy.co.za/item/44544732/Koisan_Automatic_Solar_Powered_Fish_Feeder.html; 2021 [accessed 17.07.21].

[107] FIAP Solar Feeder | Fresh by Design. Available from: https://freshbydesign.com.au/aquaponic-aquaculture-products/fish-feeders/fiap-solar-feeder/; 2021 [accessed 17.07.21].

[108] Integrated Livestock-Fish Production Systems. n.d.

[109] Premchand Mahalik N, Kim K. Aquaculture monitoring and control systems for seaweed and fish farming. World J Agric Res 2014;2:176−82. Available from: https://doi.org/10.12691/wjar-2-4-7.

[110] Water Gardends and Fish Ponds Blue Thumb. Available from: https://shopbluethumb.com/water-gardens-fish-ponds.html; 2021 [accessed 07.05.21].

[111]   Zhao Z, Xu Q, Luo L, Wang C, Li J, Wang L. Effect of feed C/N ratio promoted bio-
         flocs on water quality and production performance of bottom and filter feeder carp in
         minimum-water exchanged pond polyculture system. Aquaculture 2014;434:442−8.
         Available from: https://doi.org/10.1016/j.aquaculture.2014.09.006.
[112]   Solar Powered Submersible Garden Pond Oxygenator. Available from: https://www.
         amazon.co.uk/dp/B085QLK7Y1/ref = as_li_ss_tl?SubscriptionId = AKIAJO7E5O
         LQ67NVPFZA&ascsubtag = 488629737-311-1331268825.1620379159&tag = best_re-
         views_ uk_3-21; n.d. [accessed 07.05.21].
[113]   Short TD, Thompson P. Breaking the mould: solar water pumping-the challenges and the
         reality. Sol Energy 2003;75:1−9. Available from: https://doi.org/10.1016/S0038-092X
         (03)00233-0.
[114]   Simbeye DS. A wireless sensor network based solar powered harvesting system for aqua-
         culture. J Inf Sci Comput Technol 2018;7:733−43.
[115]   solar water pumping system - SunTech Solar India. Available from: http://www.suntech-
         solarindia.com/solar-water-pumping-system/; 2021 [accessed 28.07.21].
[116]   Solariver Solar Water Pump Kit - 360 + GPH Submersible Pump with Adjustable Flow.
         Available   from: https://www.amazon.com/Solariver-Solar-Water-Pump-Kit/dp/
         B01F4MKBBG; 2021 [accessed 28.07.21].
[117]   Firatoglu ZA, Yesilata B. New approaches on the optimization of directly coupled PV
         pumping systems. Sol Energy 2004;77:81−93. Available from: https://doi.org/10.1016/j.
         solener.2004.02.006.
[118]   Setiawan AA, Purwanto DH, Pamuji DS, Huda N. Development of a solar water
         pumping system in karsts rural area tepus, gunungkidul through student community
         services. Energy Procedia 2014;47:7−14. Available from: https://doi.org/10.1016/j.
         egypro.2014.01.190.
[119]   Koyuncu T. Economic analysis of solar powered and on-grid trout fish growing systems.
         IOP Conf Ser Earth Environ Sci 2020;464:012009. Available from: https://doi.org/
         10.1088/1755-1315/464/1/012009.
[120]   Mukuve FM, Fenner RA. The influence of water, land, energy and soil-nutrient resource
         interactions on the food system in Uganda. Food Policy 2015;51:24−37. Available from:
         https://doi.org/10.1016/j.foodpol.2014.12.001.
[121]   Love DC, Fry JP, Li X, Hill ES, Genello L, Semmens K, et al. Commercial aquaponics
         production and profitability: findings from an international survey. Aquaculture
         2015;435:67−74. Available from: https://doi.org/10.1016/j.aquaculture.2014.09.023.
[122]   Stathopoulou P, Berillis P, Levizou E, Sakellariou-Makrantonaki M, Kormas AK,
         Aggelaki A, et al. Aquaponics: A mutually beneficial relationship of fish, plants and bac-
         teria. In: Proceedings of the 3rd international congress applied ichthyology aquatic envi-
         ronment. Volos, Greece, 2018. p. 8−11.
[123]   Nagayo AM, Mendoza C, Vega E, Al Izki RKS, Jamisola RS. An automated solar-
         powered aquaponics system towards agricultural sustainability in the Sultanate of Oman.
         2017 IEEE international conference on smart grid smart cities. ICSGSC; 2017. p. 42−9.
         https://doi.org/10.1109/ICSGSC.2017.8038547.
[124]   Wongkiew S, Hu Z, Chandran K, Lee JW, Khanal SK. Nitrogen transformations in aqua-
         ponic systems: a review. Aquac Eng 2017;76:9−19. Available from: https://doi.org/
         10.1016/j.aquaeng.2017.01.004.
[125]   Project Feed 1010 I Aquaponics. Available from: http://www.projectfeed1010.com/what-
         is-aquaponics/; 2021 [accessed 27.05.21].
[126]   Allen Pattillo D, Allen D. An overview of aquaponic systems: hydroponic components part
         of the agriculture commons. NCRAC Tech Bull North Cent Reg Aquac Cent 2017;19:1−10.

[127] Engle CR. Economics of Aquaponics. Srac 2015; No. 5006:4.

[128] Liang JY, Chien YH. Effects of feeding frequency and photoperiod on water quality and crop production in a tilapia-water spinach raft aquaponics system. Int Biodeterior Biodegrad 2013;85:693−700. Available from: https://doi.org/10.1016/j.ibiod.2013.03.029.

[129] Zou Y, Hu Z, Zhang J, Xie H, Guimbaud C, Fang Y. Effects of pH on nitrogen transformations in media-based aquaponics. Bioresour Technol 2016;210:81−7. Available from: https://doi.org/10.1016/j.biortech.2015.12.079.

[130] A diagram of Solar aquaponic systems - Google Search; n.d.

[131] Solar Powered Aquaponics System | Aquaponics For Everyone. Available from: https://www.aquaponicsforeveryone.com/solar-powered-aquaponics-system/; n.d. [accessed27.05.21].

[132] Mohamad NR. Development of aquaponic system using solar powered control pump. IOSR J Electr Electron Eng 2013;8:01−6. Available from: https://doi.org/10.9790/1676-0860106.

[133] Karimanzira D, Rauschenbach T. Optimal utilization of renewable energy in aquaponic systems. Energy Power Eng 2018;10:279−300. Available from: https://doi.org/10.4236/epe.2018.106018.

[134] Tous D, De F, Sánchez R, Sánchez EM. Design of an aquaponic system run on solar power for a family business in Chad. Eur J Fam Bus 2019;9:39−48. Available from: https://doi.org/10.24310/ejfbejfb.v9i1.5220.

[135] Pantazi D, Dinu S, Voinea S. The smart aquaponics greenhouse − an interdisciplinary educational laboratory. Rom Rep Phys 2019;71:1−11.

[136] Galido E, Tolentino LK, Fortaleza B, Corvera RJ, De Guzman A, Española VJ, et al. Development of a solar-powered smart aquaponics system through internet of things (IoT). Lect Notes Res Innov Comput Eng Comput Sci 2019;31−9.

[137] Panchal H, Patel J, Chaudhary S. A comprehensive review of solar milk pasteurization system. J Sol Energy Eng Trans ASME 2018;140:1−8. Available from: https://doi.org/10.1115/1.4038505.

[138] Sain M, Sharma A, Zalpouri R. Solar energy utilisation in dairy and food processing industries - current applications and future scope. J Community Mobilization Sustain Dev 2020;15:227−34.

[139] Panchal H, Patel R, Sathyamurthy R. Investigation and performance analysis of solar milk pasteurisation system. Int J Ambient Energy 2021;42:522−9. Available from: https://doi.org/10.1080/01430750.2018.1557552.

[140] Carielo G, Calazans G, Lima G, Tiba C. Solar water pasteurizer: productivity and treatment efficiency in microbial decontamination. Renew Energy 2017;105:257−69. Available from: https://doi.org/10.1016/j.renene.2016.12.042.

[141] Bologna A, Fasano M, Bergamasco L, Morciano M, Bersani F, Asinari P, et al. Techno-economic analysis of a solar thermal plant for large-scale water pasteurization. Appl Sci 2020;10:4771. Available from: https://doi.org/10.3390/app10144771.

[142] Denkenberger D, Pearce J. Design optimization of polymer heat exchanger for automated household-scale solar water pasteurizer. Designs 2018;2:11. Available from: https://doi.org/10.3390/designs2020011.

[143] Sizirici B. Modified biosand filter coupled with a solar water pasteurizer: decontamination study. J Water Process Eng 2018;23:277−84. Available from: https://doi.org/10.1016/j.jwpe.2018.04.008.

[144] Dobrowsky PH, Carstens M, De Villiers J, Cloete TE, Khan W. Efficiency of a closed-coupled solar pasteurization system in treating roof harvested rainwater. Sci Total Environ 2015;536:206−14. Available from: https://doi.org/10.1016/j.scitotenv.2015.06.126.

Index

Note: Page numbers followed by "*f*" and "*t*" refer to figures and tables, respectively.

Printed in the United States
by Baker & Taylor Publisher Services